$38.00

ESTIMATING HOME BUILDING COSTS

Revised

By W.P. Jackson; revised by Brian E.P. Beeston

What's on the CD-ROM?

- An Excel workbook, with a worksheet for each phase of construction
- A summary sheet that totals all the worksheets and adds overhead, contingency and profit, for a complete estimate of home building costs
- Excel worksheets for quickly calculating material quantities
- All the worksheets in PDF format that you can print and fill out manually

I0808977

Craftsman Book Company
6058 Corte del Cedro / P.O. Box 6500 / Carlsbad, CA 92018

To Katherine,

whose patience and support made this rewrite possible.

With thanks and with love, Brian

Library of Congress Cataloging-in-Publication Data

Jackson, W. P.
Estimating home building costs / W.P. Jackson ; revised by Brian E.P. Beeston. — Rev. ed.
p. cm.
Includes index.
ISBN-13: 978-1-57218-205-9
ISBN-10: 1-57218-205-9
1. Building—Estimates. 2. House construction—Costs. I. Beeston, Brian E. P. II. Title.

TH435.J19 2010
690'.837--dc22 2008020483

Layout: Devona Quindoy

Contents

1 **The Building Site** 5
- The Purchase Price 8
- Recording and Legal Fees 9
- Engineering Fees 10
- Cost Estimate Worksheet 12

2 **Preliminary Costs** 13
- Architect's Fees 13
- Stock Plans 13
- Plot Plans 14
- Building Permits 16
- Water Connection 20
- Sewer Connection 21
- Temporary Services 21
- Cost Estimate Worksheet 23

3 **Site Clearing, Excavation and Fill Dirt** 25
- Site Clearing 26
- Excavation 28
- Fill Dirt 33
- Site Cleaning and Hauling 36
- Cost Estimate Worksheet 36

4 **Footings** 37
- Layout 38
- Estimating Concrete Quantity 38
- Estimating Reinforcing Rod 43
- Estimating Forms for Footings 44
- Estimating Labor Costs 48
- Cost Estimate Worksheet 51

5 **Foundations** 53
- Estimating Masonry Blocks or CMUs 53
- Blocks Displaced by Stepped Footings 60
- Estimating Mortar 65
- Basement Openings 67
- Foundation Supports and Reinforcing 68
- Waterproofing and Footing Drains 76
- Masonry Labor 79
- Cost Estimate Worksheet 80

6 **Floor Systems** 81
- Board Measure 81
- Estimating Girders and Joist Hangers 84
- Floor Joists 88
- Subfloor 96
- Labor Costs for Floor Systems 101
- Cost Estimate Worksheet 105

7 **Superstructure** 107
- Exterior and Interior Walls 107
- Studs 111
- Headers 115
- Wall Sheathing 118
- Ceiling Joists 123
- Rafters 129
- The Ridge 140
- Roof Sheathing 144
- Trusses 152
- Porch Shed Roof Framing 154
- Stair Stringers 159
- Labor Costs for the Superstructure 160
- Cost Estimate Worksheet 163

8 **Roofing** 167
- Roof Covering 167
- Flashing 169
- Asphalt and Fiberglass Shingles 174
- Putting the Estimate Together 175
- Labor Costs for Roofing 177
- Cost Estimate Worksheet 179

9 **Electrical, Plumbing, Heating and Air Conditioning** 181
- Electrical 181
- Plumbing 183
- HVAC 184
- Cost Estimate Worksheet 187

10 **Estimating Brickwork** 189
- Estimating Brick Materials 190
- Fireplaces and Chimneys 193
- Labor Costs for Brickwork 200
- Cost Estimate Worksheet 202

11 **Saving Energy** 203
- Insulation 203
- Where to Insulate 206
- Estimating Insulation Materials 211
- Cost Estimate Worksheet 212

12 **Interior Wall and Ceiling Finish** 213
- Thin-Coat Plaster 213
- Drywall 214
- Whole House Wallboard Estimate 218
- Estimating Labor 221
- Cost Estimate Worksheet 222

13 **Exterior Trim** 223
- Windows 223
- Exterior Doors 224
- Siding 224
- Fascia, Frieze and Rake Boards 226
- Soffit and Porch Ceilings 228
- Porch Column Posts 229
- Moldings 230
- Exterior Trim Labor 230
- Estimating the Sample House Exterior Trim 231
- Cost Estimate Worksheet 236

14 **Concrete Floors, Walkways and Terraces** 239
- Concrete 239
- Concrete Accessories 240
- Cold Weather Pours 245
- The Material Estimate 246
- Labor Estimate for Concrete 253
- Cost Estimate Worksheet 256

15 **Interior Trim** 257
- Floor Underlayment 258
- Wood Flooring 259
- Interior Doors 260
- Window Trim 260
- Baseboard and Base Shoe 261
- Wall Molding 262
- Paneling 264
- Kitchen Cabinets and Vanities 266
- Stairs 267
- Bathroom Trim 269
- Estimating Trim 270
- Labor 282
- Cost Estimate Worksheet 283

16 **Painting, Floor Covering and Appliances** 287
- Painting 287
- Floor Covering 288
- Appliances 290
- Labor 294
- Cost Estimate Worksheet 295

17 **Gutters, On-Site Improvements and Miscellaneous** 297
- Gutters 297
- On-Site Improvements 298
- Labor 302
- Cost Estimate Worksheet 303

18 **Overhead, Contingency and Profit** 305
- Overhead 305
- Contingency 308
- Profit 309
- Form 100 – Estimating Building Costs 311

How To Use the CD-Rom 312

Glossary 313

Index 325

Chapter 1
The Building Site

ONE OF THE BIGGEST investments most people make is in a home. Ask any person on the street, and they'll tell you — in detail — what their dream home would be. Most contractors try to put a little dream in every home they build. Some people dream of a huge country kitchen. Others visualize a cozy bedroom fireplace. What their dreams have in common is that the whole package has to be there. Not just the perfect kitchen or fireplace but, maybe the most important thing of all, the ideal setting. Even if our dreams take us to the Bahamas or Hawaii, what most of us seek is — location, location, location. So that's what we'll focus on first in this book: selecting the right site at the right price.

Site selection costs are included in this book for speculative (spec) builders: those who buy land and build houses to sell for profit, rather than those under contract to build for someone else. The cost of the building site isn't a construction cost unless you buy it and build a house for subsequent sale. In that case, the site cost is factored in. This chapter will serve as a guide to evaluating site costs, which include the purchase price of the site, recording and legal fees, engineering fees for the survey, interest, taxes, liability insurance and other expenses incurred before the house and lot are sold.

All too often, a spec builder is content to build a house without giving thought to possible disadvantages of the site. A wise builder knows what conditions add value to the finished home, and avoids any site that could reduce his profit.

When the house is finished and the property is appraised, the appraiser will look for:

1. Growth/decline of the local housing market
2. Where the property fits in the growth/decline pattern
3. The appearance and desirability of the street or area

4. Demographic and economic indicators for the area, such as population, employment opportunities, rate of growth or decline of the population and the reason why
5. Accessibility to good schools, churches, shopping centers, recreation areas and public transportation
6. Terrain
7. Adequacy of water supply
8. Adequacy of sewage disposal
9. Mandatory preservation of trees on the site (trees can add as much as 25 percent to the appraised land value, in many locations)

The Move Back to the City

The move to the suburbs slowed a bit in the early 2000s, when fuel prices skyrocketed. Urban renewal has provided desirable neighborhoods, shopping variety, nearby medical facilities, and a full range of cultural opportunities. Urban construction generally has the advantage of immediate sewer and water connections, with streets, walks and utilities already in place.

However, there may not be many large building sites available in the city. If you intend to build more than one structure, you may only find lots that are scattered, rather than adjoining. You could have a tough time finding *any* prime sites, and may be forced to reconsider a site you once passed up. You may have to take a second look in older sections of town, as well as any vacant lot that has become a catch-all for neighborhood debris. Take another look at hillside sites that you once passed over. With adequate preparation, they could be the most profitable locations.

Never pay an inflated price for property based on a rumor that new industry is coming into the area. Check first to find out if the rumor is true. Your best resource for this information is the city planning department.

Check sales of comparable lots in the area. Go to the county recorder's office to find the most recent records of sale. Taxes paid on the transfer can help you determine a fair market price. Take along the legal description of the land in question. If the legal description and the name of the owner of the land aren't available at the county recorder's office, inquire at the local tax department. It also helps to have the names and addresses of surrounding property owners. The agency responsible for property taxes has the name and address of the owner of the property on file. For a nominal fee, anyone can check these records.

The selling price of nearby sites doesn't always tell you how much you should pay for a lot. That information just helps you get a ballpark figure. A building site with the lowest price may turn out to be the most expensive to build on, in the long run. But there are also times when lots, reclaimed after on-site demolition, can be bought and developed into building sites for less than the surrounding lots. Estimate your costs to develop the property, as opposed to building on more expensive sites. The expensive sites may end up being more economical in the long run.

When looking for suitable lots in or near the city, keep in mind:

- Ideally, a single-family house shouldn't take up more than 40 percent of the lot. Lot size should be adequate for the planned house, and shaped so that the structure can be situated without violating local building codes.
- There should be easy access to (and circulation around) the building. When it's completed, can the house be maintained without trespassing on adjoining property?
- If required, there should be sufficient room for safe and sanitary installation of an individual water supply and sewage disposal system.

Zoning ordinances vary from community to community, but they're all intended to protect the health and safety — as well as the investments — of residents. Zoning rules once fit on one piece of paper, with room to spare. Not any more. Zoning ordinances are multisection, multipage legal documents that include every conceivable exception and special requirement. Fortunately, they're also readily available. Most cities and counties have planning departments or websites to help you find zoning ordinances.

Check the Site Before Buying

Your goal is to build and sell a spec house quickly, when (or even before) it's completed. It's critical that you know as much as possible about the site before you buy.

Determine beforehand the type of house you want to build and the target price range. If your house is substantially more (or less) expensive than the average house in the neighborhood, you could have a problem selling it. That narrows the prospective market, so you'll wait longer for a buyer. A good rule of thumb: price your house within 15 to 25 percent of the average price in the neighborhood.

A good neighborhood is always a plus at selling time. Buy in growing areas. Proximity to schools, shopping centers, public transportation, hospitals and recreation areas are desirable to homebuyers. Vacant houses in the area may be signaling a declining market. You may find lower land prices near the proposed site of an airport or a future highway, but keep in mind that selling any house built there could be difficult, and will usually be at a lower profit margin.

Always check with the local planning commission for zoning regulations. Zoning regulations and deed restrictions may be the reason properties are sitting vacant, so know the area inside and out before purchasing a site. Don't count on breaking the rules and getting away with it.

City staff — and your neighbors — take zoning laws very seriously. Ask about future plans for the area that could affect your decision to buy in that jurisdiction. A building moratorium may be in place, or planned. If that's the case, don't buy the land until you get confirmation that the moratorium will be lifted by a specified date. Take an option on the land if necessary. Taxes, interest, and insurance are your financial responsibility the minute that property becomes yours. Finally, buying in a no-growth area can mean financial suicide: idle land is 100 percent liability.

If you're buying land as an investment, answer the following questions before buying:

1. Can you afford the monthly payments, taxes, and other assessments on the land until a house is built and the property is sold?
2. Is the area expected to continue growing? Is that growth expected in the residential, industrial, or commercial sector?
3. Are any highways planned through this area?
4. What are the future plans for zoning in the area?

There are rarely opportunities to both select the lot(s) you want and orient the house(s) on-site for maximum energy efficiency and livability. Some of the following types of property you initially rejected may be worth a second look:

- Wooded areas are expensive to develop, but with careful planning, can become beautiful building sites. They'll have a much higher resale value, if properly designed.
- Bare land, or land with few trees is cheaper to develop and build on, but landscaping it can be expensive.
- Hillside lots can have views that take your breath away. They can be developed into beautiful sites, but expect to have more than your breath taken away. Building to code here can be twice as expensive as construction on a flat lot. Not surprisingly, some of the most exclusive houses built today are on hillside lots.

Purchase Price of Site

You've found the land you want to purchase. What steps do you take to buy it?

1. Hire a reputable attorney who specializes in land-use law.

2. Verify that a clear title to the property exists.
3. Check present and future zoning regulations for the area.
4. Do your research. Is a highway, shopping mall, or industrial park intended for nearby property in the foreseeable future? Is the site under the flight path of a present or future airport?
5. Is the site on or near a landfill or toxic dump site? Is there evidence the land could be chemically contaminated? If there's a chance of toxic contamination, thoroughly investigate the property.
6. Know the present and projected tax rate.
7. If there aren't already service lines to the property, contact the utility companies about installing them.
8. Research building codes, deed restrictions, easement rights, and any other building requirements for the land.

*You can find Blank Cost Estimate Worksheets on the CD inside the back cover of this book. Make entries only in the underlined cells.

There's a cost estimate worksheet for the building site at the end of this chapter. When completed, it can show you what the total cost of the building site will be when you sell the improved property. Any additional costs of the land, along with the purchase price, are included in your building site cost. These costs are explained in the following sections. If you build under contract for a landowner (who's paying these additional costs), don't make entries in cost lines 1.1 through 1.7 on the *Cost Estimate Worksheet*.

*Enter the **purchase price of the land** on Line 1.1 of your Cost Estimate Worksheet.* We'll use a figure of $60,000 in our Example Cost Estimate Worksheet.*

Recording and Legal Fees

In many states, a lawyer must complete any real estate transaction and prepare and record the deed. The seller (*grantor*) normally pays for preparing the deed, but the buyer (*grantee*) pays to transfer the property, including the deed of trust (if there is one), the recording fee, his share of the transfer tax, and the title check (to confirm a clear title).

The title should be free of any mortgages, mechanics liens, easements, or encumbrances that could influence your decision to buy the land.

If the building site is on a private road, determine the assessments for maintenance, snow removal, and utility lines. These fees add to the cost of the land. If the site is on a private street, make sure there's a permanent easement to the property.

*Enter the **recording and legal fees** on Line 1.2 of your Cost Estimate Worksheet. We'll use a figure of $2,000 in our Example Cost Estimate Worksheet.*

Engineering Fees

If you're unsure of property lines, a registered land surveyor can determine them for you. When the corners have been identified, the surveyor uses permanent markers (rather than wooden stakes) to indicate their locations. These markers may cost more, but they'll save you from having to resurvey if the wooden stakes are removed or broken off. Make sure the lot described in the deed is indeed the lot you intend to buy. People have built on the wrong lot. It's not as hard to make that mistake as you might think. Don't let it happen to you.

An engineer can give you a heads up on conditions that could affect construction costs. For instance, the property's topography can affect excavation costs, driveway grade, walks and drainage. Plans for outdoor living areas may need revising if the topography isn't user-friendly.

The building site should be free of hazards that could affect the structural soundness of the building, or the health and safety of the occupants. These include subsidence (excessive settling caused by unstable soil or high groundwater), flooding, and erosion. Underground springs produce hydrostatic pressure that can cause leakage in basement walls and floors. High groundwater means an additional expense to waterproof the basement, involving drain pipes, sump pumps, and waterproofing the foundation walls. Springs and high groundwater could necessitate raising the elevation of the finished floor.

Run-off can damage surrounding property. Many jurisdictions prohibit building any structure unless there's an approved erosion and sedimentation plan in place. In fact, it could be unlawful to clear, grade, or otherwise change the contour of land without these plans. You may also be required to get approval before you remove or destroy trees, shrubs, or other plant life. The city engineer can give advice for your specific situation.

Here's a little trick that could help in your situation. If you want to save money on storm sewers, use the smallest-diameter pipe permissible. A decrease in the size of the storm pipe decreases site development costs. When an on-site sewage system is required, the property must meet local sanitation requirements. Leave the job of designing the system to your engineer.

*Enter the **engineering fees** on Line 1.3 of your Cost Estimate Worksheet.*
We'll use $2,000 in our Example Cost Estimate Worksheet.

Interest

Many contractors borrow money from a bank or lending institution to purchase land. That means interest payments from the time of purchase to the time the completed property is ultimately sold. If you borrow money to buy your site, you could pay interest for over a year. And with the present cost of building land, interest payments can be substantial. You need to decide if buying and improving a property is going to work for you and your financial situation.

Guesstimate how long you'll need the loan, based on your building schedule. Also consider how long it'll take to sell the finished building. In a good market, you may be able to pre-sell the property. In a slow market, the property could be on the market for months after completion. To maximize your investment, you need answers to all the "what-ifs."

> *Enter the estimated time of the loan (months) and the **interest** rate on Line 1.4 of your Cost Estimate Worksheet. We'll use 12 months and 7 percent in our Example Cost Estimate Worksheet.*

Taxes

Contact the tax assessor to determine if you'll have to pay property tax on your building site before and during construction. Taxes are typically assessed on real property: land, improvements to land, structures, and certain equipment affixed to those structures. Property is generally appraised at 100 percent of its fair market value, according to the highest and best use of the property. Fair market value is the price a buyer is willing to pay a willing seller. Don't assume that because there's no completed structure on the site, you don't owe any taxes. The amount of property tax you'll have to pay depends on the length of time between land purchase and eventual property sale. Again, you'll have to estimate this time period. Obviously, the more time it takes, the more taxes you'll pay.

> *Enter the estimated property **taxes** (if any) on Line 1.5 of your Cost Estimate Worksheet. We'll use 12 months and $100 a month in our Example Cost Estimate Worksheet.*

Maintenance

If you can't start building immediately after purchasing the property, you'll be responsible for maintenance of the site in the interim. You can't just let the vegetation grow unchecked. In many areas, underbrush poses a fire hazard, particularly in dry, desert climate areas. This was brought home during the California wildfires. Maintenance costs generally depend on the length of time between the purchase of the land and when you start construction. You'll need to estimate that time and, again, the longer the time, the more you'll pay.

> *Enter the **maintenance** costs on Line 1.6 of your Cost Estimate Worksheet. We'll use 3 months and $100 a month in our Example Cost Estimate Worksheet.*

We won't include site clearance and preparation costs yet — they'll be addressed in Chapter 3, when we discuss site cleaning, excavation and fill.

Other Costs

You may need to include other costs, such as liability insurance. As the landowner, you could be held liable for any accidents that happen on the property. If your site is on a hillside, heavy rainfall could cause a landslide, damaging property at the bottom of the slope. Or runoff from a rainfall could cause flooding of other properties. Be sure that you have sufficient liability insurance — it's worth its cost for the peace of mind it gives. This is especially true if you aren't on site often, and don't catch accidents waiting to happen. The total cost depends on how much time passes between the land purchase and property sale. Estimate how long you expect construction to take, and remember, the longer the time, the more cost you'll incur.

*Enter any **other costs** on Line 1.7 of your Cost Estimate Worksheet. We'll use 12 months and $100 a month in our Example Cost Estimate Worksheet.*

COST ESTIMATE WORKSHEET FOR BUILDING SITES

#	Item	Time	Rate	Subtotal
1.1	**Purchase price of site**			$60,000.00
1.2	**Recording and legal fees**			$2,000.00
1.3	**Engineering fees**			$2,000.00
1.4	**Interest**	12 Months	@ 7 % annual interest rate	$4,200.00
1.5	**Taxes**	12 Months	× $100.00 per month	$1,200.00
1.6	**Maintenance** *(Clearing underbrush, etc. while lot is idle)*	3 Months	× $100.00 per month	$300.00
1.7	**Other costs** *(Liability insurance, assessments, etc.)*	12 Months	× $100.00 per month	$1,200.00
	Total cost of site (entered on line 1 of Form 100)*			**$70,900.00**

* Form 100 is the Cost Estimate Form used to compute the total cost of the house and building site — it is shown in Chapter 18 of this book. If the cost of the building site is not to be included in the estimate, enter zero for the total cost of site. This will then be transferred to line 1 of Form 100.

Chapter 2

Preliminary Costs

IF YOU HIRE AN architect to draw up blueprints for a project, his fee may or may not be included in your construction costs. You may opt to purchase stock plans, in which case they'd be part of your costs. If you're building spec houses, the cost of the plans is always included. But if you're bidding on a contract or working on a cost-plus basis, the blueprints will be provided by the owner, and shouldn't be included in your cost estimate.

Architect's Fee

Even the smallest structure would be difficult to build without drawings to help illustrate the building instructions. These drawings are known as *blueprints*, named after the way they were originally made. Before the 1940s, blueprinting was the only way to copy drawings. The drawing was traced in India ink onto paper or cloth. When exposed to sunlight, a blue image of that illustration appeared through the tracing material. The material was then taken indoors, washed, and dried. A clear copy of the drawing appeared as white lines on a dark blue background; hence "blueprint."

The process has gone through a series of improvements since the 1940s. But even though plans today are processed on flatbed printers linked to computers, the term *blueprint* is still used.

Stock Plans

An architect isn't usually hired to draw plans for residential construction unless large tracts or custom homes are being built. Small scale builders can buy standard plans quite reasonably from companies that specialize in home building-plan services. Many of these companies have Internet sites, such as www.allplans.com. There are companies offering complete blueprints for houses of virtually any size and design. Many people who've decided to build a house find a plan that, with a few minor changes, is compatible with their needs and taste. If the house is factory-built (modular), the manufacturer will furnish all necessary blueprints as part of the package price.

However, few standard plans are so complete that absolutely no changes are needed. Unless you're an expert in drawing house plans, I recommend hiring a professional to do it. Amateurs often think that by drawing their own plans, they'll save money. Big mistake. One error in the blueprints could be costly enough to halt construction.

You generally need a copy of the blueprints to show when you apply for financing, and later when you request building permits. Architectural and engineering exhibits are commonly submitted when applying for a construction loan or mortgage. These exhibits describe in detail the size and location of the house, and the types of the materials to be used. An appraisal (based on your exhibits) is made of the proposed structure before the loan can be finalized. If you have complete plans and are using quality material, the proposed structure should appraise at full value.

> *Enter the* ***architect's fee*** *or the* ***cost of stock plans*** *on Line 2.1 of your Cost Estimate Worksheet. We'll use $5,000 in our Example Cost Estimate Worksheet.*

Plot Plans

A plot plan is an essential part of the working drawings for house construction. It's needed to help you accurately estimate construction costs. Plot plans are usually prepared by the builder or an engineer. If you hire an engineer, his fee is included in the engineering costs of the building site, as explained in Chapter 1. Plot plans can also be paid for by the property owner. If the job is open to bid, this fee isn't part of the builder's costs, and would be billed and paid separately.

The plot plan, as seen in Figure 2-1, must have the following minimum information:

- Compass direction indicating north
- Metes and bounds (see Glossary) of the property lines and their distances
- Lot corners
- Description of the lot and section number
- Elevations of each floor level
- Location and dimensions of easements
- Situation of the house on the lot
- Grade elevations
- Location and dimensions of water and sewer lines
- Location of electric, gas, telephone, and TV cable service

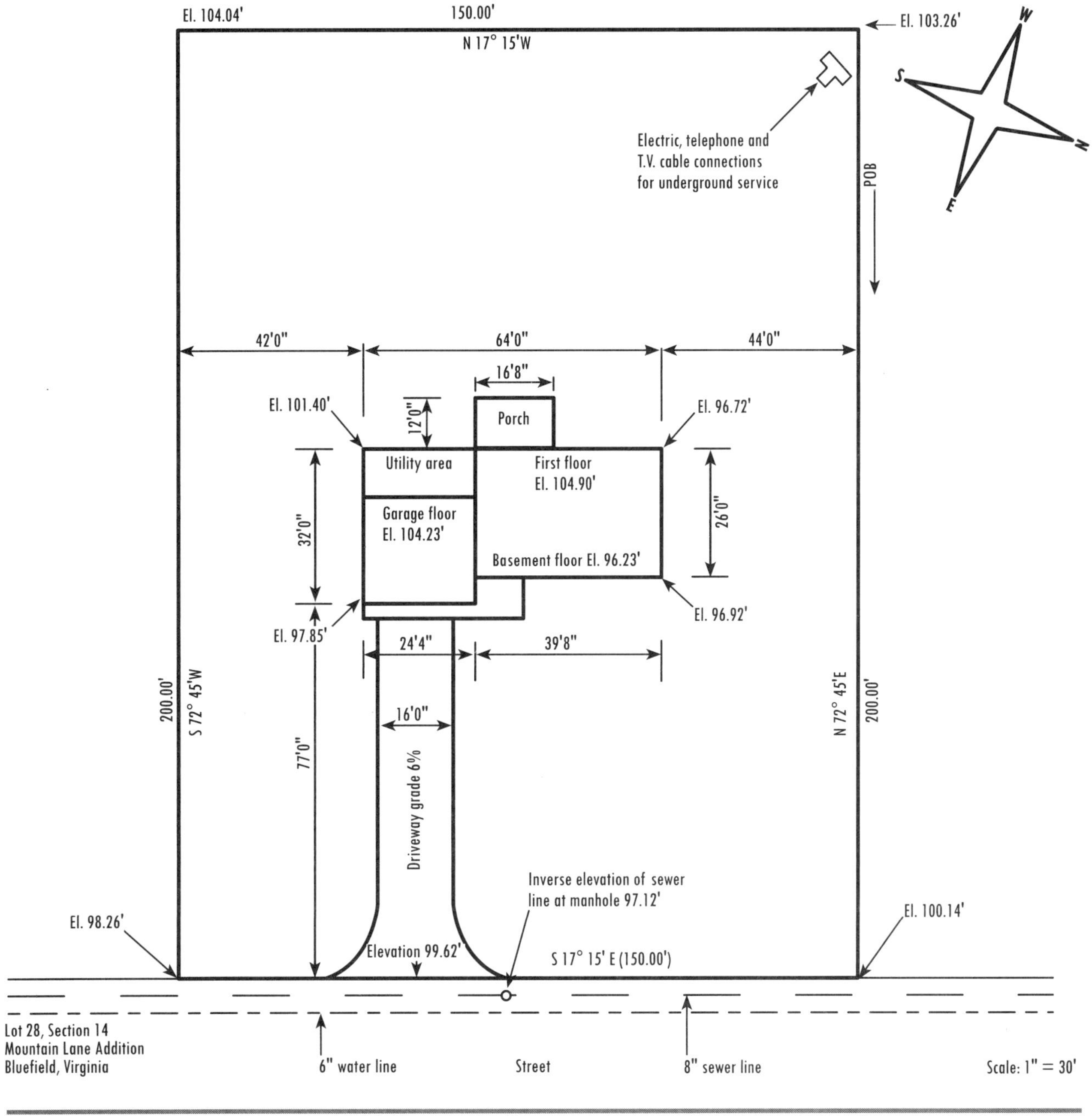

Figure 2-1
Plot plan

- Finished elevations showing driveway to street connection
- Finished elevations at each corner of the dwelling
- Dimensions of the house, garage, carport, and other buildings
- Location and dimensions of walks and driveways
- Location of steps, terraces, and porches
- Scale of the drawing
- Location of existing trees (necessary in some jurisdictions)

The elevations can be given in relation to sea level (if known), or above or below a *bench mark* (BM), an arbitrary elevation based on a fixed point. The bench mark is normally assigned an elevation of 100 feet. All other elevations on the plot plan refer to this point. A bench mark may or may not be shown on the plot plan.

Estimating Home Building Costs contains sample estimates for a 2,238-square-foot two-story house, with a basement, a porch, and a two-car garage. All references, unless specifically noted, refer to this house, beginning with the plot plan in Figure 2-1. From the information given, you can make the following construction cost calculations:

1. Using the metes and bounds of the lot, you can calculate the interior angles of the lot corners. A book entitled *Building Layout,* available from Craftsman Book Company, shows how to convert these bearings into interior angles. This information is extremely important for the builder to know, because lot corners tend to get lost. Or there could be obstructions (such as trees or rocks) that prevent sighting with a transit from one lot corner to another. Corner angles aren't always 90 degrees, so don't miscalculate the property line. That could prove very costly.
2. Finished grade and floor elevations indicate how much excavation or fill dirt will be required. In our plot plan, the elevations shown require an estimated 1,540 cubic yards of compacted fill dirt to be imported to the jobsite. The basement floor elevation (96.23 feet) is lower than the elevation of the sewer line at the manhole (97.12 feet) where the sewer connection will be made. So, even though no bathroom or laundry is planned in the basement, a sump pump is required to remove any water that could seep in.
3. Location of the building and the dimensions of the driveway and walks determine their cost.
4. Location of the utility connections show how far the service lines must be run.
5. Easements shown on the plot plan may alter the plans for a particular area, such as the location of an accessory building or patio.

*Enter your cost of **plot plans** on Line 2.2 of your Cost Estimate Worksheet.*
We'll use $750 in our Example Cost Estimate Worksheet.

Building Permits

A building permit regulates the location, size and type of structure that can be built on a particular plot. A plans examiner uses the blueprints and plot plan you supply to make sure that you won't be violating any zoning ordinances. If, after checking the blueprints and plot plan, the examiner is satisfied that the building is in compliance with code, a building permit will be issued. If a minor violation is found, the plans examiner could present the permit application to the Planning Commission to be approved or disapproved.

In addition to the building permit, plumbing and electrical permits must be obtained. Because the plumbing and electrical work is usually done by subcontractors, the responsibility and expense of obtaining these permits is theirs. The costs will be included in the total construction bid, under their respective headings.

The cost of permits varies, depending on the area of the country. One way cost can be determined is based on the anticipated value of the structure. There's a set fee (for the first determined valuation), with increments for each additional thousand dollars. For example, assuming our house's estimated value is $300,000, the building permit might be $500.00 for the first $100,000 of valuation, and $2.50 for each additional $1,000. Our permit would cost $1,000 [$500 + (200 x $2.50) = $1,000].

The value of any property can be debated. Disputes over the building's worth frequently arise between the property owner and the department issuing the permit. Consequently, many jurisdictions have developed a simpler process for establishing property value. An example is shown below.

Habitable space	$100 per square foot
Garage	$50 per square foot
Unfinished basement/storage	$40 per square foot
Deck/patio/porch	$25 per square foot
Carport	$25 per square foot
Fence	$10 per linear foot

Using these costs per square foot, the permitted value of the property for the permit in Figure 2-2 is:

Habitable space		
First floor	1,198 square feet @ $100 per square foot	$119,800
Second floor	1,040 square feet @ $100 per square foot	104,000
Garage	610 square feet @ $50 per square foot	30,500
Basement	1,031 square feet @ $40 per square foot	41,240
Porch	200 square feet @ $25 per square foot	5,000
Total value:		$300,540

The jurisdiction where our sample house is located uses a sliding scale of permit costs for property valuation or for remodeling work, shown here:

$1 to $500	$30	
$501 to $2,000	$30 for 1st $500	$1.50 each additional $100
$2,001 to $25,000	$50 for 1st $2,000	$6.50 each additional $1,000
$25,001 to $100,000	$200 for 1st $25,000	$4.00 each additional $1,000
$100,000 and up	$500 for 1st $100,000	$2.50 each additional $1,000

BUILDING PERMIT

ALL SPACES MUST BE COMPLETED

Date ______________

A BUILDING PERMIT MUST BE ISSUED BEFORE CONSTRUCTION IS STARTED.

Application for a Building Permit must be made to the Building Officials.
Application is hereby made for a Building Permit in accordance with the description and for the purpose hereinafter set forth. This application is made subject to all local and State laws and ordinances and which are hereby agreed to by the undersigned and which shall be deemed a condition entering into the exercise of this permit.

Name of Owner *W. P. Jackson* Address *109 Fincastle Lane*

Name of Contractor/Builder *Same* Address *Same*

Certified State Contractor's No. *15474* Zone Classification* *R-1*

If for Alteration or Repairs, state in detail *None*

____

* Zone classification must be specified.

Street Name *Mountain Lane* Lot No. *28* Section of *Mt. View* Subdivision

Size of Lot *150' × 200'* If purchased within the past two years from *N/A*

Date ____________ I hereby certify that on January 1 the land described above is listed in the

name of *W. P. Jackson* .

NOTE: Permit for septic tank and approval of location and of well must be obtained from the County Health Department after the lot has been cleared and building has been staked out, but before construction has been started.

Plot Plan ______________ Construction Plans ______________

Estimated Date of Completion ______________

I hereby certify that I have the authority to make the foregoing application, that the information given is correct and that the construction will conform with the regulations in the Building Code, Zoning Ordinances and private building restrictions, if any, which may be imposed upon the above property by deed. I also agree to be responsible for any and all damage to any and all property, public or private, caused by above construction/repair.

Signature of Owner or Authorized Agent ______________________

Telephone______________ Email Address ______________________

Figure 2-2
Typical building permit

I, the undersigned, hereby make application for a permit to erect a *Two-story*

building to be used for *Residence* (Commercial/Residence) on my property at *219 Mountain Lane*

Lot No. *28* Section No. *14* of Subdivision *Mt. View*

The general shape of the lot and the location of the proposed improvements are accurately set forth in the plot plan.

Front yard available *77'0"* ft — Type of heating *Electric*

Side yards available *42'0" and 44'0"* ft — No. of flues *None*

Rear yard available *91'0"* ft — Size of flues *N/A*

Note: All flues must be lined.

Type of roof *Asphalt shingles* — Ceiling joists and rafters on *16-inch* centers

Expected amount of total electrical load *200 amps*

Type of floors *Wall-to-wall carpet over subfloor and floor underlayment* — Joists on *16-inch* centers

No. of rooms *Seven* — Type of construction *Drywall and panel*

Number of baths *Three* — Type of construction *Drywall*

Basement size *39'8" × 26'0* — Type of construction *Concrete block with concrete basement floor*

Garage/carport *24'4" × 22'0" + 7'6" × 10'0"* — Square feet *610*

First floor square feet *1,198* — Second floor square feet *1,040*

NOTE: Soil bearing test for other than one-family dwelling shall be required. Soil test report *N/A*

I, the undersigned, do affirm that all of the foregoing figures and statements are true, full and correct to the best of my knowledge and belief, and all sanitary, safety and building ordinances of the Town of Bluefield, Virginia will be complied with in said construction. I tender with this application the sum of $ ________ covering permit, with $ ________ covering sewer tap and $ _______ covering water tap.

A TOTAL AMOUNT OF: $ ________________

Building Permit Approved ____________________ Applicant ________________________

Disapproved ____________________ Code reference ____________________

By: Building Inspector ____________________

Engineer ________________________ Manager ________________________

Figure 2-2
Typical building permit (continued)

Our permit, based on a property value of $300,540, would fall into the *$100,000 and up* category for a permit based on valuation. The permit cost would be:

The 1st $100,000	$500.00
$200,540 @ $2.50 for each $1,000	<u>$501.35</u>
[(200,540/1000) x $2.50 = 501.35]	
Permit cost:	**$1,001.35**

There may also be a state surcharge fee (ours adds 7 percent), and a plan check fee, which is around 65 percent of the permit amount. This levies additional costs of:

Permit	$1,001.35
Surcharge fee (7 percent of $1001.35)	$70.09
Plan check fee (65 percent of $1001.35)	<u>$650.88</u>
Total Fees:	**$1,722.32 (rounded to $1,720)**

*Complete line 2.3 of your Cost Estimate Worksheet to obtain the total **permit** fee. We'll enter these figures on our Example Cost Estimate Worksheet.*

Figure 2-2 is a typical building permit application for a two-story house. This application would also include the plot plan from Figure 2-1. It's a good idea to keep a duplicate copy of the blueprints, in case the copy submitted with the permit request isn't returned.

Water Connection

To complete your estimate, you must also figure the cost of getting water to the building site. A fee is charged for connecting to the water supply if it's supplied by a public utility. If the service line isn't brought to the building site, a line must be run from the main source to the site. Before you submit an estimate, find out if this applies to your site.

If there's no public water supply, you'll have options for alternative water sources. One choice could be to dig a well; another is to have water piped from nearby springs (which may serve more than one family). If the water supply is privately owned, be sure you know who's responsible for maintaining it. Finally, ask the local Health Department to check the water for purity. It's hard to sell a house that doesn't have water you can drink.

*Enter the **water connection** fee or cost on Line 2.4 of your Cost Estimate Worksheet. We'll enter $500 on our Example Cost Estimate Worksheet.*

Sewer Connection

As with the water supply, sewage disposal may be a public system. In that case, you'll be charged for the sewer tap. A private sewage disposal system generally consists of a septic tank or a privately-owned sewage disposal system that serves more than one family. When you apply for a permit to install a septic tank, an agent from the Health Department will come to the site and design a sewage disposal system for you.

Some of the factors used to determine the size of the tank include results from percolation tests, the size and shape of the building site, the size of the house, and whether or not a garbage disposal will be installed. It may come as a surprise that the material a garbage disposal adds to the septic tank can overtax its capabilities. In fact, in some jurisdictions, they aren't allowed at all.

After the system is designed, the Health Department will issue a permit to build it. An inspector will check the tank's installation for code compliance before final approval is given to bury the system.

Sewer system expenses are part of construction costs, to be included in your cost estimate.

*Enter the **sewer connection** fee on Line 2.5 of your Cost Estimate Worksheet. We'll enter $500 on our Example Cost Estimate Worksheet.*

Temporary Water Service

If water is supplied by a public utility, determine the cost of the water you anticipate using during construction, and add it to the cost estimate. In most cases, there's a minimum charge per month for the water. Except for brief periods, like when brick masons or drywall subcontractors are working, the minimum monthly charge will generally cover the cost of the water used. The minimum monthly water charge multiplied by the estimated number of months to complete the house gives a ballpark cost for temporary water service. The cost estimate worksheet at the end of this chapter can help you figure this cost.

*Complete Line 2.6 of your Cost Estimate Worksheet for the **temporary water service.** We'll use $50 a month for 9 months in our Example Cost Estimate Worksheet.*

Temporary Electric Service

Because permanent electric service can't be connected until the house is completed, you'll need a temporary meter for service during construction. Utility companies charge for installing these meters, and bill monthly for the electricity used. As with temporary water service, the monthly charge for electricity multiplied by the estimated number of months to complete the house gives a reliable estimate for anticipated electricity cost. If you'll be working in cold weather, be sure to include an allowance for the electricity needed to run heating units. Again, the cost estimate worksheet at the end of this chapter shows how to calculate the total cost of electric service.

Complete Line 2.7 of your Cost Estimate Worksheet for the ***temporary electric service****. We'll use a $500 installation fee plus $100 per month for 9 months in our Example Cost Estimate Worksheet.*

COST ESTIMATE WORKSHEET FOR PRELIMINARY COSTS

#		Amount		Cost					Subtotal
2.1	**Architect's fee or cost of plans**								$5,000.00
2.2	**Plot plans**								$750.00
2.3	**Building permit**								
	First floor	1,198 sf	@	$100.00 per sf			=	$119,800	
	Second floor	1,040 sf	@	$100.00 per sf			=	$104,000	
	Garage	610 sf	@	$50.00 per sf			=	$30,500	
	Basement	1,031 sf	@	$40.00 per sf			=	$41,240	
	Porch	200 sf	@	$25.00 per sf				$5,000	
						Total valuation		$300,540	
	If the cost of the permit is based on valuation								
		$100,000	@	$500.00 per		$100,000	=	$500.00	
		$200,540	@	$2.50 per		$1,000	=	$501.35	
			@	per			=		
								$1,001.35	
	State Surcharge Fee			7 **%**	of	$1,001.35	=	$70.09	
	Plan Check Fee			65 **%**	of	$1,001.35	=	$650.88	
						Total permit fee		$1,722.32	$1,720.00
2.4	**Water connection**								$500.00
2.5	**Sewer connection**								$500.00
2.6	**Temporary water service**								
	(a) Minimum cost per month			$50.00					
	(b) Estimated months to complete			9					
			(a)	$50.00 ×	(b)	9	=	$450.00	$450.00
2.7	**Temporary electric service**								
	(a) Charge for temporary meter			$500.00					
	(b) Minimum cost per month			$100.00					
	(c) Estimated months to complete			9					
			(a)				=	$500.00	
			(b)	$100.00 ×	(c)	9	=	$900.00	
						Total		$1,400.00	$1,400.00
	Total preliminary costs (entered on line 2 of Form 100)								**$10,320.00**

Chapter 3

Site Clearing, Excavation and Fill Dirt

RESIDENTIAL CONSTRUCTION IS A highly competitive business. The success or failure of any project begins with the accuracy of your cost estimates. This is true whether you're estimating as a spec builder or bidding on a contract. You arrange financing based on the estimated cost of the house. If you underestimate the job, and as a result underestimate the loan taken out at the bank, you may need to borrow additional funds to complete it. This will lower your profit, at best. And, worst case scenario, you could end up *losing* money on the project.

Bad estimates cost money. If your estimate is too low, you'll lose money. If your estimate is too high, you probably won't get the contract. Don't get careless by rushing your estimate and omitting something vital. Once a bid is submitted and a contract is signed, it's binding. Unfortunately, estimators and builders *do* overlook items, sometimes major ones (like the cost of brick on a brick-faced house), and often the error isn't discovered until it's too late.

Never make an estimate using a square-foot or cubic-foot method. It's just not accurate enough. For example, a house with eight exterior corners will cost significantly more to build than a house of the same size with a simple rectangular design and only four exterior corners. The square-foot method should only be used to get a ballpark figure, or to check a properly-done estimate in case you've overlooked an item, such as the brick in the instance described above.

The type and quality of the materials used in a house — or even the pitch of the roof — can change the cost by thousands of dollars without adding one square foot of living space. The same house plan, built by the same crew on adjacent lots, can also vary in cost because of the individual lot requirements. Each house should be estimated individually.

Costs may vary significantly for identical work performed months apart. Labor productivity for each tradesman can change from day to day. Inflation, especially in the cost of some building materials, often increases from month to month. Material theft or weather extremes resulting in lost time have to be addressed in any construction estimate. Always make allowances in your estimate for conditions that vary from the norm. Experience and accurate cost records from previous jobs are essential.

A day lost during construction extends the time it takes to complete the house. Every extra day adds to the interest you'll pay, which is based on the full amount of the construction loan. Delays happen, but by keeping them to a minimum, you can limit unforeseen expenses. To help avoid delays, schedule the work on a day-to-day basis. Enlist the cooperation of your suppliers and subcontractors.

On any residential construction job, your estimated costs should be within 5 percent of the actual cost. If you use a step-by-step cost estimating system to keep from forgetting important costs, you should be pretty close to on the money.

However, the accuracy of your estimate depends on the accuracy of the individual costs you use — for both construction materials and labor. Cost records of similar jobs are the most reliable source of cost data. They take into account geographical cost differences, weather-related variations and labor productivity. That's another good reason to keep accurate records and compare actual versus estimated costs once each job is completed. If those records aren't available, you'll need a source of labor and material costs.

A good reference for both material and labor costs is the *National Construction Estimator*, published by Craftsman Book Company. It gives costs for every aspect of residential, commercial and industrial construction. In addition, you'll find adjustment factors to apply for different states and counties, for different weather conditions, and for any construction situations that can affect productivity, like job size or working-height above grade. The order form bound into the back of this book tells you how to order the book, or a digital version.

The *National Construction Estimator* is where we've gotten most of the costs used in the Cost Estimate Worksheets at the end of each chapter in this book. The cost examples used come from the edition of the *National Construction Estimator* current at the time of this writing. They won't be valid if you're reading this book in a later year, but the method shown will always be good.

Site Clearing

The building site must be cleared and prepared before construction can begin. *Always* walk the site before making your cost estimate. Don't assume anything or take someone else's word for the condition of the site where you plan to build. Another person's idea of a cleared site may not meet your standards. Make sure you identify any potential problems that could mean additional costs.

Some building sites are wooded or have old structures that must be removed. Fences and poles, debris (like old construction materials), brush and tree limbs must be removed. Wood that's buried underground attracts termites, and must be removed from the building site.

Small trees and stumps can be removed with a dozer or loader, but larger trees have to be cut down by experienced tree-removal contractors. That can be expensive, especially if the trees have extensive root systems. A loader can demolish old fences, poles, and paved areas, but be careful tearing down any buildings on the site. You could destroy otherwise reusable material. Keep an eye out for any toxic materials on the site, like asbestos in old buildings or chemical spills resulting from industrial activity or dumping. Asbestos removal must be performed by specialists, at considerable additional cost. And any evidence of chemical spills could be reason enough to avoid the site completely. It's difficult to determine site-clearing costs, especially if extensive work is involved. Get bids from at least two contractors who specialize in the type of work you need to have done.

Site Clearing Cost Guidelines

If you need a preliminary estimate, use the following guidelines.

Removal of tree branches and trunk:	
Diameter from 12 to 36 inches –	$100 to $200 each
Stump removal:	
Diameter from 12 to 36 inches –	$100 to $200 each
Removal of paved areas:	
Bituminous paving –	$2.00 to $4.00 per square yard
Concrete paving –	$3.00 to $9.00 per square yard
Removal of old buildings:	
Depending on size of building –	$3.00 to $6.00 per square yard
Removal of old fences:	
Depending on type of fence –	$1.00 to $2.00 per linear foot

Removal of debris from site:

Depends on the quantity of debris and the distance to be hauled. Check with a contractor who specializes in this work for a firm price.

These costs are only guidelines, based on prevailing costs at the time this book was written. They shouldn't be used as firm cost estimates. If you don't have costs from past jobs, refer to the current edition of the *National Construction Estimator*. When site-clearing expenses are expected to be sizable, get a firm bid on the work before completing your final estimate.

*Enter the calculated or quoted **site clearing** cost on Line 3.1 of your Cost Estimate Worksheet. We'll use a figure of $3,000 in our Example Cost Estimate Worksheet.*

Excavation

Estimating excavation isn't easy. More errors are made here than in any other phase of residential construction. The construction of any foundation inevitably involves removal of earth. The quantity to be removed depends on the design of the house, the floor elevations, the location of the excavation on the building site, and if the house has a basement, post-and-beam, slab, or a combination of these types of foundations.

You may have some knowledge of determining elevations, calculating floor elevations and computing cut and fill. If you don't, *Building Layout*, published by Craftsman Book Company, is a good place to start. However, you still may need to consult an engineer.

Before you do an excavation estimate, answer the following questions:

- Is there an area on the building site large enough to store topsoil during construction, or must it be hauled away? If it has to be hauled away, how far is the dump site? Does the soil need to be hauled back?
- Will shoring be required?
- Will special safety precautions be needed around the excavated area (some cities' requirements)?
- Will blasting be necessary?
- Has this area been a landfill, necessitating additional excavation?

There's a difference between a general contractor and a residential builder. The general contractor has the right equipment for constructing commercial and industrial buildings. He has the earthmoving equipment needed for this kind of work. By comparison, the small-volume residential builder rarely owns excavating equipment, and usually needs to hire an excavation subcontractor.

When estimating excavation, take into consideration the soil conditions, whether or not there's rock present, and the cost of hauling equipment to the building site. Here again, accurate cost records from previous jobs are very helpful. Machine excavation where there's no rock should cost between $4.00 and $6.00 per cubic yard, depending on the soil conditions and location. If blasting is required, it can increase the excavation costs by 200 percent or more, depending on the amount of blasting necessary (and proximity of surrounding buildings and utilities).

Weather is another factor to be reckoned with. Cold temperatures can increase excavation costs by 15 percent or more. Frozen earth is much harder to excavate and haul, and can damage earthmoving equipment. Wet-dry and freeze-thaw cycles are also troublesome. The topsoil can be soft and muddy, while the earth underneath remains hard. Builders usually try to get rough-grading completed before winter sets in. However, that also adds to the interest they'll pay on loans until work can start again in the spring.

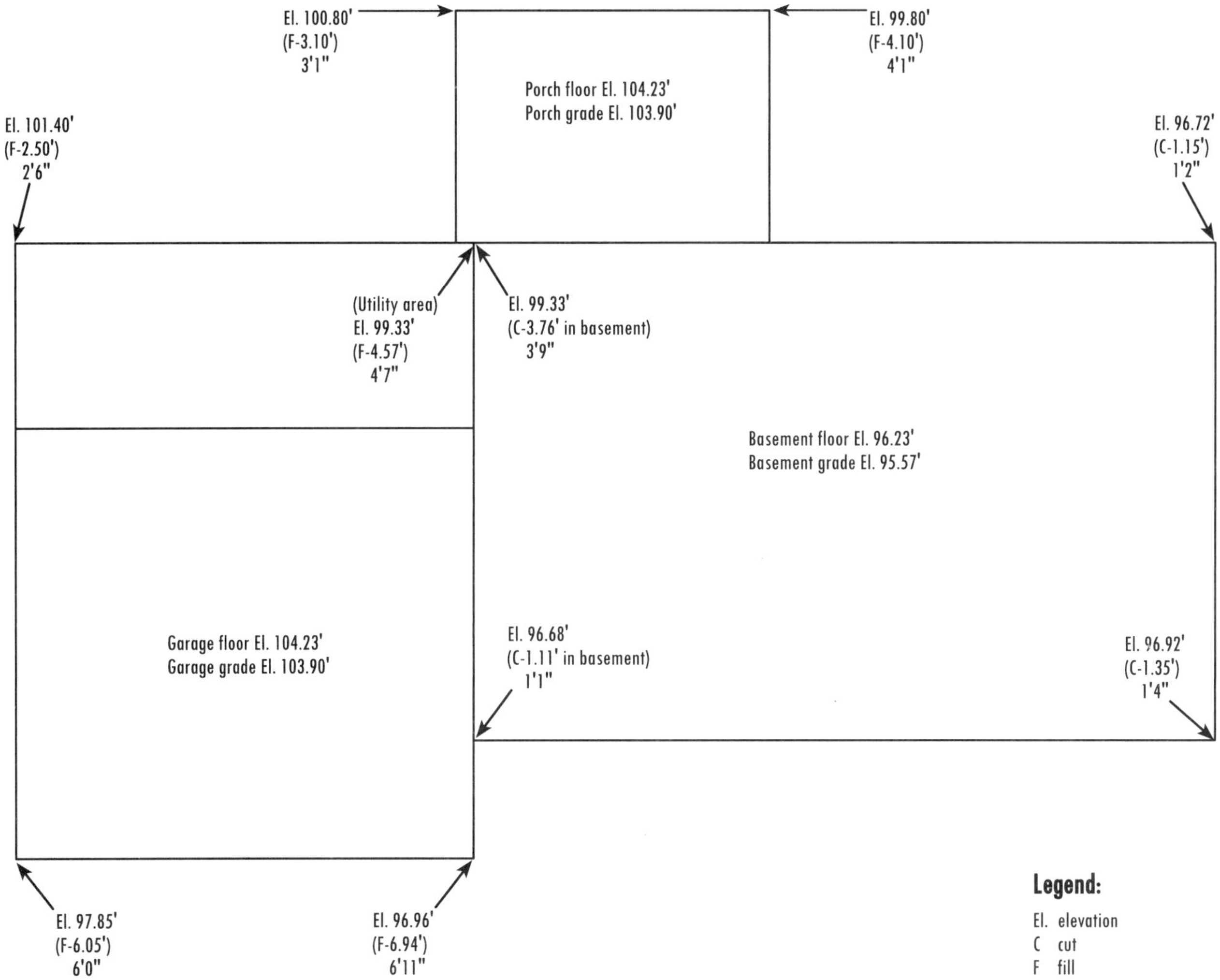

Figure 3-1
Foundation plan

Access roads should be graded, compacted and graveled, so trucks can deliver without getting stuck in the mud. Don't overlook these costs in your estimate. It's always a good idea to rough-grade the site before starting any footing excavation, so surface water can flow away from the footing area. Be prepared to store and protect building materials, and don't forget to account for the storage costs on your estimate.

Figure 3-1 is the foundation plan of a house staked out with dimensions from a plot plan. The existing grade elevations at each corner of the house have been taken, and the floor and grade elevations have been determined and recorded.

From this information and the building dimensions given in Figure 3-2, you can calculate the amount of earth to be excavated.

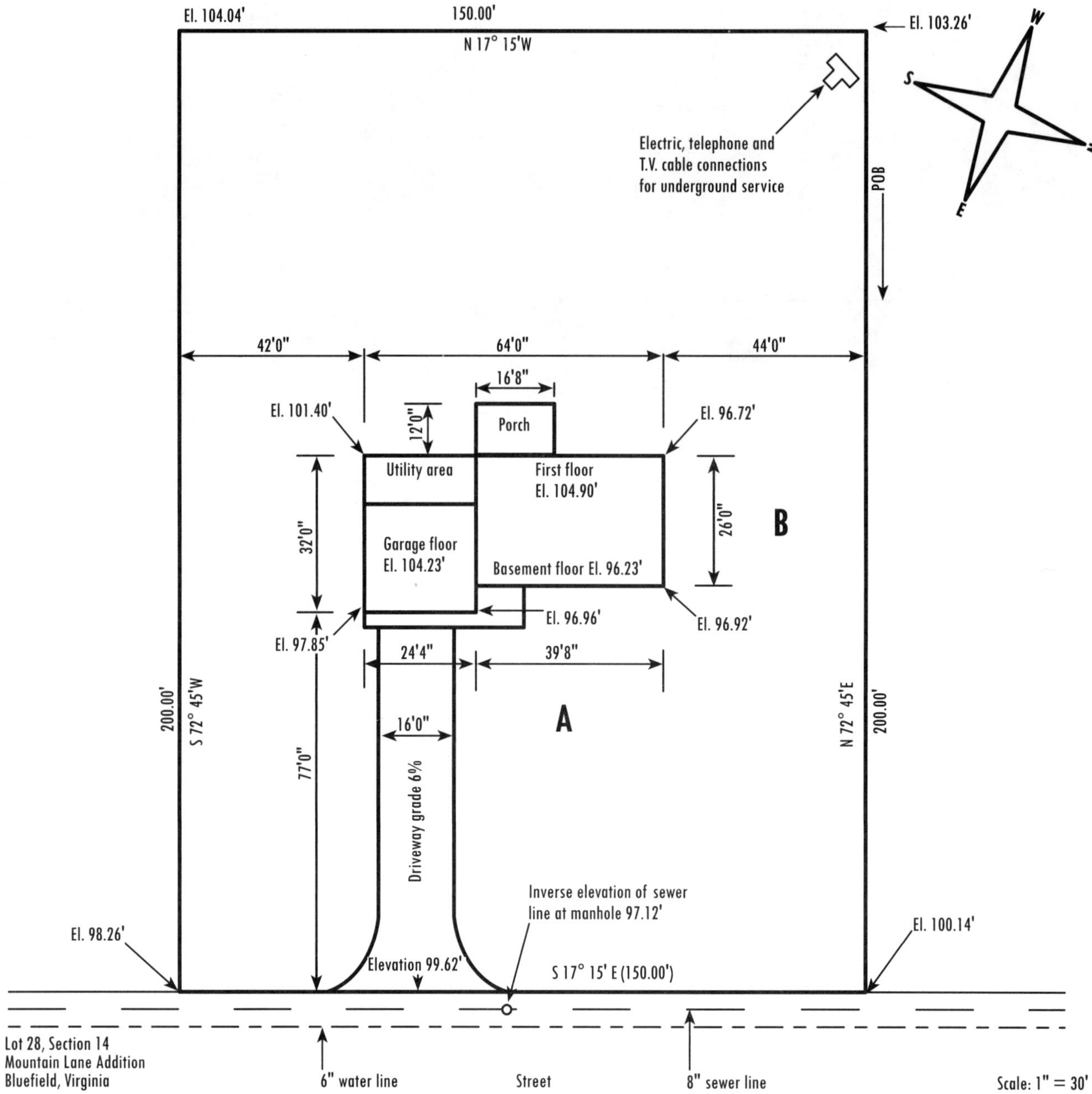

Figure 3-2
Grade and floor elevations

Always add an extra 2 feet for work space around the foundation perimeter. Calculate the number of cubic yards of earth to be excavated as follows:

1. Basement foundation wall dimensions from the blueprints, and shown on the plot plan in Figure 3-2, are 39′8″ × 26′0″ (indicated as A and B in the figure). Allowing 2 additional feet for working room around the foundation, basement excavation dimensions are:

$$[39'8'' + (2'0'' \times 2)] \times [26'0'' + (2'0'' \times 2)] = 43'8'' \times 30'0''$$

Whole Inches	Fractional Parts of an Inch							
	0	1/8	1/4	3/8	1/2	5/8	3/4	7/8
0	0.00	0.01	0.02	0.03	0.04	0.05	0.06	0.07
1	0.08	0.09	0.10	0.11	0.13	0.14	0.15	0.16
2	0.17	0.18	0.19	0.20	0.21	0.22	0.23	0.24
3	0.25	0.26	0.27	0.28	0.29	0.30	0.31	0.32
4	0.33	0.34	0.35	0.36	0.38	0.39	0.40	0.41
5	0.42	0.43	0.44	0.45	0.46	0.47	0.48	0.49
6	0.50	0.51	0.52	0.53	0.54	0.55	0.56	0.57
7	0.58	0.59	0.60	0.61	0.62	0.64	0.65	0.66
8	0.67	0.68	0.69	0.70	0.71	0.72	0.73	0.74
9	0.75	0.76	0.77	0.78	0.79	0.80	0.81	0.82
10	0.83	0.84	0.85	0.86	0.88	0.89	0.90	0.91
11	0.92	0.93	0.94	0.95	0.96	0.97	0.98	0.99

Figure 3-3
Decimal to fraction conversion table

2. In Figure 3-1, basement grade elevation is calculated as 95.57 feet. Grade elevation at the right back corner of the basement foundation is 96.72 feet. A cut of 1.15 feet is required here. A minus sign in the calculation result means that a *cut* is required.

   95.57' − 96.72' = −1.15' (1'1¾" rounded to 1'2")

Figure 3-3 shows decimals (hundredths) of a foot and equivalents in inches and fractions of an inch. You can use this table to convert 1.15 feet to inches and fractions of an inch. In this case, the decimal to be converted is 0.15. Find 0.15 in the body of the table. The heading above 0.15 shows that it's in the ¾-inch column. Now look to the left and you'll see 1. That tells you that 0.15 of a foot equals 1¾ inches. So 1.15 feet is equivalent to 1 foot 1¾ inches, rounded to 1 foot 2 inches, in our figure.

3. Grade elevation at the right front corner of the basement foundation is 96.92 feet. A cut of 1.35 feet (1′4¼″) is required.

   95.57' − 96.92' = −1.35' (1'4¼" rounded to 1'4")

4. The left front corner of the foundation adjacent to the garage area has a grade elevation of 96.68 feet. Here, a cut of 1.11 feet (1′1⅜″) will be needed.

   95.57' − 96.68' = −1.11' (1'1⅜" rounded to 1'1")

5. The left back corner of the foundation adjacent to the utility area has a grade elevation of 99.33 feet. This requires a cut of 3.76 feet (3′9⅛″).

   95.57' − 99.33' = −3.76' (3'9⅛" rounded to 3'9")

Excavation is only required for the basement area of this foundation. The other areas of the foundation (the garage section and utility area) need fill dirt, which will be discussed later in this chapter. To determine the cubic yards of earth to excavate for the foundation shown in Figure 3-1, do the following:

1. Average the depth of the four cuts for the basement area.

   $$\frac{1.15' + 1.35' + 1.11' + 3.76'}{4} = 1.84'\ (1'10\tfrac{1}{8}''\text{ rounded to }1'10'')$$

   These cuts average 1.84 feet (1′10″).

2. Using the conversion table in Figure 3-3, convert the length (43′8″) and width (30′0″) of the basement foundation measurements into their decimal equivalents of a foot. The results are 43.67 feet and 30.00 feet, respectively. We'll use those decimals for our calculations.

3. Multiply the basement excavation dimensions by the average cut depth and divide by 27 to convert cubic feet to cubic yards:

   $$L\ (\text{ft}) \times W\ (\text{ft}) \times D\ (\text{ft}) \div 27 = \text{cubic yards (cy)}$$

   $$\frac{43.67' \times 30.00' \times 1.84'}{27} = 89.28\text{ cy}$$

Estimate how much excavation is necessary to provide an access road and an area for storing the material. In this case, we'll need to excavate an additional 65 cubic yards of earth. We're going to round the basement excavation up to 90 cubic yards, so the total excavation will be:

Basement	90
All other	65
Total:	155 cubic yards

We can store the earth we excavate on the building site, so there's no hauling expense involved. No blasting allowance is necessary either, since there's no rock present. Finally, the house is scheduled to be completed before winter, so we don't need an allowance for cold weather.

If records of the excavation per cubic yard costs are available, they provide the most reliable cost estimate. If not, there's a cost for excavation with heavy equipment in the *National Construction Estimator.* We'll use a backhoe, an operator and one laborer, working in average soil, for our excavation.

*Enter your estimate for **excavation** and cost per cubic yard on Line 3.2 of your Cost Estimate Worksheet. We'll use 155 cubic yards on our Example Cost Estimate Worksheet.*

Fill Dirt

It's easy to make errors when estimating fill dirt, too. The quantity required may be even more difficult to figure than estimating excavation. Compaction can add as much as 25 percent to the yards needed, depending on soil-type and moisture content. When estimating fill dirt quantities, be sure to take into consideration present and finished grade elevations and the location of the structure on the building site. Elevations shown on the plot plan provide the essential information you need to make accurate calculations.

Never use organic material such as tree roots or stumps as fill material — settling could continue for years. It'd be a monumental mistake if you used it as fill under a concrete slab.

Compaction of fill dirt must be done in layers. On deep fills, compaction should be continuous, as the fill dirt is imported to the area. Fills that support concrete slabs should be compacted in 4- to 6-inch layers.

From grade and floor elevations, the dimensions shown in Figure 3-2, and the foundation plan in Figure 3-1 showing the cuts and fills, you can calculate the fill material you'll need.

Garage and Utility Area

Dimensions of the garage and utility area are 24′4″ × 32′0″. A fill will be required for the entire area, and must be compacted in 4- to 6-inch layers, since it'll support the concrete floor.

Garage and utility *floor* elevation has been calculated as 104.23 feet, as shown in Figure 3-2. The specifications call for a 4-inch concrete slab over the fill dirt. So the garage and utility area *grade* elevation will be 103.90 feet.

104.23' − 0.33' (4") = 103.90'

This area will have to be filled to reach the required elevation.

1. The elevation at the right front corner of the garage is 96.96 feet (Figure 3-1). The fill required will be 6.94 feet:

   103.90' − 96.96' = 6.94'

2. The elevation at the left front corner of the garage is 97.85 feet. The fill required here will be 6.05 feet:

   103.90' − 97.85' = 6.05'

3. The elevation at the left back corner of the utility area is 101.40 feet. The fill required here will be 2.50 feet:

   103.90' − 101.40' = 2.50'

4. The elevation at the right back corner of the utility area is 99.33 feet. The fill required will be 4.57 feet:

   103.90' − 99.33' = 4.57'

Add these four numbers and divide by 4 to find the average depth of fill in the garage and utility area:

$$\frac{6.94' + 6.05' + 2.50' + 4.57'}{4} = 5.02' \ (5'\tfrac{1}{4}")$$

The number of cubic yards of fill dirt is calculated by multiplying the area of the garage and utility area by the depth of fill, and dividing by 27 to convert cubic feet to cubic yards.

$$\frac{24.33' \ (24'4") \times 32.00' \ (32'0") \times 5.02' \ (5'\tfrac{1}{4}")}{27} = 144.75 \text{ cy}$$

Front Yard

The plans call for a walk from the front of the garage to the front porch. The elevation of this walk and the porch will be the same as the garage and utility floor elevation (104.23 feet), shown in Figure 3-2. The finished grade from the left front corner of the garage to the right front corner of the house will be the same, and will taper to the left and right front corners of the lot at the street. We can determine the amount of fill dirt required in the front yard, as follows:

1. Elevation at the left front corner of the *garage* is 97.85 feet. The fill required here will be 6.38 feet:

   $$104.23' - 97.85' = 6.38'$$

2. In Figure 3-1, you'll see that elevation at the right front corner of the *house* is 96.92 feet. The fill required here will be 7.31 feet:

   $$104.23' - 96.92' = 7.31'$$

3. In Figure 3-2, the elevation at the left front corner of the *lot* is 98.26 feet. No fill is required at this point.

4. The elevation at the right front corner of the *lot* is 100.14 feet. No fill is required at this point.

5. The average depth of fill in the front yard will be:

   $$\frac{6.38' + 7.31' + 0.00' + 0.00'}{4} = 3.42'$$

6. The front yard fill will taper from the front corners of the house to the front corners of the lot (in the shape of a trapezoid), so the average width of fill will be:

   $$\frac{\text{Width of house} + \text{Width of lot}}{2}$$

   $$\frac{64.00' + 150.00'}{2} = 107.00'$$

7. The average depth of the front yard from the house to the lot frontage is 80 feet. This is calculated as follows.

   The depth from left front of garage to the lot frontage is 77 feet. The depth from right front of the house to the lot frontage (77.00′ + 6.00′) is 83 feet.

$$\frac{77.00' + 83.00'}{2} = 80.00'$$

8. The volume of fill dirt needed for the front yard is calculated by multiplying the average width of the fill (107 feet) by the average depth of the front yard from the house to the lot frontage (80 feet) and the average depth of fill (3.42 feet), and dividing by 27 (to convert cubic feet to cubic yards).

$$\frac{107.00' \times 80.00' \times 3.42'}{27} = 1{,}084.27 \text{ cy}$$

9. The fill dirt for the rear porch and the side and back yards was calculated, and the amount of soil from excavation plus rough-grading in the back yard will take care of these areas.

10. The amount of fill dirt to be imported to the building site:

Garage and utility area	144.75 cubic yards
Front yard	1,084.27 cubic yards
Side and back yards	0.00 cubic yards
Subtotal:	1,229.02 cubic yards
Allow 25 percent more for compaction	307.26 cubic yards
Total amount of fill dirt:	1,536.28 cubic yards
	(rounded to 1,540 cubic yards)

The cost of fill dirt depends on the location and trucking distance from the source. (You may even be able to find a nearby job where they want to get rid of *excess* dirt. Then you should get the fill dirt free and only have to pay the trucking charge.) Your records will again be the best source of fill dirt cost in your local area. Always get a hauling quote before submitting your estimate, particularly if a large amount of fill dirt is required.

*Enter your estimate for **fill dirt** and cost per cubic yard on Line 3.3 of your Cost Estimate Worksheet. We'll use 1,540 cubic yards in our Example Cost Estimate Worksheet.*

The cost of spreading and shaping the dirt from loose piles delivered and dumped on site must also be estimated. Again, previous records are your best source of costs. If they're not available, use a published cost reference for an appropriate cost.

*Enter your cubic yards of fill dirt and cost per cubic yard **to spread** on Line 3.3.1 of your Cost Estimate Worksheet. We'll use 1,540 cubic yards in our Example Cost Estimate Worksheet.*

Site Clearing and Hauling

There's expense involved in disposing of debris, both during and after every construction project. A clean work site is a safe work site, so don't let waste material pile up and expect to remove it at the completion of the job. Biodegradable material such as scrap lumber, scrap roofing, and cement bags must be hauled to an authorized dump site. Toxic or nonbiodegradable material like waste from gypsum board and empty paint cans must be removed from the site and dealt with according to local regulations. Burning waste is illegal in most areas. Other items to be hauled away from the building site at the conclusion of the construction are equipment, scaffolding, and any leftover reusable material.

The expenses of cleaning the area and hauling these materials away are part of construction costs that shouldn't be overlooked when you make a cost estimate. The expense depends on the overall size and cost of the project. You may have to complete your entire cost estimate before you can estimate the site cleanup costs. Remember, experience and previous records help you estimate. If you need an additional source for information, monthly jobsite cleanup and debris removal costs can be found in the *National Construction Estimator*.

Enter your project ***months*** *on Line 3.4 of your Cost Estimate Worksheet.*
We'll use 9 months in our Example Cost Estimate Worksheet.

COST ESTIMATE WORKSHEET FOR SITE CLEARING, EXCAVATION AND FILL DIRT

#	Item	Qty		Cost	Per		Subtotal
3.1	**Site clearing**						$3,000.00
3.2	**Excavation**	155 cy	×	$5.01	cy	=	$780.00
3.3	**Fill dirt**	1,540 cy	×	$5.00	cy	=	$7,700.00
3.3.1	**Fill dirt spreading**	1,540 cy	×	$1.07	cy	=	$1,650.00
3.4	**Site cleaning and hauling**	9 months	×	$300.00	month	=	$2,700.00
	Total site clearing costs (entered on line 3 of Form 100)						**$15,830.00**

Chapter 4
Footings

FOOTINGS ARE MEANT TO support a house as long as it stands, so they must be designed and built for a minimum of settling. In residential construction, soil bearing analysis is seldom required, but conditions vary, so determine the type of soil present on your site before you estimate footings. Follow the footing design on the blueprint and adhere to local building code requirements. The minimum width of a footing must be whatever width will safely support the total design load without significant settling.

An architect who draws plans for a house on a particular building site also designs the footings for that specific site. Standard plans from a house building-plan service may show footings that would support the house under normal soil conditions. But local zoning codes may require wider and deeper footings due to soil conditions that aren't the norm. For this reason, many ready-made blueprints don't give footing specifics, leaving it to the builder to meet local codes.

Footings must be poured on a solid surface. Pouring on soft or wet soil will cause cracks in the foundation wall. Footings for detached garages or porches may be poured later, but your estimate should still include the cost.

Footing construction estimates affect the cost of a house. If stepped footings are required, the cost increases. Adverse soil conditions can affect labor and material costs. This is another reason why square foot and cubic foot cost estimates can only be approximate. Conduit, under or through the footing, may be required for the water, sewer, electricity, gas, telephone and cable utility service lines. These are installed before pouring the footings. Don't forget these costs in your estimate.

Construction during the winter can also increase the cost of footings. In inclement weather, a concrete truck may not be able to get to the site, so a front end loader would need to haul the concrete from the truck to the site. In freezing weather, the concrete should be poured as soon as possible after excavation, and be protected against freezing until cured. Calcium chloride, a water-absorbing chemical, can be added to concrete to accelerate the curing. Be careful not to add too much calcium chloride or you'll weaken the concrete. In freezing weather, use approximately two pounds of calcium chloride per bag of cement. In extremely cold temperatures, further adjustments may be necessary.

Layout

Once excavation is completed, lay out the foundation to determine the permanent corners before you start forming up the footings. Even if you've previously marked a rough layout of the foundation for excavation, reference stakes rarely stay in place.

The cost of laying out the corners (setting the forms and grade stakes and leveling them) will vary from one job to the next. The cost can be determined more accurately if the job is contracted out, or when you have previous job records on which to base your estimate.

> *Enter your estimate for **layout** on Line 4.1 of your Cost Estimate Worksheet. We'll use $3,000 in our Example Cost Estimate Worksheet.*

Estimating Concrete Quantity

Estimating concrete for footings before construction begins is easy on paper, but the quantity calculated from the blueprints and what you actually need could be different. True, the dimensions of the foundation won't change, but it's easy to forget the following:

- Excavation is frequently wider than the specified footing width (the footing should never be less than what's specified). Concrete forms can be kept to a true dimension, but accuracy is difficult when you're excavating, whether by machine or by hand.
- Excavating to the exact depth is also difficult. The depth of concrete should never be less than the specifications call for. If excavation ends up deeper than required, it must be filled with concrete, not loose dirt.
- When stepped footings are required, the amount of concrete used in the vertical rise can easily be misjudged.

The factors just mentioned are regularly overlooked in estimating concrete for footings. Another problem area that's sometimes miscalculated or not considered is concrete spills, either from the truck during unloading or from wheelbarrows carrying the concrete to its final location. Some of this concrete can be salvaged, but much of it is wasted. It's generally cheaper to pay for a little extra concrete in the beginning than pay the extra cost for a small batch that's specially prepared to complete the job.

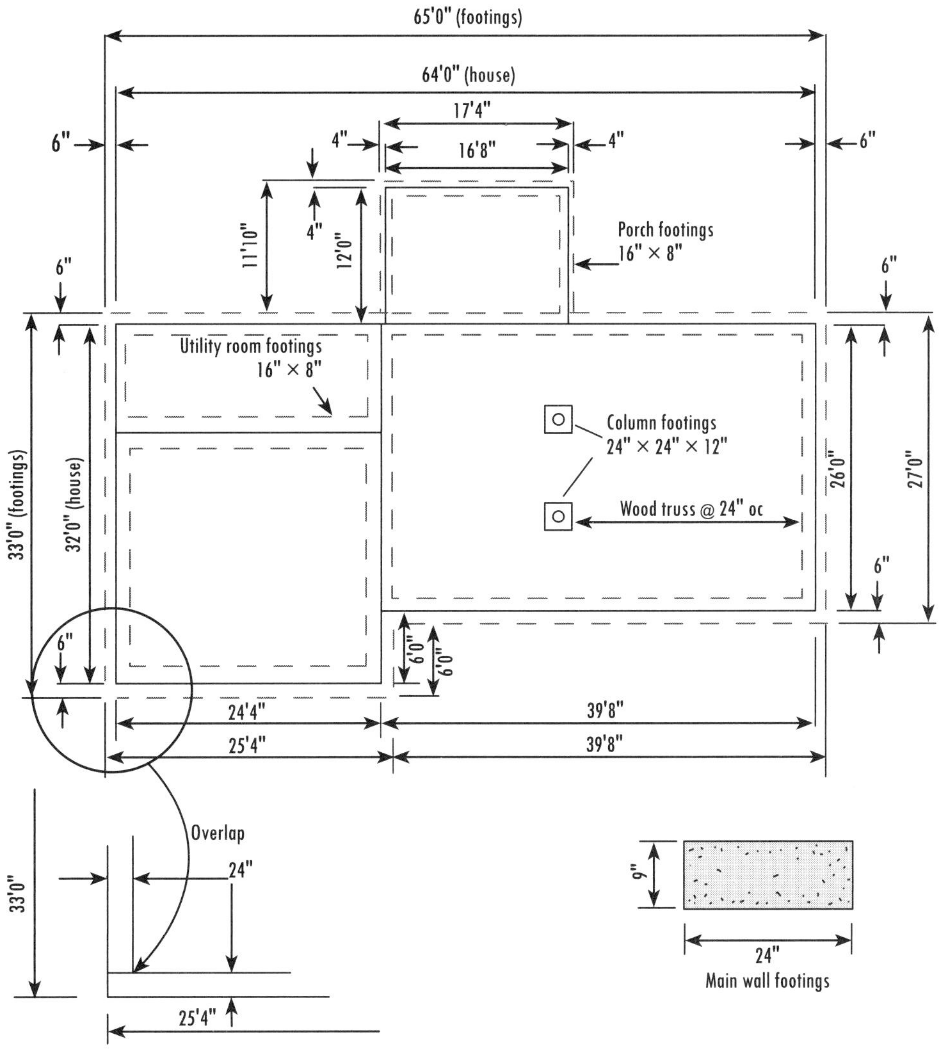

Figure 4-1
Foundation footing plan

Most ready-mix concrete companies have a minimum delivery charge. A small amount of concrete in a special delivery can cost as much as three or four cubic yards brought during the regular delivery. Consequently, estimators have gotten creative in order to offset waste. One common solution is to double the amount of concrete needed for your corners. Rather than measuring the length of a foundation from the *inside* point of the foundation at a right angle corner, measure from the *outside* point. That way, you'll have enough concrete to fill the corner twice over.

House Footing Calculations

The amount of concrete required for our house footings can be estimated using Figures 4-1, 4-2, and 4-3. The dimensions of the footings, except in the special areas indicated (porch, utility room and column footings), are

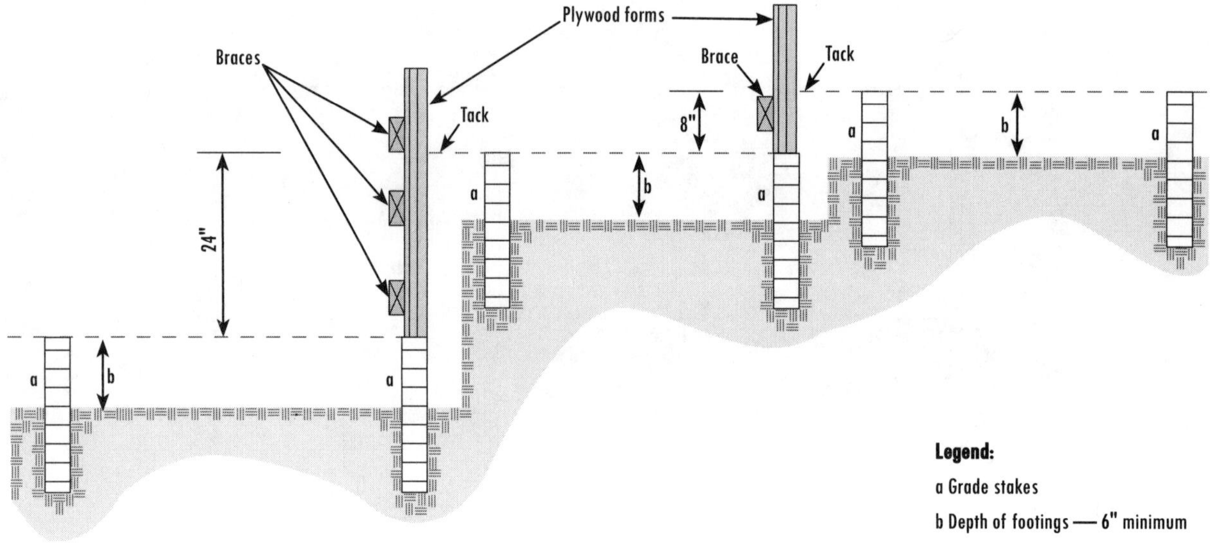

Figure 4-2
Construction of stepped footing form

24 inches wide and 9 inches in depth. The total linear feet of the main wall footings (disregarding the corner overlaps), as shown in Figure 4-1, is:

Left side of basement	27′0″
Left side, garage and utility room	33′0″
Back	65′0″
Right side (27′0″ + 6′0″ of garage offset)	33′0″
Front	65′0″
Total:	223′0″

The amount of concrete to be estimated is:

$$\frac{223.00'(223'0'')\times 2.00'(24'')\times 0.75'(9'')}{27(\text{cf to cy})} = 12.39\text{cy}$$

The footings for the back porch and utility area are 16 inches wide × 8 inches deep. The concrete required for these footings will be:

Back porch (11′10″ + 17′4″ + 11′10″)	41′0″
Utility area (front wall)	24′4″
Total:	65′4″

$$\frac{65.33'(65'4'')\times 1.33'(16'')\times 0.67'(8'')}{27} = 2.16\text{cy}$$

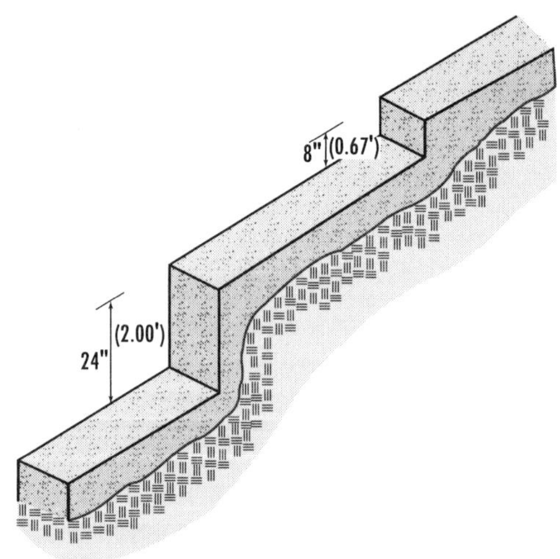

Figure 4-3
Stepped footing

The blueprints show two column footings in the basement, each with dimensions of 24″ × 24″ × 12″. The concrete for these will be:

$$\frac{2.00'(24'')\times 2.00'(24'')\times 1.00'(12'')\times 2(\text{columns})}{27} = 0.30\text{cy}$$

Stepped Footing Calculations

Before completing your estimate of the quantity of concrete required for the footings, check the grade elevations on the foundation plan (see Figure 3-1, Chapter 3) to determine if stepped footings will be required. Figures 4-2 and 4-3 show examples of stepped footing construction. If stepped footings are needed, you can estimate the concrete required in the vertical rise as follows:

- In Figure 3-1, the basement grade elevation is shown as 95.57 feet and the grade elevation at the left rear corner of the utility area is 101.40 feet. The garage and utility areas require fill dirt, and the footing here will start from the present grade elevation. This is a rise of:

  $$101.40' - 95.57' = 5.83'\ (5'10''\ \text{or}\ 70'')$$

- Several steps, in multiples of 8 inches, will have to be constructed from the left front corner of the basement (adjacent to the garage area) around the perimeter of the front and left side of the garage, to the left rear corner of the utility area. The sum of the vertical rise for these steps, in multiples of 8 inches, is 72 inches (since 70 inches is not a multiple of 8 inches).

The required amount of concrete for the vertical rise portion of the stepped footing will be 24 inches (width of footings) × 6 inches (thickness of footings) × 72 inches (vertical rise) × 2 (front and back perimeter) divided by 27 (to convert cubic feet to cubic yards):

$$\frac{2.00'(24'') \times 0.50'(6'') \times 6.00'(72'') \times 2}{27} = 0.44\text{cy}$$

This isn't a lot of concrete, but it could be costly if you didn't estimate it in the beginning, and the concrete truck has to make a special trip to bring it to the jobsite.

So the total amount of concrete required for the footings will be:

Main walls	12.39 cubic yards
Porch and utility area	2.16 cubic yards
Column footings	0.30 cubic yard
Vertical rise in stepped footings	0.44 cubic yard
Subtotal	15.29 cubic yards
Add 5 percent for waste	0.76 cubic yard
Total:	16.05 (rounded to 16*) cubic yards of concrete

*As a rule, decimals of material are rounded up to insure adequate coverage. In this situation, .05 is so small, we'll round down.

House Footing Estimates Using Table Factors

You can also use table factors as an alternative to doing the math. Figure 4-4 provides information to estimate concrete quantities for various footing depths and widths. By multiplying the factor in the footing width column (where it intersects with the footing depth row) by the total linear feet of the footings, you can figure the cubic yards of concrete required. Let's look at a few examples.

Factors for Concrete Volume										
Footing Depth	Footing Width									
	12"	14"	16"	18"	20"	22"	24"	26"	28"	30"
6"	0.01852	0.02160	0.02469	0.02778	0.03086	0.03395	0.03704	0.04012	0.04321	0.04630
7"	0.02160	0.02521	0.02881	0.03241	0.03601	0.03961	0.04321	0.04861	0.05041	0.05401
8"	0.02469	0.02881	0.03292	0.03704	0.04115	0.04627	0.04938	0.05350	0.05761	0.06173
9"	0.02778	0.03241	0.03704	0.04167	0.04630	0.05093	0.05556	0.06019	0.06481	0.06944
10"	0.03086	0.03601	0.04115	0.04630	0.05144	0.05658	0.06173	0.06687	0.07202	0.07716
11"	0.03395	0.03961	0.04527	0.05093	0.05658	0.06224	0.06790	0.07356	0.07922	0.08488
12"	0.03704	0.04321	0.04938	0.05556	0.06173	0.06790	0.07407	0.08025	0.08642	0.09259
13"	0.04102	0.04681	0.05350	0.06019	0.06687	0.07356	0.08025	0.08693	0.09362	0.10031
14"	0.04321	0.05041	0.05761	0.06481	0.07202	0.07922	0.08642	0.09362	0.10082	0.10802
15"	0.04630	0.05401	0.06173	0.06944	0.07716	0.08488	0.09259	0.10031	0.10802	0.11574

Figure 4-4
Factors for concrete volume

The total linear feet for the main wall footings in our example is 223 feet. The footings are 24 inches wide × 9 inches deep. In Figure 4-4, go across the footing width line to the 24″ column. In the 24″ column, go down to the 9″ footing depth row. Where the column and row intersect, you'll find 0.05556 (solid outlined). This is the factor you'll use for 24-inch-wide × 9-inch-deep footings. So, the number of cubic yards of concrete required for the main wall footings will be:

223' × 0.05556 = 12.39 cy

The total linear feet for the back porch and utility area footings is 65.33 feet (65′4″). The footings are 16 inches wide × 8 inches deep. In Figure 4-4, the factor for 16″ wide × 8″ deep is 0.03292 (shown with dashed outline). Multiply this factor by the total linear feet to get the number of cubic yards of concrete required:

0.03292 × 65.33' (65'4") = 2.15 cy

The two column footings are each 24 inches wide, 12 inches in depth and 24 inches in length. How much concrete will we need for these footings? First find the amount needed for one, then multiply it by 2. The factor for 24″ wide × 12″ deep footings is 0.07407 (double outlined in Figure 4-4). So the concrete needed for two column footings is:

0.07407 × 2.00' (24") = 0.15 cy each

0.15 × 2 = 0.30 cy for both

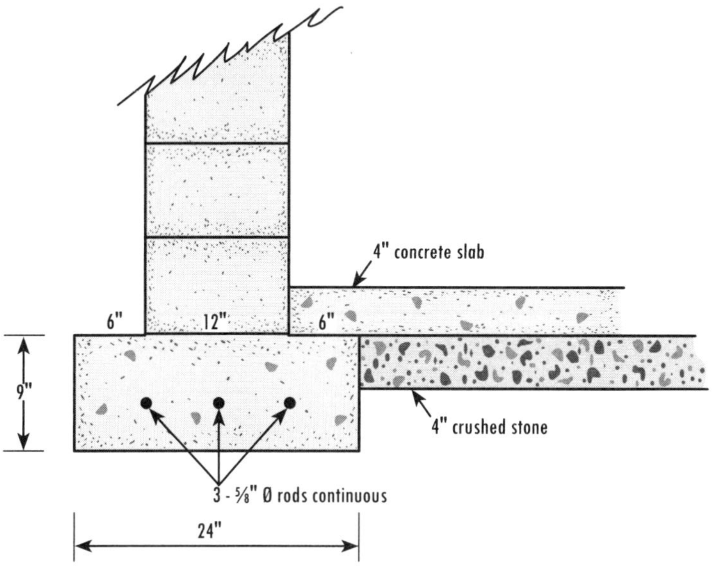

Figure 4-5
Continuous reinforcing rods in footing

Footings Wider Than 30 Inches

If footings are wider than the maximum width (30 inches) shown in Figure 4-4, a chimney footing for instance, two factors can be added together to make the width you need. Let's calculate the concrete we'd need for a chimney footing.

The chimney's footings are 3 feet (36 inches) wide, 6 feet long, and 12 inches deep. For the 3-foot (36-inch) width footing factor, simply add the factors for 20-inch and 16-inch footings:

The factor for 20″ × 12″ is 0.06173
The factor for 16″ × 12″ is 0.04938
The factor for 36″ × 12″ is 0.11111
Multiply by the 6-foot footing length:

$$0.11111 \times 6.00' = 0.67 \text{ cy}$$

> **Note:** When calculating large quantities of concrete, there may be a slight difference between the result you get when using factors and using the formula shown in our last calculation. I recommend using the formula to calculate cubic yards. Figure 4-4 should only be used to get a ballpark figure for larger quantities.

Check your work by multiplying the dimensions and dividing by 27 to get cubic yards:

$$\frac{3.00' (36") \times 6.00' \times 1.00' (12")}{27} = 0.67 \text{ cy}$$

If you've kept records of concrete costs on your jobs, they'll usually give you the most accurate numbers. If not, the costs of concrete footings are listed in the current *National Construction Estimator.*

> *Enter the **footing** dimensions on line 4.2 of the Cost Estimate Worksheet. We'll use the footing dimensions we've just calculated on our Example Cost Estimate Worksheet.*

Estimating Reinforcing Rod

The size and number of reinforcing rods for our footings is shown in Figure 4-5. Local zoning regulations can require the size and number of rods be increased in adverse soil conditions. The greater of the two requirements — blueprint or zoning code — must be used.

Reinforcing rods are sized by their diameter in inches (e.g. ½-inch diameter rod), or by numerals that identify them (e.g. #4 rod). The bar numbers are based on 1/8-inch increments in the diameter of the rod. So a #4 diameter rod is 4 × 1/8 inch or ½ inch in diameter, and a #3 diameter rod is 3 × 1/8 inch or 3/8 inch in diameter (Figure 4-6). Usually, #5 (5/8-inch diameter) reinforcing rods are used in footing construction.

Figure the total linear feet of reinforcing rods needed for the footings in our sample house, then add 10 percent to allow for overlap.

Reinforcing Rods

Rod Size		Weight
Diameter in Inches	**Rod Number**	**Pounds per Foot**
¼"	2	0.167
⅜"	3	0.376
½"	4	0.668
⅝"	5	1.043
¾"	6	1.502
⅞"	7	2.044
1"	8	2.670

Note: If the reinforcing rods are sold by weight, compute the total weight and cost as follows:

1. Multiply the total number of linear feet by the weight per foot

Example: The weight of 940 linear feet of ½" diameter rod is 940 x 0.668 = 627.92 lbs

2. Total weight (rounded off) multiplied by the rate = cost

Figure 4-6
Table of rod numbers and weights

Figure 4-1 shows the foundation footing plan for the house. Add the dimensions to get the total linear feet of the foundation perimeter:

Left	33′0″
Back	65′0″
Right (27′0″ + 6′0″ offset)	33′0″
Front:	65′0″
Utility room footing	24′4″
Back porch [(2 × 11′10″) + 17′4″]	41′0″
Column footing (2 × 24″)	4′0″
Vertical rise in steps (2 × 72″)	12′0″
Total:	277′4″

> **Note:** The *National Construction Estimator* gives costs that include tie wire for splicing, with 10 percent already added for overlapping, so a 10 percent deduction has been made to the cost of reinforcing rods in our Example Cost Estimate Worksheet.

Since we know that three continuous rods are required, multiply the total linear feet by 3, then add in the additional 10 percent for overlap to find the total number of reinforcing rods required:

3 × 277.33' (277'4") = 831.99'
831.99' × 0.10 = 83.20'
915.19' (rounded to 920) linear feet

*Enter your estimate of the **linear feet and size of reinforcing rod** required on line 4.3.1 of the Cost Estimate Worksheet. We'll use 920 linear feet of reinforcing rod #5 (5/8-inch diameter) in our Example Cost Estimate Worksheet.*

Estimating Forms for Footings

Build the footing forms for the basement area as shown in Figures 4-7 and 4-8. The top of the forms should be at the required footing elevation. Using 2 × 4 framing to form the footings, the crushed stone under the concrete floor will be level with the top of the footings (see Figure 4-5 and Figure 4-9).

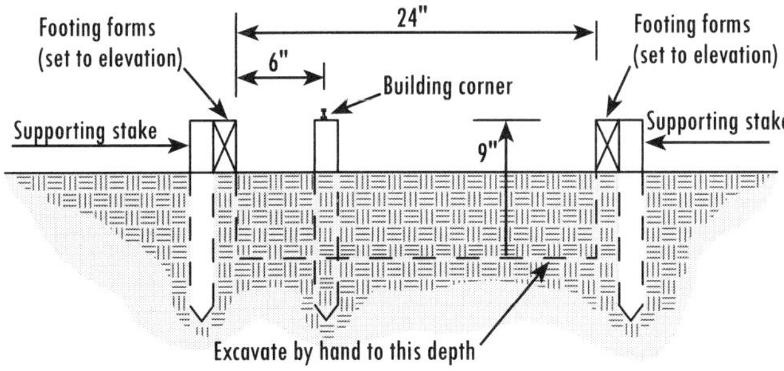

Figure 4-7
Forms for footing before excavation

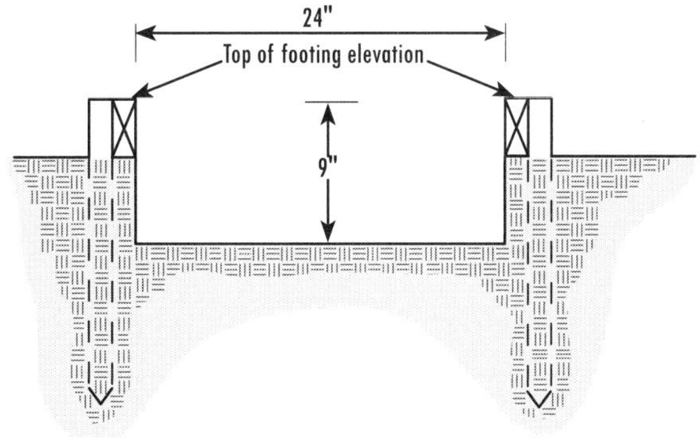

Figure 4-8
Forms for footing after excavation

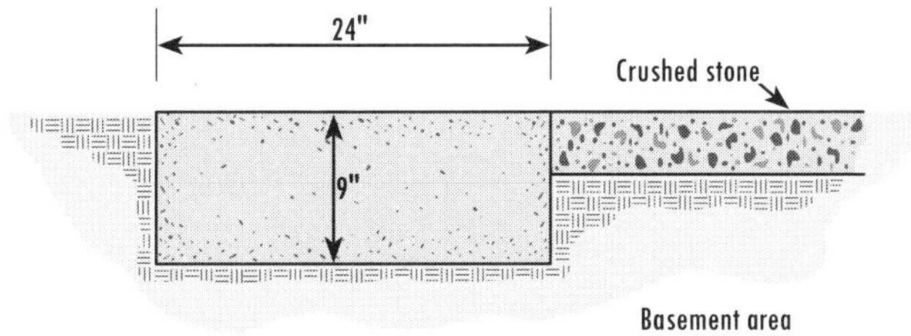

Figure 4-9
Crushed stone and footing

The perimeter of the basement footing from the plans and Figure 4-1 is:

Front	39′8″
Left side	27′0″
Back	39′8″
Right side	27′0″
Total perimeter of the basement footing	133′4″ (133.33′)
2 column footers [2 × (4 × 24″)]	16′0″
Total:	149′4″
	(149.33 rounded up to 150 lf)

Forms are estimated by the square foot, so multiply the linear feet we just calculated by 2 (for the two sides of the footing), then multiply that by the height of the framing material (4 inches). The square footage of framing material required is:

$$150.00' \times 2 \times 0.33' (4") = 99 \text{ (rounded to 100) sf}$$

Forms can be reused on other jobs, so only about 30 percent of the cost of form material is charged to any one job. We'll use a third of the cost (33.33 percent) of the forms in our estimate.

*Enter your estimate of **linear feet and height of basement footing** on line 4.4.1 of the Cost Estimate Worksheet. We'll use 150 feet of 4-inch high material on our Example Cost Estimate Worksheet.*

Estimating Supporting Stakes for Forms

Use 20-inch-long reinforcing rods as supporting stakes for the forms. Place them every 5 feet, plus one starter stake for each end and one for each column face. Here's how you calculate the number of supporting stakes.

The perimeter of the basement footing is 133.33 feet. Multiply by 2 for both sides of the footing, then divide by 5 (one stake every 5 feet):

$$133.33' \times 2 = 266.67' \text{ of forms}$$

$$\frac{266.67'}{5'} = 53.33 \text{ supporting stakes}$$

Now add one stake for each starting point (4), and one stake for each column face (2 × 4 = 8):

$$53.33 + 4 + 8 = 65.33 \text{ (rounded to 66) supporting stakes}$$

Each stake is 20 inches long, so the total length of reinforcing rod required will be:

$$\frac{66 \times 20''}{12} = 110 \text{ lf}$$

Stakes can also be reused, so only 30 percent or so of the total cost of the stakes is usually charged to any one job. We'll divide the linear feet by 3 for the linear feet of stakes to be charged to this job:

$$\frac{110 \text{ lf}}{3} = 36.67 \text{ lf}$$

*Enter your estimate of the **linear feet of reinforcing rod** required for stakes on line 4.3.2 of Cost Estimate Worksheet. We'll use 36.67 linear feet in our Example Cost Estimate Worksheet.*

Estimating Grade Stakes

When the forms for the footings are constructed, the top of the forms should be at the calculated footing elevation. When there are no forms, grade stakes cut 20 inches long from reinforcing rods can be used (Figure 4-10). The tops of the stakes should be set to the footing elevation. We'll use grade stakes in the garage and utility area of our house. The perimeter of the footings was determined from the plans, and can be seen in Figure 4-1.

Front	25′4″
Left side	33′0″
Back	25′4″
Right side (basement)	0′0″
Total:	83′8″ (83.67′)

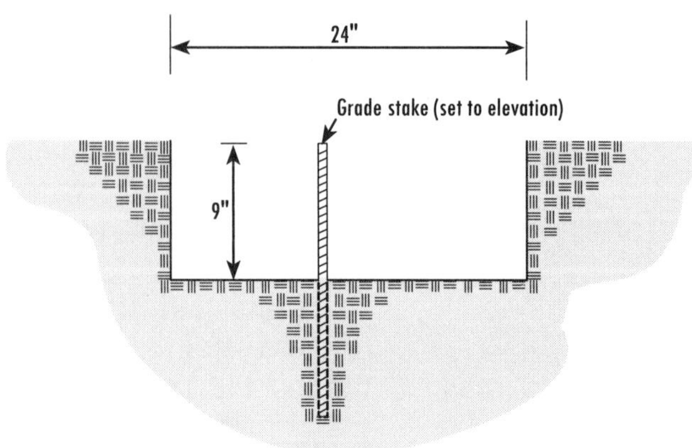

Figure 4-10
Grade stakes when forms aren't used

We need one 20-inch grade stake for every 5 feet of footing, plus one starter on each side of three sides of the garage and utility area. (The one side in common with the basement doesn't need grade stakes.)

The perimeter of the basement footing is 83.67 feet. One stake every 5 feet requires:

$$\frac{83.67'}{5} = 16.73 \text{ stakes}$$

Add in one starter stake for each side:

$$16.73 + 3 = 19.73 \text{ (rounded to 20) grade stakes}$$

If each stake is 20 inches long, the total length of rod required is:

$$\frac{20 \times 20''}{12} = 33.33 \text{ lf}$$

Note: Grade stakes are embedded in the concrete, so they can't be reused like the supporting stakes we discussed. As a result, the full cost of grade stakes must be charged to each job.

*Enter your estimate of the **linear feet of reinforcing rod required for grade stakes** on line 4.3.3 of Cost Estimate Worksheet. We'll use a figure of 33.33 feet in our Example Cost Estimate Worksheet.*

Estimating Particleboard for Stepped Footings

Particleboard is called out for the stepped footings (Figure 4-2). Estimate 4 sheets of ½″ × 4 × 8 for each cubic yard of concrete needed for the vertical rise of the stepped footings, which we determined to be 0.44 cubic yard. We'll estimate 2 pieces of ½″ × 4 × 8 particleboard. These sheets are cut to size for the particular job, so they can seldom be reused. Consequently, you'll charge the full amount to each job for these stepped footing sheets.

*Enter your estimate of the **number of ½″ × 4 × 8 sheets** required on line 4.5 of Cost Estimate Worksheet. We'll use 2 sheets of ½″ × 4 × 8 in our Example Cost Estimate Worksheet.*

Estimating Bracing for Stepped Footings

Braces are needed to support the vertical rise in the stepped footings around the garage and utility areas. You'll usually be able to use 2 × 4 × 12′ forms from previous jobs, so charge only a portion of the cost to any one job. We'll again use 33.33 percent of the cost in our estimate. Twelve 2 × 4 × 12′ forms are needed. At 12 × 12′, we'll need 144 linear feet.

*Enter your estimate of the **linear feet of form material** required on line 4.4.2 of Cost Estimate Worksheet. We'll use 144 linear feet in our Example Cost Estimate Worksheet.*

Estimating Labor Costs

Material can be estimated pretty accurately. But labor isn't as easily determined, and a serious labor estimating error can be the difference between a contractor making and losing money on a project. Labor productivity varies with every craftsman and from day to day. Guidelines are published for estimating labor costs on each phase of building a house, but they're only guidelines. The most accurate method of estimating labor costs, as I've stressed, is using your payroll sheets from previous jobs for estimating labor costs on similar jobs. Allowances must be made in any cost estimate for labor in less than normal conditions. Winter weather or temperatures above 95 degrees can increase labor costs from 10 to 15 percent. That's where accurate cost records from previous jobs come in handy.

First, calculate how much labor is needed to excavate the footings. This is relatively simple, since the amount of dirt excavated is equal to the amount of concrete that takes its place. We calculated 16 cubic yards of concrete. If you need to hand-excavate footings, you'll find typical labor costs in the current *National Construction Estimator*.

Enter your estimate of the ***cubic yards to be excavated*** *on line 4.6.1 of Cost Estimate Worksheet. We'll use 16 cubic yards in our Example Cost Estimate Worksheet.*

Next, calculate the cost of labor to build footing forms, pour the concrete, and strip the forms after the material sets up, using the dimensions shown in Figure 4-1. If you have records from previous jobs, refer to time sheets of recent similar construction, as in Figures 4-11 and 4-12.

The daily log at the bottom of these time sheets shows that form work started on Thursday, August 19th, and was completed on Wednesday, August 25th. The concrete was poured on Thursday, August 26th (an extra man was needed because some of the concrete had to be placed with wheelbarrows). On Friday, August 27th, two men stripped the forms from the footings.

A total of 145 manhours, beginning Thursday, August 19th, through Friday, August 27th, was required to form, hand-excavate, pour and strip the forms for 13.91 cubic yards of concrete (rounded to 14), including the stepped footings.

$$\frac{\text{Total manhours}}{\text{cy}} = \text{Manhour factor}$$

$$\frac{\text{145 manhours}}{\text{14cy}} = 10.36\ \text{(manhour factor)}$$

Weekly Time Sheet

For Period Ending: 8/21 **Job:** Brown Family

	Name	Exemptions	Month: August							Hours Worked		Rate	Total Earnings
			Day	M	T	W	T	F	S				
			Date	16	17	18	19	20	21	Regular Hours	Overtime Hours		
			Hours										
1	Doug White			X	X	X	8.00	8.00	X	16.00			
2	Robert Kidd			X	X	X	8.00	8.00	X	16.00			
3	Ray Farlow			X	X	X	8.00	8.00	X	16.00			
4													
5													
6													
7													
8													
9													
10													
11													
12													
13													
14													
15													
16													
17													
18													
19													
20													

Daily Log

Monday -------

Tuesday -------

Wednesday -------

Thursday Forms for footings in basement area

Friday Forms for footings in basement area

Saturday ------

Figure 4-11
Time sheet for basement footings

Use the following formula to compute labor costs:

1. 16 cubic yards of concrete are needed for the footings shown in Figure 4-1. The manhours estimated to build the forms, pour the concrete, and strip the forms are:

    cubic yards × manhour factor = total manhours

    16 cy × 10.36 (manhour factor) = 165.76 manhours

Weekly Time Sheet

For Period Ending: 8/28 Job: Brown Family

			Month: August							Hours Worked				
	Name	Exemptions	Day	M	T	W	T	F	S	Regular Hours	Overtime Hours	Rate	Total Earnings	
			Date	23	24	25	26	27	28					
			Hours											
1	Doug White			8.00	8.00	6.00	6.50	2.50	X	31.00				
2	Robert Kidd			8.00	8.00	6.00	6.50	2.50	X	31.00				
3	Ray Farlow			8.00	8.00	6.00	6.50	X	X	28.50				
4	Nick Noel			X	X	X	6.50	X	X	6.50				
5														
6														
7														
8														
9														
10														
11														
12														
13														
14														
15														
16														
17														
18														
19														
20														

Daily Log

Monday Forms for stepped footings in garage area

Tuesday Forms for stepped footings in garage area

Wednesday Finished stepped footings - ready for concrete

Thursday Poured concrete for footings (14 cu yds)

Friday Stripped forms from footings

Saturday ------

Figure 4-12
Time sheet for garage footings

2. Three workmen will be assigned to this job, a cement mason and two laborers. The time allotted for each man is:

$$\frac{\text{Total manhours}}{\text{Number of workmen}} = \text{time per man}$$

$$\frac{165.76\text{ hrs}}{3} = 55.25\text{ (rounded to 56) hours per worker}$$

> **Note:** Labor rates include FICA, FUTA, workers' comp, and liability insurance costs.

> *Enter your estimate of the* ***labor hours*** *and* ***labor rates*** *on line 4.6.2 of Cost Estimate Worksheet. We'll use 56 hours for a cement mason, CF, and two laborers, BL, in our Example Cost Estimate Worksheet.*

COST ESTIMATE WORKSHEET FOR FOOTINGS

#	Item	Qty / Size / Cost / Per	Subtotal
4.1	**Layout**		$3,000.00
4.2	**Concrete**		
4.2.1	**Wall footings**		
	Size:	223.00 lf × 2.00 width (ft) × 0.75 depth (ft) = 334.50 cf	
	Size:	65.33 lf × 1.33 width (ft) × 0.67 depth (ft) = 58.22 cf	
4.2.2	**Column footings**	number of columns 2	
	Size:	2.00 width (ft) × 2.00 length (ft) × 1.00 depth (ft) = 8.00 cf	
4.2.3	**Vertical rise for stepped footings**	number of rises 2	
	Size:	2.00 width (ft) × 0.50 thick (ft) × 6.00 rise (ft) = 12.00 cf	
4.2.4	**Chimney footings**	number of chimneys 1	
	Size:	3.00 width (ft) × 6.00 length (ft) × 1.00 depth (ft) = 18.00 cf	
		Total 430.72 cf	
		Total cf 430.72 ÷ 27 = 15.95 cy	
		Add 5% for waste 0.80 cy	
		Total concrete required = 16.75 cy @ $104.00 cy =	$1,742.00
4.3	**Reinforcing rods**		
4.3.1		920.00 lf for footings	
4.3.2		36.67 lf for supporting forms — Assume 3 uses	
4.3.3		33.33 lf for grade stakes	
	Total	990.00 lf — # 5 size rods @ $0.45 lf =	$445.50
4.4	**Forms for footings (2")**		
4.4.1		150 lf — 2 sides of footings = 300 lf	
		300 lf — 4 " framing height = 100.00 sf	
4.4.2		144 lf — 4 " framing height = 48.00 sf	
		Assume 3 uses — 148.00 sf @ $1.01 sf =	$149.48
4.5	**Particleboard for stepped footings**		
		2 ½" × 4' × 8' sheets @ $16.10 sheet =	$32.20
		Subtotal	$2,369.18
		Sales tax @ 7.75%	$183.61
		Total material cost for footings	$2,552.79
4.6	**Labor**		
4.6.1	**Footings excavation**	16.00 cy of footings @ $45.29 cy =	$724.64
4.6.2	**Forms, concrete pouring and stripping**		
		56.00 manhours - cement mason, CF @ $30.44 hour =	$1,704.64
		56.00 manhours - laborer, BL @ $26.64 hour =	$1,491.84
		56.00 manhours - laborer, BL @ $26.64 hour =	$1,491.84
		Total labor cost for footings	$5,412.96
		Cost of footings (entered on line 4 of Form 100)	**$10,965.75**

Chapter 5

Foundations

WHEN PLANS SPECIFY CRUSHED stone under a concrete basement floor, it's more economical to haul the stone in and spread it before the foundation masonry blocks are installed. Trucks hauling the stone can then get as close to the foundation area as concrete trucks do. Otherwise, the stone would need to be hauled in by wheelbarrow. It's also easier to spread the stone before the foundation walls are constructed. And, not only does it save you money to spread the stone early, it also provides a safer work area for the masons and their helpers.

Estimating Crushed Stone Fill

Crushed stone is usually sold by weight, and priced by the ton. Occasionally, it's sold by volume and priced by the cubic yard. To figure the volume and weight of the crushed stone required for the basement area, refer back to Chapter 4, Figure 4-1, and use the following information:

- Outside dimensions of the basement area are 39′8″ × 26′0″.
- Inside dimensions are 38′8″ × 24′0″ [39′8″ minus the thickness of the block wall on the right side (12″), and 26′0″ minus the thickness of the block wall in the front and back (12″ + 12″)].
- The basement footing was formed so that when the 4 inches of crushed stone specified is added, the material will be level with the top of the footing (Figure 5-1).
- The amount of stone needed for the basement area will be:

$$\frac{38.67'(38'8'') \times 24.00'(24'0'') \times 0.33'(4'')}{27} = 11.34 \text{ cy}$$

Divide by 27 to convert cubic feet into cubic yards.

- The weight of crushed stone is 2,800 pounds per cubic yard.

$$\frac{11.34\text{ cy} \times 2{,}800\text{ lbs}}{2{,}000\text{ (lbs per ton)}} = 15.88\text{ tons}$$

Add 15 percent for extra stone needed to level uneven grade:

$$15.88 \times 1.15 = 18.26\text{ (rounded to 18) tons of crushed stone}$$

> *Enter the number of **tons of crushed stone** on line 5.1 of your Cost Estimate Worksheet. We'll use 18 tons in our Example Cost Estimate Worksheet.*

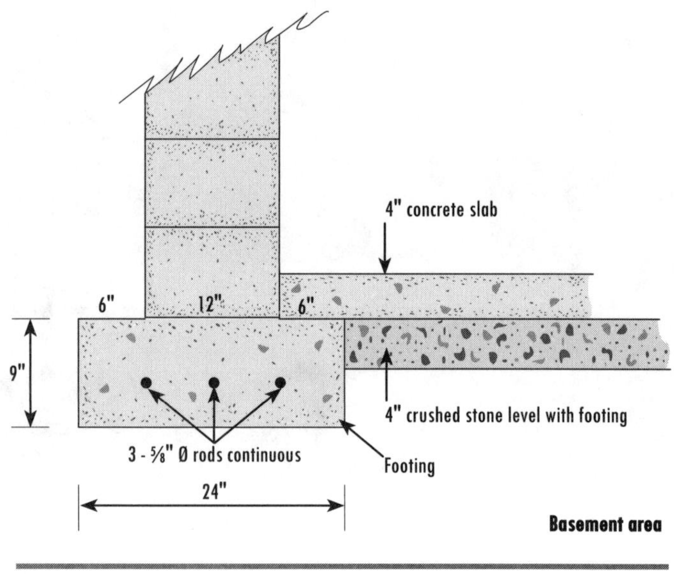

Figure 5-1
Concrete slab over crushed stone

The payroll log from previous jobs shows that an average of 0.65 manhour is required to spread one ton of crushed stone. The labor needed to spread the stone is calculated as follows:

$$18\text{ tons} \times 0.65 = 11.7\text{ (rounded to 12) manhours}$$

> *Enter the number of **manhours to spread crushed stone** on line 5.9.1 of your Cost Estimate Worksheet. We'll use 12 hours at the National Construction Estimator hourly laborer rate, BL, in our Example Cost Estimate Worksheet.*

Estimating Masonry Blocks or Concrete Masonry Units (CMUs)

Always use the foundation plan, the specs, and the plot plan with grade elevations to help you estimate the masonry block required for the house foundation. The foundation plan shows the dimensions of the foundation, and whether or not there'll be a basement, crawl space, or both. It gives the height of the foundation walls, the size and location of windows and doors, and the number and size of piers and pilasters. The specs will show if any external finish brick is to start below grade. With this information, you'll be able to identify the course of blocks on which the brick will start. This will be the same course where the 8 × 8 × 16 block must start to support the bricks to be laid on the exposed outer 4 inches of the 12 × 8 × 16 block.

When the number of block courses doesn't vary, estimating is fairly easy. But when the footings are stepped, an accurate estimate becomes more difficult. Use the plot plan grade elevations (see Chapter 2, Figure 2-1) to calculate stepped footings when they're necessary.

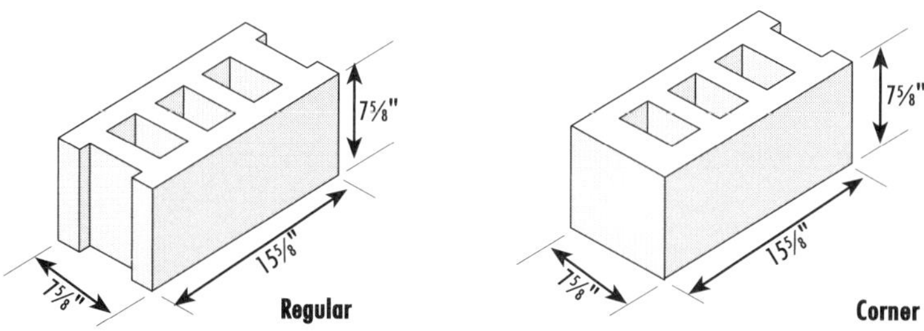

Figure 5-2
Standard masonry block dimensions

When estimating block, keep in mind that the actual size of the block isn't the same as the nominal size. The actual size of masonry block is reduced from the nominal size enough so that the block plus one mortar joint (usually ⅜ inch) will equal the nominal size. The actual size of 8 × 8 × 16 masonry block is shown in Figure 5-2. One 7⅝" × 7⅝" × 15⅝" regular or corner block plus one ⅜-inch mortar joint on three faces is:

- 8 inches wide
- 8 inches high
- 16 inches long

Nominal Block Size	Actual Block Size
Width × Height × Length	Width × Height × Length
4 × 8 × 16	3⅝" × 7⅝" × 15⅝"
6 × 8 × 16	5⅝" × 7⅝" × 15⅝"
8 × 4 × 16	7⅝" × 3⅝" × 15⅝"
8 × 8 × 16	7⅝" × 7⅝" × 15⅝"

Figure 5-3
Nominal and actual block sizes

Figure 5-3 shows the nominal and actual sizes for commonly used block.

Number of Blocks per Course (Course Factor)

Calculate the number of blocks required per course by multiplying the length of the wall in feet and inches by a 0.75 course factor. Why 0.75? Imagine a course of block 10 feet long. How much block was used? Each block, plus mortar joint, is 16 inches long. So the number of blocks is:

$$\frac{10'0''(120'')}{16''} = 7.5 \text{ blocks}$$

A 10-foot course requires 7.5 blocks, so each foot of each course requires 0.75 block, giving us the 0.75 course factor. A 4-foot-long wall will need:

$$4.00' \times 0.75 \text{ (course factor)} = 3 \text{ blocks per course}$$

How many blocks would we need for each course in a wall 39'8" long?

$$39.67' \text{ (39'8'')} \times 0.75 \text{ (course factor)} = 29.75 \text{ blocks per course}$$

Calculating Block per Wall Area

There are two methods you can use to calculate the number of blocks you'll need for a wall area. The first is to multiply the number of blocks per course by the number of courses; and the second is estimating by the square foot of wall area, using a *square foot factor*. Let's compare the two methods. We'll estimate the number of 8-inch-high by 16-inch-long blocks we'll need for a wall that's 40 feet long and 8 feet high (320 square feet).

Multiplying Blocks per Course by Number of Courses

- First convert the 8-foot wall height into inches (96 inches) and divide that by the block height (8 inches) to determine the number of courses.

$$\frac{96''\ (8'0'')}{8''\ \text{per course}} = 12\ \text{courses}$$

- Then multiply wall length (40 feet) by the 0.75 course factor and the number of courses to find the number of blocks we need.

$$40'0'' \times 0.75 \times 12\ \text{courses} = 360\ \text{blocks}$$

Using a Square Foot Factor

- We know that the wall area is 320 square feet and the block we're using is 8 inches high and 16 inches long.
- Multiply the square feet by a square foot factor of 1.125 to find the number of blocks we need.

$$320\ \text{sf} \times 1.125 = 360\ \text{blocks}$$

The square foot factor is a known factor for that size block. Using 8-inch-high by 16-inch-long blocks, you need 112.5 blocks per 100 square feet of wall space:

$$\frac{112.5}{100} = 1.125$$

How do we come up with this factor? You have to calculate it out using the first method. Let's try it on another example.

To keep things simple, let's say our wall is 25 feet long and 4 feet high. That's 100 square feet of wall area. Again, we'll use 8-inch-high, 16-inch-long blocks. Start by converting the wall height to inches:

$$12'' \times 4' = 48''$$

$$\frac{48''}{8''} = 6\ \text{courses}$$

$$25'0'' \times 0.75 \times 6\ \text{courses} = 112.5\ \text{blocks for 100 square feet of wall}$$

Using the square foot factor:

$$100\ \text{sf} \times 1.125 = 112.5\ \text{blocks}$$

What if we use 4-inch-high, 16-inch-long blocks for the same wall? Then our square foot factor will be 2.25:

$$12'' \times 4' = 48''$$

$$\frac{48''}{4''} = 12\ \text{courses}$$

$$25'0'' \times 0.75 \times 12\ \text{courses} = 225\ \text{blocks for 100 square feet of wall}$$

Using the square foot factor:

$$100\ \text{sf} \times 2.25 = 225\ \text{blocks}$$

Make a chart for yourself with a square foot factor for all the block sizes you use. Once you've done that, you'll have a quick and easy way to estimate block. Many estimators use square foot factors for estimating block for walls that don't vary in height. However, when you have a stepped foundation, or a wall with different size blocks for different courses, it's more accurate to estimate your block by the course, using the first method we discussed.

Calculating Block for the Basement Walls

The basement of the house in our example is 39.67 feet long by 26 feet wide. The three external walls of the basement will have a brick finish, which will start below grade. The brick will start on top of the eighth course, so the top four courses of regular block must be 4 inches narrower to allow for the width of the brick.

The plans for the external walls show 12 courses of regular block and one course of 4-inch-high solid cap block:

8 courses of 12 × 8 × 16 block
4 courses of 8 × 8 × 16 block
1 course of 8 × 4 × 16 solid block

To calculate the number of blocks needed for the three external walls, start by figuring their total linear feet:

39.67' (front) + 26' (side) + 39.67' (rear) = 105.34'

Then, multiply the linear feet by the 0.75 course factor and the number of courses to find how many of each size block you need.

Note: The extra block at overlapping corners will be applied against waste.

12 × 8 × 16 block:
105.34' × 0.75 course factor × 8 courses = 632 blocks

8 × 8 × 16 block:
105.34' × 0.75 course factor × 4 courses = 316 blocks

8 × 4 × 16 block:
105.34' × 0.75 course factor × 1 course = 79 blocks

The fourth basement wall is the 26-foot internal wall backing up to the garage, so it doesn't require the external brick finish. For this one, you need 12 courses of 12 × 8 ×16 block and one course of 8 × 4 × 16 block.

12 × 8 × 16 block:
26′ × 0.75 course factor × 12 courses = 234 blocks

8 × 4 × 16 block:
26′ × 0.75 course factor × 1 course = 19.5 (rounded to 20) blocks

The total number of each size block required for the basement walls is:

12 × 8 × 16 block:
632 + 234 = 866

8 × 8 × 16 block:
316

8 × 4 × 16 block:
79 + 20 = 99

Calculating Block for the Garage and Utility Area

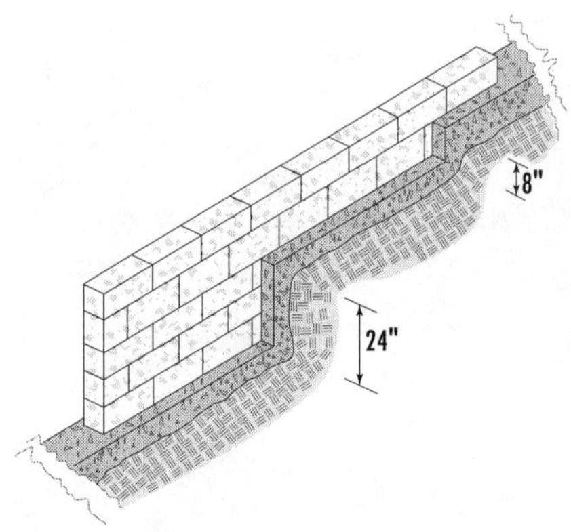

Figure 5-4
Blocks on a stepped footing

Stepped footings (shown in Figure 5-4) are necessary in many foundations, and estimating masonry block for these footings is more difficult. Unlike the basement, which will have 8-foot-high walls on all four sides, the garage and utility room will need stepped footings because of the different grade elevations under those areas. Figure 5-5 shows the layout for the stepped footings.

Many estimators guesstimate the amount of block displaced by a stepped footing. They deduct (or add) 50 percent, 25 percent, etc. But that's only a guess. It may be close, or it may not even be in the ballpark. A more accurate way to estimate masonry block for stepped footings is to calculate using a scaled layout like the one in Figure 5-5, or to use scaled block wall drawings, like those in Figures 5-6 through 5-10.

Estimating the Stepped Footings

Calculate the height and length of the stepped footings for the garage and utility area as follows:

- The garage floor elevation in the plot plan (Chapter 2, Figure 2-1) is 104.23 feet, indicating that a fill is required in this area.
- The 6-foot offset from the basement wall to the right front corner of the garage shows a rise of 1.06 feet. Using the elevations in Figure 5-5, calculate this rise by subtracting the baseline footing elevation in the basement (95.90 feet) from the grade elevation at the right front corner of the garage (96.96 feet):

  96.96' − 95.90' = 1.06' (or 13", rounded)

  One 8-inch stepped footing will be built here, 3 feet from the garage corner toward the basement wall.
- The grade elevation rises 0.89 foot from the right front corner of the garage to the left front corner of the garage:

  97.85' − 96.96' = 0.89' (or 11", rounded)

  One 16-inch stepped footing will be constructed here, at approximately the halfway mark.
- From the left front corner of the garage to the left back corner of the utility area, the grade elevation rises 3.55 feet:

  101.40' − 97.85' = 3.55' (or 43", rounded)

  A four-step footing will be constructed for this left side wall. The first step will be 16 inches high, approximately 8 feet from the left garage front corner. The second step will also be 16 inches high, 8 feet from the first step. The third and fourth steps will each be 8 inches in height; the third step 6 feet from the second step, and the fourth step 6 feet from

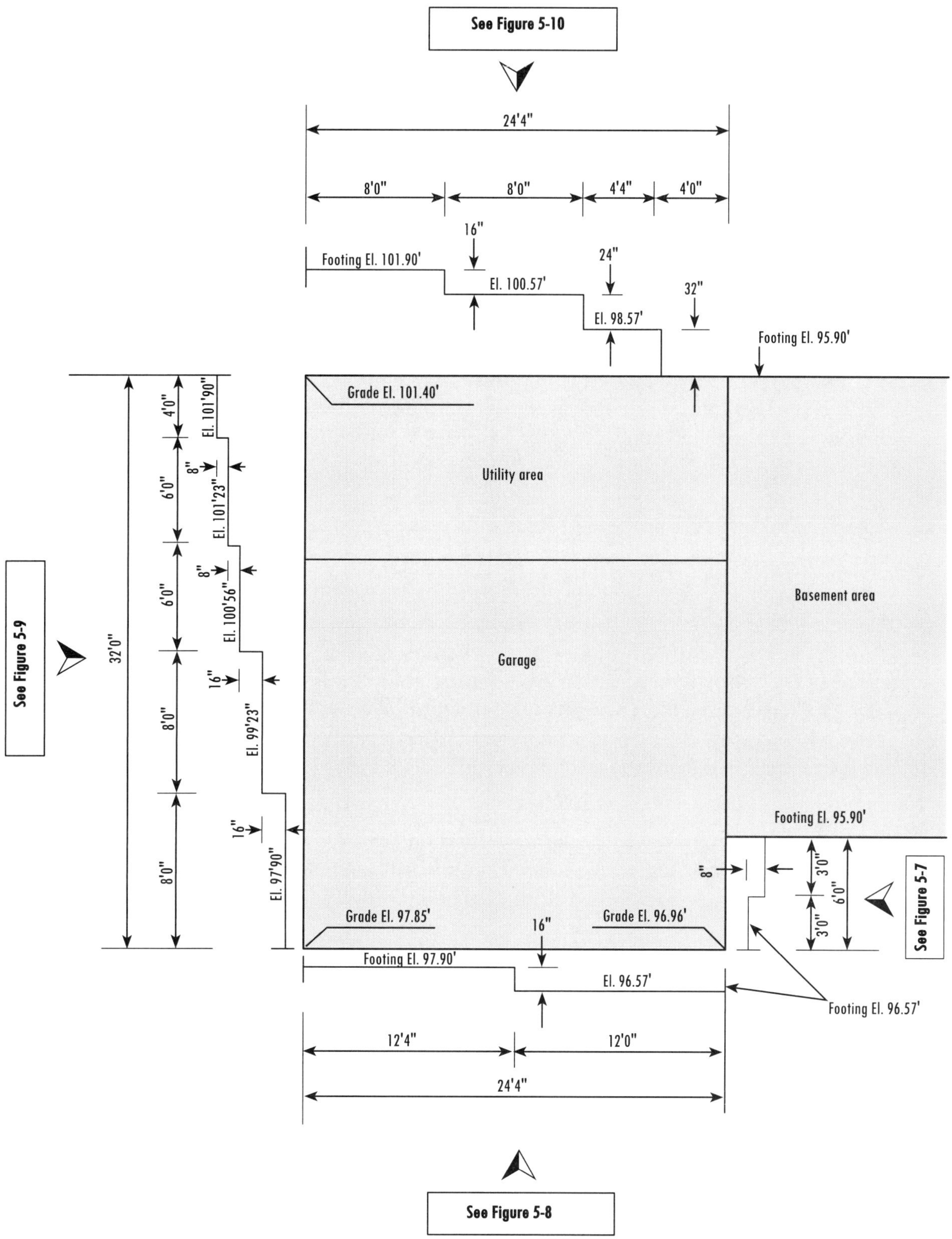

Figure 5-5
Layout of stepped footings

Figure 5-6
Drawing to scale used for computing displaced blocks

the third step and 4 feet from the left back corner of the utility area. This will be a total rise of 4 feet for the footings on the left side. The top of the footings at the left back corner of the foundation will be 0.50 feet higher than the present grade.

101.90' – 97.90' = 4.00' (or 48")

101.90' (footing) – 101.40' (grade) = 0.50' (or 6")

This won't be a problem, because the finish grade elevation will be above the footing.

- From the left back corner of the utility area to the right back corner of the utility area next to the basement, there's a fall of 6.00 feet in elevation. We've established the footing elevation at the left back corner of the utility area at 101.90 feet and the basement footing elevation is 95.90 feet.

  101.90' – 95.90' = 6.00' (or 72")

- The footing across the back of the utility area will fall 72 inches in 24 feet 4 inches, so we'll need higher, but fewer footing steps on this side of the building. The first footing step will be 16 inches high, approximately 8 feet from the left back corner. The second step will be 24 inches in height, 8 feet from the first step. And the third step will be 32 inches high, 4.33 feet from the second step, and 4 feet from the basement wall.

Blocks Displaced by Stepped Footings

Figure 5-6 shows a masonry block wall drawn to scale. Using a scale drawing like this is very helpful in estimating masonry block for walls where stepped footings are needed. We'll use a scale drawing to calculate the blocks displaced by the step footings on each section of the foundation wall for the garage and utility area.

Right Offset, Garage Wall We've estimated that we're going to need one 8-inch stepped footing in the 6-foot offset from the basement wall to the right front corner of the garage, as shown in Figure 5-5. The scale drawing in Figure 5-7 shows this stepped footing, with the displaced blocks

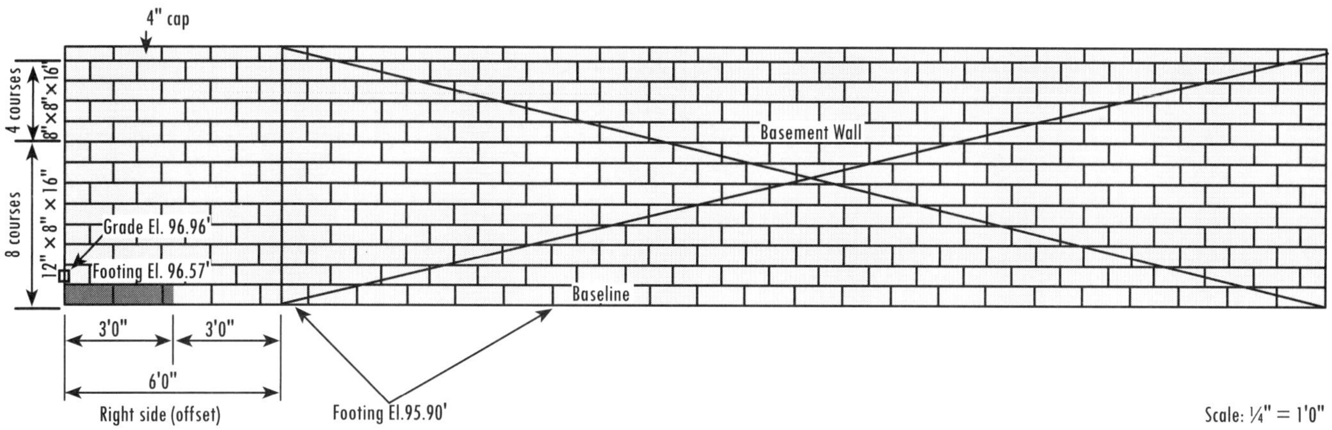

Figure 5-7
Blocks displaced in offset on right foundation wall

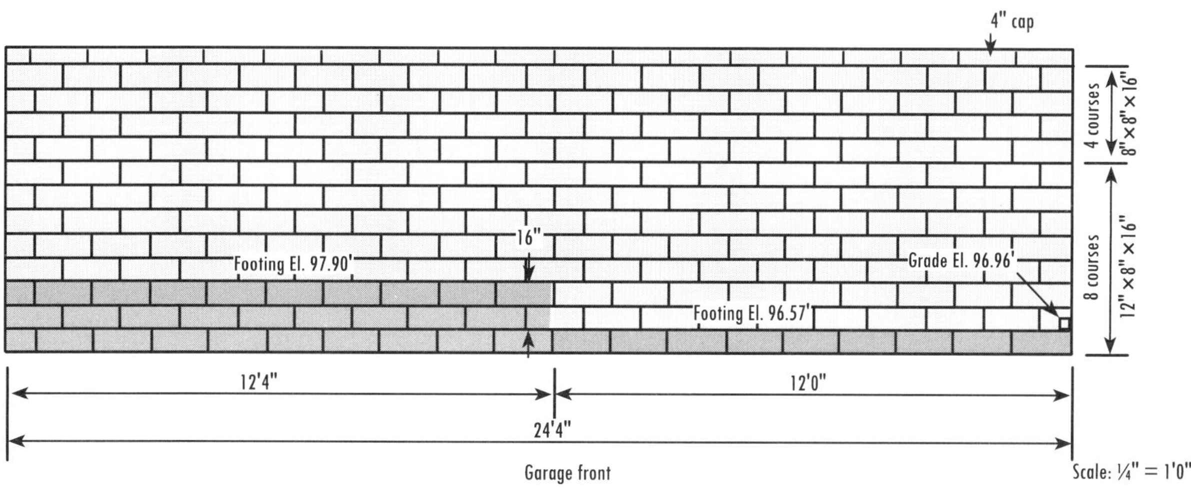

Figure 5-8
Blocks displaced at garage front foundation wall

indicated by the shaded area. The following calculation indicates that a total of 2¼ (2.25) blocks will be displaced by the stepped footing for this wall.

Number of 12 × 8 × 16 blocks displaced (shaded area):

One course, 3′0″ long, 3.00′ × 0.75 (course factor) =	2.25
Total 12 × 8 × 16 blocks displaced:	2.25 blocks

Garage Front Wall One 16-inch stepped footing is required across the front of the garage (see Figure 5-5). The scale drawing in Figure 5-8 indicates the blocks displaced (shaded area) in this footing section. The calculations show that a total of 36¾ blocks will be displaced from this wall by the stepped footing.

Number of 12 × 8 × 16 blocks displaced (shaded area):

One course, 24′4″ long, 24.33′ × 0.75 (course factor) × 1 =	18.25
Two courses, 12′4″ long, 12.33′ × 0.75 (course factor) × 2 =	18.50
Total 12 × 8 × 16 blocks displaced:	36.75 blocks

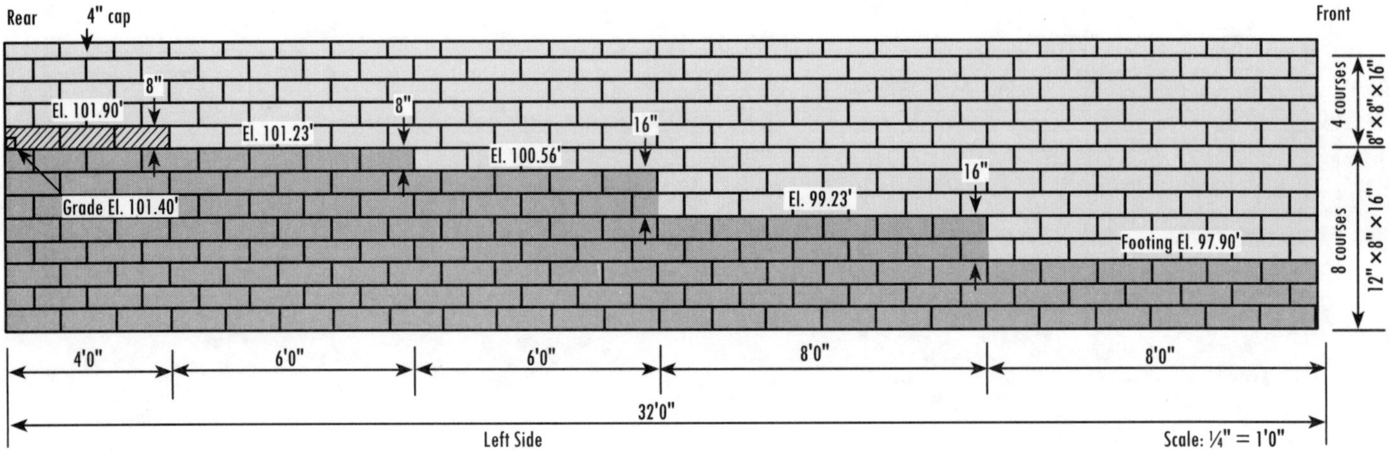

Figure 5-9
Blocks displaced on left foundation wall

Garage/Utility Area, Left Wall Four stepped footings are required along the left side of the garage/utility area (Figure 5-5). The scale drawing of this wall is in Figure 5-9. The calculations show that a total of 142½ blocks will be displaced by the stepped footing (shaded area) for this wall: 139½ are 12 × 8 × 16 blocks and 3 are 8 × 8 × 16 blocks.

Number of 12 × 8 × 16 blocks displaced (shaded area):	
Three courses, 32′0″ long, 32.00′ × 0.75 (course factor) × 3 =	72
Two courses, 24′0″ long, 24.00′ × 0.75 (course factor) × 2 =	36
Two courses, 16′0″ long, 16.00′ × 0.75 (course factor) × 2 =	24
One course, 10′0″ long, 10.00′ × 0.75 (course factor) × 1 =	7.5
Total 12 × 8 × 16 blocks displaced:	139.5 blocks

Number of 8 × 8 × 16 blocks displaced (hatched area):	
One course, 4′0″ long, 4.00′ × 0.75 (course factor) × 1 =	3
Total 8 × 8 × 16 blocks displaced:	3 blocks

Utility Back Wall Three steps are required in the footing along the back wall of the utility area from the left corner over to the basement wall (see Figure 5-5). Figure 5-10 is a scale drawing of this wall, viewed from the front. The calculations show that this stepped footing will displace a total of 109 blocks from the back wall: 103 are 12 × 8 × 16 blocks, and 6 are 8 × 8 × 16 blocks.

Number of 12 × 8 × 16 blocks displaced (shaded area):	
Four courses, 20′4″ long, 20.33′ × 0.75 (course factor) × 4 =	61
Three courses, 16′0″ long, 16.00′ × 0.75 (course factor) × 3 =	36
One course, 8′0″ long, 8.00′ × 0.75 (course factor) × 1 =	6
Total 12 × 8 × 16 blocks displaced:	103 blocks

Number of 8 × 8 × 16 blocks displaced (hatched area):	
One course, 8′ long, 8.00′ × 0.75 (course factor) × 1 =	6
Total 8 × 8 × 16 blocks displaced:	6 blocks

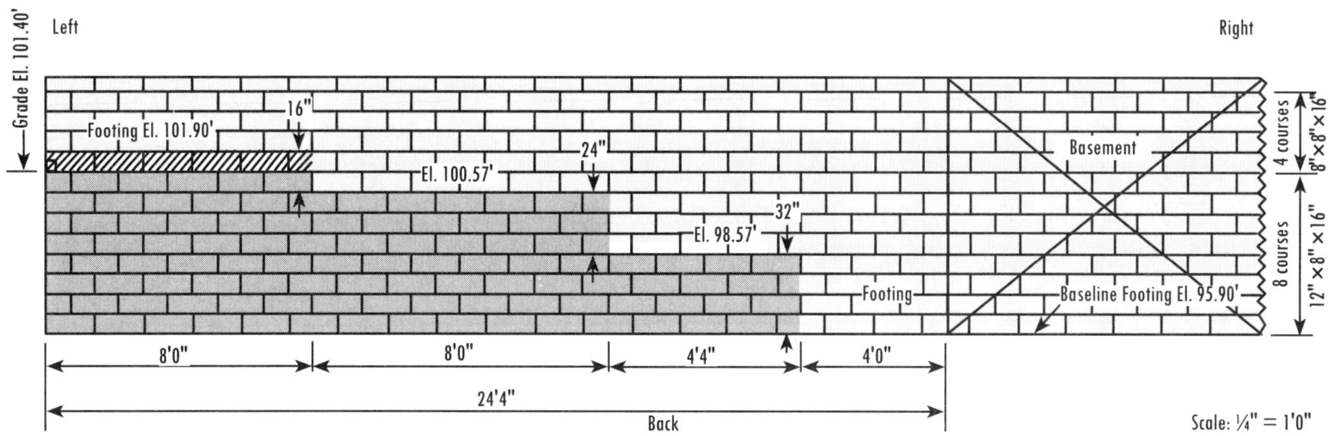

Figure 5-10
Blocks displaced on rear foundation wall

Block Required for Garage and Utility Area Foundation

You can follow the steps below to estimate the masonry block needed for the garage and utility area with reasonable accuracy. First, calculate the quantity of block required if there were no stepped footings. Then, using the calculations from the scale drawings in Figures 5-7 through 5-10, deduct the block displaced by each of the stepped footing from the total number calculated.

1. Total the block required without stepped footings.

   Add the linear feet of masonry foundation walls in the garage and utility area requiring stepped footing (see in Figure 5-5):

   6'0" (right) + 24'4" (front) + 32'0" (left) + 24'4"(rear) = 86'8" (86.67')

   Calculate the number of each size block needed to complete a wall 8.33 feet high and 86.67 long:

   12 × 8 × 16 block (8 courses)
   86.67′ × 0.75 (course factor) × 8 courses = 520 blocks

   8 × 8 × 16 block (4 courses)
   86.67′ × 0.75 (course factor) × 4 courses = 260 blocks

   8 × 4 × 16 solid block (1 course)
   86.67′ × 0.75 (course factor) × 1 course = 65 blocks

2. Total the block displaced by stepped footings.
   Calculate the total number of each size block displaced:

   12 × 8 × 16 block

Right offset, garage wall	2.25
Garage front wall	36.75
Garage/utility area, left wall	139.50
Utility back wall	103.00
Total 12 × 8 × 16 block displaced:	281.50 (rounded to 281)

8 × 8 × 16 block

Garage/utility area, left wall	3.00
Utility back wall	6.00
Total 8 × 8 × 16 block displaced:	9.00

3. Total each block type required for foundation wall. Subtract the number of displaced block from the total block needed for each of the three block sizes:

12 × 8 × 16: 520 – 281 = 239 blocks
8 × 8 × 16: 260 – 9 = 251 blocks
8 × 4 × 16: 65 – 0 = 65 blocks

You now have an estimate of the block required for the garage and utility area foundation wall.

Notes: The stepped footing (Figures 5-5, 5-7, 5-8, 5-9, and 5-10) rises above the baseline footing elevation. The foundation height is calculated, then the blocks displaced are deducted from the totals. If the stepped footing drops below the baseline footing (e.g. a basement area on a sloping lot that requires fill for part of the basement), the block for the stepped footings would be *added* to the total quantity of masonry units.

Total Blocks Required for the Foundation

Add the total for each size block that's been calculated for the basement foundation and the garage and utility area foundation.

Block	Basement		Garage/Utility	Total
12 × 8 × 16	866	+	239	1,105 blocks
8 × 8 × 16	316	+	251	567 blocks
8 × 4 × 16*	99	+	65	164 blocks

*Solid block

Notes: Window and door openings have been calculated as solid, and no allowance has been made for block overlapping the corners. These displaced blocks allow for waste.

This gives you the number of blocks required for the foundation, but there'll be additional block requirements as well.

Additional Requirements for Blocks

A total of 74 linear feet of 8 × 8 × 16 block will be needed as filler around the floor trusses.

74.00' × 0.75 = 55.50 (rounded to 56) 8 × 8 × 16 block

- The 24'4" supporting wall under the partition between the utility and garage area will need one course of 8 × 8 × 16 block and one course of 8 × 4 × 16 solid block:

  24.33' (24'4") × 0.75 × 1 =18.25 (rounded up to 19) 8 × 8 × 16 block

  24.33' (24'4") × 0.75 × 1 =18.25 (rounded up to 19) 8 × 4 × 16 block

- The retaining wall for the outside steps leading to the basement will be constructed with 12 × 8 × 16 block. We'll estimate 130 additional 12 × 8 × 16 blocks.

There's always some amount of waste involved in laying masonry block. Some block may be broken during unloading from the trucks, some broken by the helpers, and some will be broken by the masons working with them. There's no guideline to allow for this waste. Again, cost records from previous jobs are very helpful in making this allowance.

Notes: Corner and sash block are normally included in large orders by the supplier, so it's not necessary to specify them. About 10 percent of the order supplied will be corner and sash block.

To help offset waste, many builders and estimators don't deduct for window and door openings, or for any overlapping block at the corners. Waste may be anywhere from 4 to 8 percent or more, depending on the conditions and the workmen. Just remember to allow for waste, in one way or another, when you estimate masonry block. We'll make no deduction for the windows and door in the basement (as noted), and add 4 percent for waste.

Total number of blocks required:

	12 × 8 × 16	**8 × 8 × 16**	**8 × 4 × 16***
Basement/Garage/Utility	1,105	567	164
Trusses	0	56	0
Partition Support	0	19	19
Retaining Walls	130	0	0
Subtotal	1,235	642	183
Add 4 percent (waste)	50	26	7
Total:	1,285	668	190

*Solid block

*Enter the **number of masonry blocks** required on line 5.2 of your Cost Estimate Worksheet. We'll use these figures on our Example Cost Estimate Worksheet.*

Estimating Mortar

There's a high percentage of waste in mortar, too. It can be as much as 25 percent, though a lower percentage is usually estimated. Cost records from previous jobs can give you an idea of how much to allow for waste. The amount of mortar needed also depends on how thick the masons make their mortar joints. A ⅜-inch joint is most commonly used, but not always.

Mortar is made from masonry cement (purchased by the bag) and sand (purchased by the cubic yard). Pre-blended mortar mixes are also used. These combine precise amounts of masonry cement and dried sand, so there's no guesswork involved. Pre-blended mortar mixes (such as Quikrete® or Spec Mix®) are more expensive than purchasing masonry cement and sand separately. However, the additional cost is offset by the consistent quality, reduced mixing time, and ease of use.

You'll need different amounts of mortar to lay different-sized block. If most of the blocks to be laid are the standard 8 × 8 × 16 size, many estimators will use the mortar factor for that block size as the average figure for all block. But if you're laying various sizes of block, you'll need to use the correct mortar factor for each size, as shown in Figure 5-11.

Width × Height × Length	Core	Weight per Block* (lbs)	Bags per Block* (40 lb bag)	Bags per Block* (60 lb bag)	Bags per Block* (80 lb bag)	Bags per Block* (94 lb bag)
4 × 8 × 16	2	4.50	0.113	0.075	0.056	0.05
6 × 8 × 16	2	6.00	0.150	0.100	0.075	0.06
8 × 4 × 16	2	5.50	0.138	0.092	0.069	0.06
8 × 4 × 16	Solid	10.00	0.250	0.167	0.125	0.11
8 × 8 × 16	2	7.50	0.188	0.125	0.094	0.08
12 × 8 × 16	2	10.50	0.263	0.175	0.131	0.11

* Includes 10 percent waste allowance

Figure 5-11
Pre-blended mortar mix factors (pounds per block and bags per block)

Mortar Mix for the House Foundation and Basement

We calculated the number of blocks required for the foundation and basement, etc. of our sample house (see Chapter 4, Figure 4-1) as:

12 × 8 × 16	1,285
8 × 8 × 16	668
8 × 4 × 16 solid block	190
Total:	2,143 blocks

So the amount of bulk mortar mix we'll need is:

12 × 8 × 16	1,285 × 10.50* =	13,493 lbs
8 × 8 × 16	668 × 7.50* =	5,010 lbs
8 × 4 × 16 solid block	190 × 10.00* =	1,900 lbs
Total:		20,403 lbs (or 10.2 tons)

*Mortar factor in pounds from Figure 5-11.

The mortar mix comes in 60-pound bags, so the number of bags we'll need is:

12 × 8 × 16 block	1,285 × 0.175** =	224.88 bags
8 × 8 × 16 block	668 × 0.125** =	83.50 bags
8 × 4 × 16 solid block	190 × 0.167** =	31.73 bags
Total:		340.11 bags (rounded up to 341)

**Mortar factor, in 60-pound bags, from Figure 5-11.

*Enter the number of **pounds of mortar mix** required on line 5.3 of your Cost Estimate Worksheet. We'll use 20,403 pounds or 341 60-pound bags in our Example Cost Estimate Worksheet.*

Basement Openings

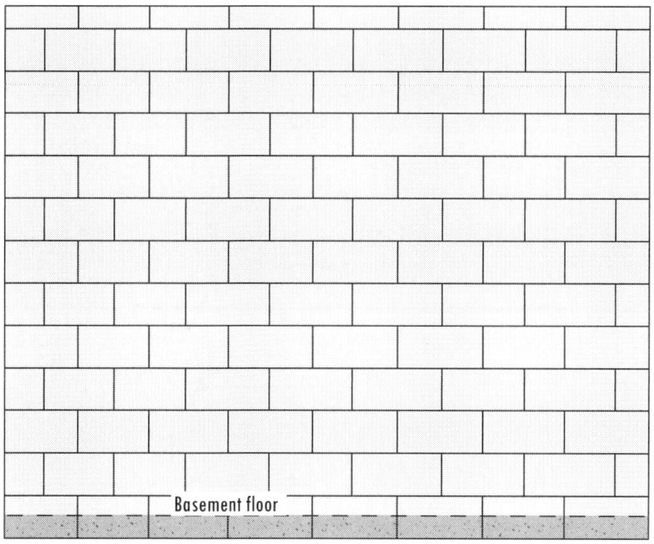

Figure 5-12
Scale drawing of foundation wall

Figure 5-12 is a scale drawing of a basement masonry wall. This drawing can help you plan openings for doors, windows and vents in the basement wall.

Foundation Vents

Natural light and ventilation must be provided in all basements. In crawl spaces, a minimum of four foundation wall vents should be provided, one at each corner of the crawl space. Since there's no crawl space in our sample house, we have no foundation vents to estimate.

Windows

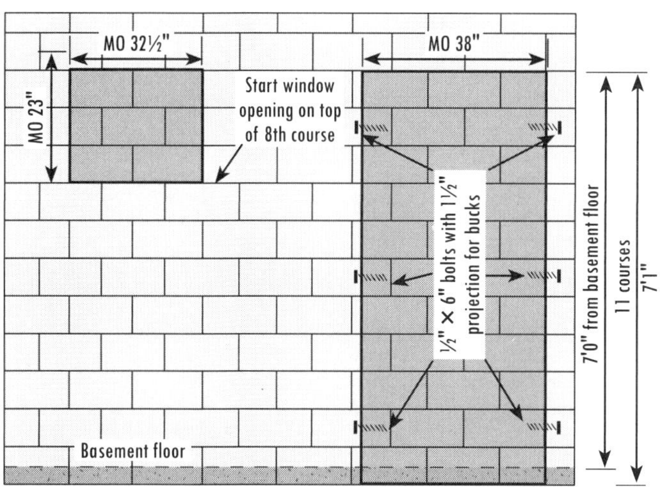

Figure 5-13
Plotting block courses for basement windows and doors

Most basement window frames are metal and made with a flange so they'll fit into the sash block the mason installs around the window opening. The window is installed before the lintel is placed above the opening. If the interior walls of the basement will be finished, regular house-type windows are usually installed. Special provisions must be made for their installation. Many builders allow an extra 3 inches in width and 1½ to 3 inches in height in the masonry opening for the bucks onto which the finish jamb and head will attach. The bucks are secured to the masonry wall by bolts placed in the block during the construction of the foundation wall. (You can see the bolt locations for the door in Figure 5-13.)

The height of the lintel for most windows and doors in a basement is the top of the 11th block course. Once you determine the location for the window opening in the masonry, measure down from the lintel the height of the window to get a quick visual picture of the window opening. You'll be able to see at which course to start the window opening, and how much space you'll need for the window sill (Figure 5-13). If you need more space than you have remaining for the sill, you can leave out one course of block to increase the opening.

Doors

A basement should have an outside entrance. Build the basement door a minimum of 2 feet 8 inches wide to accommodate materials and supplies that'll be moved to and from the basement area. As with windows, some builders set the door frame during the construction of the foundation, while others allow extra room in the masonry opening for bucks (see Figure 5-13). Let's look at an example of a masonry opening for a basement door, 2 feet 8 inches wide and 6 feet 8 inches high.

Masonry opening width	
Door	2′8″
Door frame (jamb thickness 1¼″ × 2 sides)	2½″
Bucks (1½″ × 2 sides)	3″
Clearance	½″
Total opening width:	3′2″ (or 38″)
Masonry opening height	
Door	6′8″
Door frame (jamb thickness 1¼″ × 1 side)	1¼″
Buck (1½″ × 1)	1½″
Clearance	½″
Total opening height:	6′11¼″ (or 83¼″)

The blueprints for the house in our example specify the following masonry openings in the foundation wall:

Two 36″ × 24″ basement windows
One 2′8″ × 6′8″ basement door
(No foundation vents)

*Enter your estimate for **basement windows, doors** and **vents** on line 5.4 of your Cost Estimate Worksheet. We'll use two 36″ × 24″ windows, one 2′8″ × 6′8″ door, and zero vents in our Example Cost Estimate Worksheet*

Estimating Foundation Supports and Reinforcing

The blueprints will show what type and size of supports and reinforcing will be required for the house foundation. We'll look at each of these items separately, and then put together a material take-off for our sample house.

Lintels

The lintels for masonry block come in two forms — precast reinforced concrete and flat steel plate. The type of lintel

Note: Always order lintels a minimum of 8 inches longer than the masonry opening to allow a 4-inch bearing on each end.

you'll use depends on the load it's intended to carry, and the span of the opening it'll cover. A flat steel plate lintel is usually adequate for residential construction. There's a big difference in the price of the two lintels. A 48-inch precast reinforced concrete lintel costs more than three times as much as a flat steel plate lintel. If a precast lintel is specified, use it; otherwise, use the flat steel plate lintel. When there are both a basement and a crawl space in one foundation, there'll probably be an access door from the basement to the crawl space. A lintel is required for this access door, as well.

Beams

Beams or girders support loads. There are several types of support beams: laminated, box, built-up, or steel beams. In residential construction, the only support beams usually required are the ones that support the floor system and, occasionally, the opening for a door that leads from a garage to the basement area.

The foundation plan included with the blueprints will show what types of beams are required. These beams are designed to support the specific load they must carry over the span between the column posts or piers. If a built-up beam assembled from more than one framing member is called for, it will be indicated on the foundation plan as a 3 – 2″ × 10″ wood girder (three pieces of 2″ × 10″ wood, fixed together) or a 3 – 2″ × 12″ wood girder (three pieces of 2″ × 12″ wood). See Figure 5-14. If the plans call for a steel beam, it'll be indicated by the size of the beam and the weight per foot, as indicated in Figure 5-15. If the beam is supported by column posts, as in a basement, the plans will show the size of the posts (Figures 5-14 and 5-15). If the beam is supported by piers, as in a crawl space, the plans will specify the size of the piers (Figure 5-16).

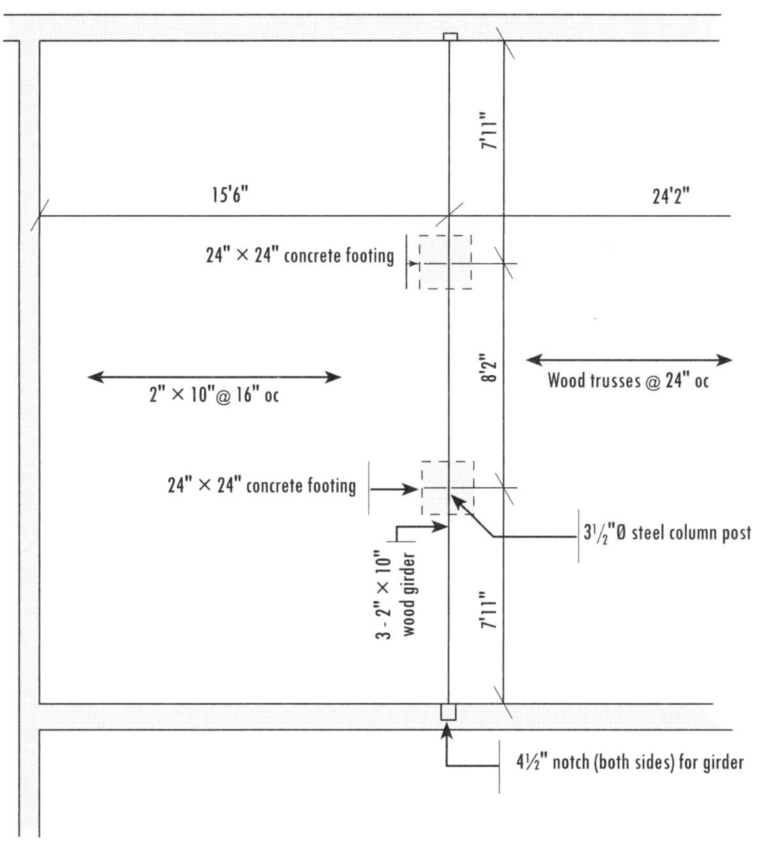

Figure 5-14
Plan indicating wood girder

Anchor Bolts

Steel anchor bolts secure the house to the foundation. The size and placement of the bolts in the foundation wall isn't left to guesswork. Each bolt should

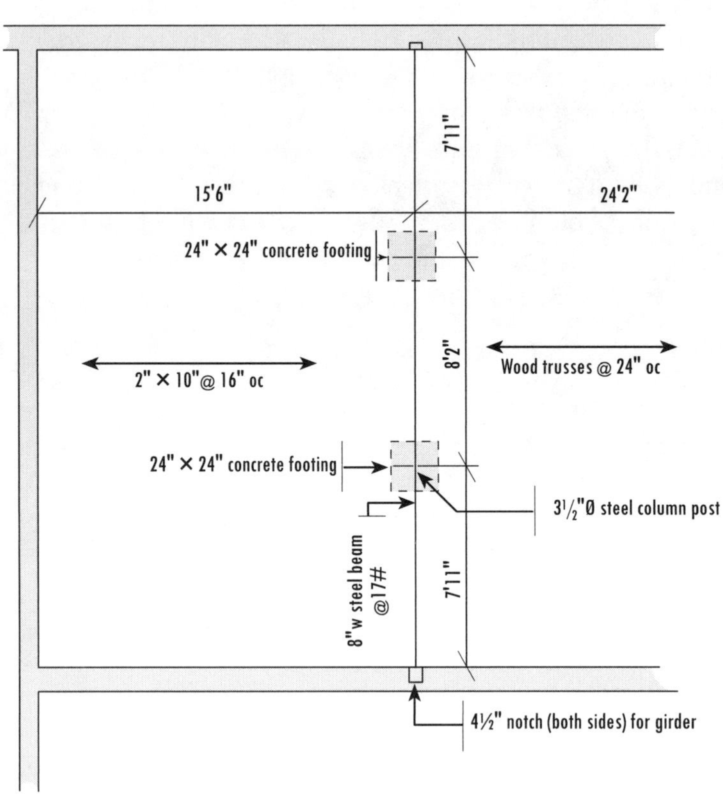

Figure 5-15
Plan indicating steel beam with steel column posts

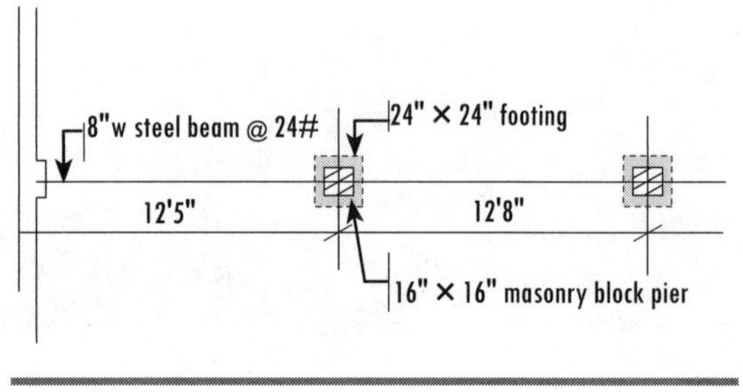

Figure 5-16
Plan indicating steel beam with masonry block piers

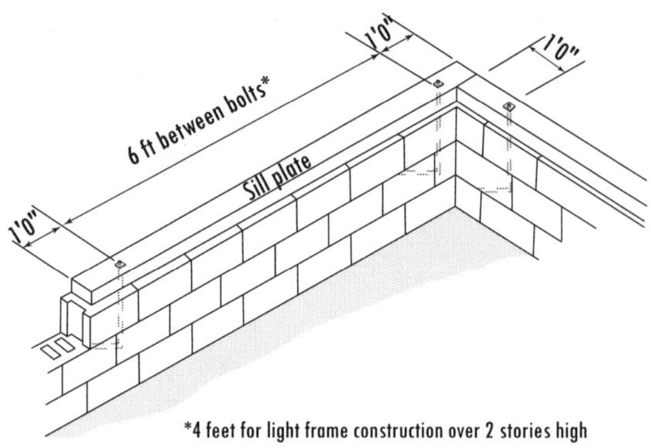

Figure 5-17
Spacing anchor bolts

be ½ inch by 18 inches, with a 3-inch square washer at least 0.229-inch thick. The spacing between bolts can't exceed 6 feet. For wood light-frame construction over two stories high, the bolt spacing shouldn't exceed 4 feet. Each sill plate needs at least two bolts, and the end bolts can't be more than 1 foot, or less than 3½ inches, from the end of the sill (Figure 5-17). In other words, an 8-foot sill plate in a two-story house would need two anchor bolts, but a 9-foot piece would need three.

To estimate the number of anchor bolts for a foundation wall:

- Calculate the perimeter of the foundation wall.
- Divide the perimeter of the foundation wall by 6 (if 6-foot bolt spacing is to be used). If the answer is not a whole number, add an additional bolt.
- Add a bolt for each section of foundation wall length longer than the maximum length of sill plate lumber to be used. Where a joint is necessary, an extra bolt is required. For example, a 32-foot section of foundation wall requires

two pieces of 16-foot sill plate, so a joint is necessary and an additional anchor bolt is required. If you use pieces of differing lengths for your sill plate and have more joints, add bolts accordingly.

- Add one bolt for each exterior and interior corner in the masonry wall. For example, if there are 6 corners on the house, you'll need 6 additional anchor bolts.
- Total the number of anchor bolts required.

Let's look at an example. Calculate the perimeter of our sample house foundation wall (refer back to Chapter 4, Figure 4-1):

Garage front	24'4"
Left side	32'0"
Rear	64'0"
Right side	26'0"
Basement front	39'8"
Garage offset, right side	6'0"
Total:	192'0"

Divide the perimeter (192 feet) by 6, if the house is 2 stories high or less. If the house was 3 stories or more, you would divide the perimeter number by 4.

$$\frac{192}{6} = 32 \text{ anchor bolts}$$

The length of our sill plate lumber is 16 feet. If you were using a different length of sill plate, use that length instead to determine the number needed. We'll need the following number and size of sill plates:

		16-foot lengths	Number of joints
Garage front	24'4"	2	1
Left side	32'0"	2	1
Rear	64'0"	4	3
Right side	26'0"	2	1
Basement front	39'8"	3	2
Garage offset, right side	6'0"	1	0
Total:			8

There are 6 exterior and interior corners, so you'll need 6 additional anchor bolts. The total number of anchor bolts needed for this foundation will be:

Perimeter	32
Joints	8
Corners	6
Total:	46

Material Take-Off

The total material take-off for lintels, beams/girders, column posts, anchor bolts and reinforcing steel for this house will be:

> **Note:** There are no crawl space access doors.

- Lintels
  Three 48-inch block lintels for two basement windows with masonry openings of 40 inches each, and one door opening of 38 inches.
- Beams/Girders
  We won't go into detail about girders here. Information about spans in the basement areas where they support floor joists will be included in Chapter 6.
- Column Posts
  Two 8-foot × 3½ -inch-diameter steel posts. The number and size of the column posts are shown on the foundation plan, and shown in Figures 5-14 and 5-15.
- Anchor Bolts
  46 (as calculated) ½-inch × 18-inch bolts with washers.
- Reinforcing Steel
  None required. In some areas, reinforcement of masonry walls is required. Local building codes, blueprints, or both will detail the size and spacing of the necessary reinforcement. Our particular house requires none.

*Enter your estimates on line 5.5 of your Cost Estimate Worksheet. We'll use the numbers we just determined for **lintels, column posts,** and **anchor bolts** on our Example Cost Estimate Worksheet*

Waterproofing and Footing Drains

Basements and habitable spaces below grade must be protected against moisture. Waterproofing keeps water from penetrating the foundation walls.

Parging is a commonly used method of waterproofing in residential construction. It involves applying coats of cement and bituminous material to masonry walls. We'll parge the foundation walls below grade in our sample house with ½-inch of masonry mortar and bituminous coating (Figure 5-18).

In any location with a high water table or where surface drainage is a problem, a membrane of hot tar and built-up fabric should be used. Special waterproofing/drainage materials may also be used. This type of work, which is seldom required in residential construction, is normally done by subcontractors.

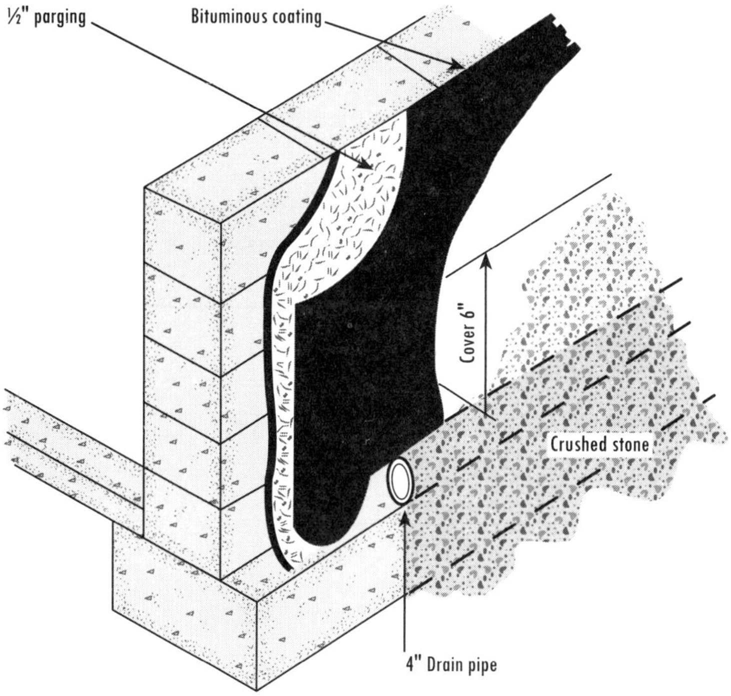

Figure 5-18
Foundation waterproofing

Parging

Let's estimate the masonry cement we need for ½-inch parging on the foundation wall of our sample house. First we need to figure how much material is needed per cubic yard of wall to be covered. We'll call that our parging factor. Once we have that, we'll multiply the total square feet of the foundation area to be parged by our parging factor.

Calculating the Parging Factor

- Multiply by 0.042 feet (the decimal conversion of ½ inch to feet) to get the number of cubic feet of mortar mix required. Then multiply the result by 1.05 to allow 5 percent for waste.
- Divide the result by 27 to convert cubic feet to cubic yards.
- Allow 3,000 pounds of mortar mix per cubic yard of finished mortar.
- This gives a parging factor of:

$$\frac{0.042 \times 1.05 \times 3{,}000}{27} = 4.90 \text{ lbs per sf}$$

Calculating the Area to be Parged

We'll parge the foundation walls of our house with a ½-inch-thick coating up to the top of the 8th course of block (64 inches or 5.33 feet) and overlap on the footings. Allowing for the footing overlap, the parging will extend to a height of 6 feet. We'll start by calculating the parging area for the basement foundation walls. Then we'll figure the parging required for the garage and utility perimeter foundation areas. These have to be figured separately in order to make adjustments for the stepped footings.

Basement External Walls The three external sides of the basement wall total 105.33 feet in length (39′8″ + 26′0″ + 39′8″ = 105′4″). The parging will extend 6 feet up the wall. To offset waste, no deduction is made for the basement openings. The total square feet to be parged:

$$6.00' \times 105.33' = 631.98 \text{ sf}$$

Garage, Right Side Offset Wall (Figure 5-7) The offset wall on the right side of the garage is 6 feet and we'll parge it up to a height of 6 feet.

$$6.00' \times 6.00' = 36.00 \text{ sf}$$

We previously figured that 2.25 blocks were displaced by the stepped footings. Each 8-inch-high, 16-inch-long block is 0.89 square foot. So, we'll need to multiply the number of displaced blocks by 0.89 and deduct that from the wall area to find the total area to be parged:

$$2.25 \times 0.89 = 2.00^{*} \text{ sf}$$

$$36.00 - 2.00^{*} = 34.00 \text{ sf to be parged}$$

*Displaced by stepped footings

Garage Front Wall (Figure 5-8) The front garage wall is 24.33 feet (24′4″).

$$6.00' \times 24.33' = 145.98 \text{ sf}$$

We previously figured that 36.75 blocks were displaced by the stepped footings.

$$36.75 \times 0.89 = 32.70 \text{ sf}$$

$$145.98 - 32.70 = 113.28 \text{ sf to be parged}$$

Left Side Wall (Figure 5-9) The left side garage/utility wall is 32 feet (32′0″).

$$6.00' \times 32.00' = 192.00 \text{ sf}$$

We previously figured that 142.5 blocks were displaced by the stepped footings.

$$142.5 \times 0.89 = 126.83 \text{ sf}$$

$$192.00 - 126.83 = 65.17 \text{ sf to be parged}$$

Rear Utility Wall The rear utility wall is 24.33 feet (24′4″).

$$6.00' \times 24.33' = 145.98 \text{ sf}$$

We previously figured that 109 blocks were displaced by the stepped footings.

$$109 \times 0.89 = 97.01 \text{ sf}$$

$$145.98 - 97.01 = 48.97 \text{ sf to be parged}$$

Foundation Wall Parging Area Now add up the total square feet of foundation wall to be parged:

Basement external walls	631.98 square feet
Right side offset wall	34.00 square feet
Garage front wall	113.28 square feet
Left side wall	65.17 square feet
Rear Utility Area Wall	48.97 square feet
Total:	893.40 square feet

Calculating the Mortar

The parging factor we calculated earlier was 4.90 pounds of mortar mix per square foot of wall area. So, to parge the foundation and basement walls (893.40 square feet), we'll need:

$$893.40 \text{ sf} \times 4.90 \text{ lbs per sf} = 4{,}377.66 \text{ lbs (rounded to 4,400 lbs, or 2.2 tons)}$$

If purchased in 60-pound bags, we'll need the following bags of mortar:

$$\frac{4{,}400}{60\,\text{lbs per bag}} = 73.33\,(\text{rounded to } 74)\text{ bags}$$

*Enter your estimated **number of pounds** and **bag size of mortar mix** on line 5.6.1 of your Cost Estimate Worksheet. We'll use 4,400 pounds of mortar mix, in 60-pound bags, in our Example Cost Estimate Worksheet.*

Parging Labor

Estimate the labor for parging the foundation and basement walls (893.40 square feet) at 2 manhours per 100 square feet:

$$\frac{893.40\,\text{sf} \times 2\,\text{manhours}}{100\,\text{sf}} = 17.87\,(\text{rounded to } 18)\text{ manhours}$$

*Enter the number of hours of **parging labor** on line 5.9.2 of your Cost Estimate Worksheet. We'll enter 18 hours at the National Construction Estimator hourly laborer rate, BL, in our Example Cost Estimate Worksheet.*

Bituminous Coating

Note: Bituminous coating is normally sold in 5-gallon cans, so in this example, we'll need two cans.

Estimate one gallon of bituminous coating per 100 square feet. Use the same area of foundation and basement walls, as we just calculated. This is 893.40 square feet, in our example.

$$\frac{893.40\,\text{sf} \times 1\,\text{gallon}}{100\,\text{sf}} = 8.93\text{ gallons}$$

*Enter your estimated **number of 5-gallon cans of bituminous coating** on line 5.6.2 of your Cost Estimate Worksheet. We'll enter two 5-gallon cans in our Example Cost Estimate Worksheet.*

Bituminous Coating Labor

The labor needed to brush on the bituminous coating can be estimated at 1½ manhours per 100 square feet.

$$\frac{893.40\,\text{sf} \times 1.5\,\text{manhours}}{100\,\text{sf}} = 13.40\,(\text{rounded to } 14)\text{ manhours}$$

*On line 5.9.3 of your Cost Estimate Worksheet, enter the **number of hours of labor to install bituminous coating**. We'll enter 14 hours, again at the National Construction Estimator hourly laborer rate, BL, in our Example Cost Estimate Worksheet.*

Footing Drain

Install a footing drain around a foundation enclosing a basement, or any other habitable space below grade, to relieve hydrostatic pressure against the foundation wall and basement floor. The drain can discharge either by gravity or by pump to an outfall, such as a drainage ditch or sump pit. From there, it can drain or be pumped into a storm or sanitary drainage system.

Pipe

Footing drains are usually made of 4-inch perforated plastic pipe, typically in 10-foot lengths*. However, 4-inch farm tile (12 inches long) or 4-inch perforated bituminous fiber can also be used. Place a minimum of 2 inches of crushed stone under the pipe, and cover it with at least 6 inches of crushed stone (Figure 5-18).

*If you use a different length of 4-inch perforated plastic pipe, substitute the new length in your calculation.

In the plot plan (Chapter 2, Figure 2-1), you can see that the grade elevation at the lowest corner of the lot (left front corner) is 98.26 feet. The basement footing elevation is 95.90 feet (Figures 5-7 and 5-10). The basement footing where the drain pipe will be laid is 2.36 feet (2′4″) lower than the lowest grade elevation on the lot. Consequently, discharge from the drain pipe will have to be pumped into the sanitary drainage system. We'll install a sump pit and pump for the basement floor drain, and the drain pipe can discharge into this sump pit.

Our estimate will be based on using 4-inch perforated plastic pipe. The material (pipe and couplings) required for the drain pipe around the basement footings will be:

Right side	39′8″
Front side	26′0″
Rear	39′8″
Total:	105′4″ (105.33′ linear feet)

$$\frac{105.33'}{10'(\text{length of pipe*})} = 10.53\,(\text{round to }11)\text{ pieces of pipe}$$

*Enter your estimate of 10-foot **perforated plastic pipes** on line 5.6.3 of your Cost Estimate Worksheet. We'll use 11 10-foot lengths in our Example Cost Estimate Worksheet.*

Couplings

The number of couplings required will always be one less than the number of pieces of piping to be joined. So we'll need 10 couplings for 11 pieces of 10-foot* perforated plastic pipe.

*Recalculate the number of couplings if the pipe lengths are other than 10 feet.

*Enter your estimate of **couplings** on line 5.6.4 of your Cost Estimate Worksheet. We'll use 10 couplings on our Example Cost Estimate Worksheet.*

Corner Ells

A corner ell is needed each time the drain makes a 90-degree turn. In our example, the drain pipe makes two 90-degree turns around the right rear and right front of the footings. So we need two 90-degree ells.

> *Enter your estimate of **90-degree ells** on line 5.6.5 of your Cost Estimate Worksheet. We'll use 2 90-degree ells in our Example Cost Estimate Worksheet.*

Crushed Stone

We need to estimate the cubic yards of crushed stone needed to fill around the drain pipe in an area that's 105.33 linear feet long, 18 inches wide and 12 inches deep. The factor for calculating the cubic yards of crushed stone needed to cover a 4-inch pipe is:

$$0.05232 \times \text{linear feet} = \text{cubic yards}$$

$$0.05232 \times 105.33' = 5.51 \text{ cubic yards}$$

You can check your answer by first calculating the cubic yards of stone for the area, and then deducting the volume of the pipe to determine the total cubic yards of stone needed.

$$105.33' \times 1.5'\,(18'') \times 1'\,(12'') = 158 \text{ cubic feet}$$

The formula for calculating the volume of the drain pipe is:

$$\pi \times \left[\frac{\text{diameter of pipe}''}{2 \times 12}\right]^2 \times \text{length (ft)}$$

$$3.142 \times \left[\frac{4''}{2 \times 12}\right]^2 \times 105.33' = 9.19 \text{ cubic feet}$$

Now subtract the pipe volume from the cubic feet of crushed stone:

$$158.00 \text{ cf} - 9.19 \text{ cf} = 148.81 \text{ cubic feet}$$

Divide your answer by 27 to convert cubic feet to cubic yards:

$$\frac{148.81 \text{ cf}}{27} = 5.51 \text{cubic yards}$$

As you can see, the answer, 5.51 cubic yards, is the same as it was using the factor above. Whichever method you use, you'll need to covert the cubic yards to pounds or tons for your supplier.

$$1 \text{ cubic yard of crushed stone} = 2{,}800 \text{ lbs}$$

$$5.51 \text{ cubic yards} \times 2{,}800 \text{ lbs} = 15{,}428 \text{ lbs or } 7.71 \text{ (rounded to 8) tons}$$

> *Enter your estimated **tons of crushed stone** on line 5.6.6 of your Cost Estimate Worksheet. We'll use 8 tons of crushed stone in our Example Cost Estimate Worksheet.*

Footing Drain Labor

The labor to install the footing drain, including the crushed stone, can be estimated at 10 manhours per 100 linear feet.

$$\frac{105.33\text{ lf}}{100\text{ lf}} \times 10\text{ manhours} = 10.53\text{ (rounded to 11) manhours}$$

*Enter the **manhours to install a footing drain** under labor on line 5.9.4 of your Cost Estimate Worksheet. We'll use 11 hours at the current National Construction Estimator hourly laborer rate, BL, in our Example Cost Estimate Worksheet.*

Estimating Miscellaneous Materials

There are some necessary foundation materials that tend to get overlooked. These materials are:

- Wall ties
  If brick is to be used and will start below grade, you need wall ties in the foundation walls.
- Polyethylene
  This film forms a vapor retarder, covering masonry cement and other materials.
- Bucks/or basement windows and doors
  2 × 4 or 2 × 6 framing lumber is most commonly used for buck material.
- Bolts
  Remember to add ½″ × 6″ bolts with washers to secure the bucks to the masonry walls (Figure 5-13).
- Door frame/ basement door
  If the door frame is to be set before the masonry block is laid, it may be included here; otherwise, include it later with exterior finish material.
- Areaways
  Windows and vents below grade need to be protected by areaways. (Areaways are enclosed spaces below grade adjacent to windows and vents that allow ventilation and light to enter.)
- Dryer vents
  If the laundry area will be in the basement, install the dryer vent during construction of the foundation.
- Mortar antifreeze
  May be required during cold weather.
- Termite protection
  Use metal shields or chemical treatment.
- Brushes for the bituminous coating (if needed).
- Scaffold rental (if needed).

Note: If brick areaways will be required for the basement windows, include the cost with your brick estimate.

We'll include the following miscellaneous materials for the foundation in our sample house cost estimate:

- 1 box wall ties (500)

- 1 roll polyethylene (12′0″ × 100′0″)
- 1 2″ × 6″ × 12′0″ (door buck)
- 1 2″ × 6″ × 8′0″ (door buck)
- 6 ½″ × 6″ bolts with washers

> *Make your estimate for the above **miscellaneous items** and enter the total on line 5.7 of your Cost Estimate Worksheet. We'll use costs from the current National Construction Estimator in our Example Cost Estimate Worksheet.*

Masonry Labor

Masons may charge an hourly rate, or they may subcontract by job or by the block. If an hourly wage is paid to the masons and their helpers, the builder has the additional expense of paying FICA, FUTA, workers' compensation, and liability insurance as well as taxes on their wages. The amount paid is a percentage of each employee's earnings, and is typically included in the hourly rate estimated. There will also be management and administrative expenses added as overhead in the final estimate. See Chapter 18 for more information on these costs.

A masonry contractor hires his own crew, so the responsibility of the payroll records, taxes and insurance is his. He'll contract for laying the masonry block either by job or by the block.

> **Note:** If the masonry work is let as a subcontract, get a copy of the masonry contractor's workers' compensation policy before any work begins.

Certain factors can limit how many blocks a mason can lay in one day. Among them are the size of the block, the height of walls and number of openings, and labor efficiency. Some masons can lay 200 blocks or more in an 8-hour shift, while others average only half that number. If you have the option of paying masons by the hour or by contract, weigh these factors before making a final decision on your payment method.

In our cost estimate, the masonry work was done at an hourly rate for a bricklayer and a bricklayer's helper. We'll estimate they can lay an average of 150 blocks of all sizes per 8-hour shift, or 9.375 blocks per manhour.

$$\frac{150 \div 8}{2} = 9.375$$

The mason will lay the following block for our sample house:

12″ × 8″ × 16″	1,285
8″ × 8″ × 16″	668
8″ × 4″ × 16″ (solid block)	190
Total:	2,143 masonry blocks

> *Enter the **number of blocks** and the total **manhours** on line 5.8 of your Cost Estimate Worksheet. At a rate of 9.375 blocks per manhour, we'll estimate it will take a two-man crew* 228.59 manhours to lay the 2,143 blocks. We'll use crew code B9 from the National Construction Estimator for the wage in our Example Cost Estimate Worksheet.*

*When two or more craftsmen or their helpers are involved in a task, the hourly rate in the *National Construction Estimator* is an average of the hourly rates for each worker.

COST ESTIMATE WORKSHEET FOR FOUNDATIONS

#	Qty		Size		Cost	Per			Subtotal
5.1	**Stone fill under concrete slab**								
	18	tons, stone		@	$17.20	ton	=	$309.60	$309.60
5.2	**Masonry blocks**								
	1,285	12" × 8" × 16"		@	$3.16	block	=	$4,060.60	
	668	8" × 8" × 16"		@	$2.01	block	=	$1,342.68	
	190	8" × 4" × 16" solid		@	$1.64	block	=	$311.60	
	2,143	Total						$5,714.88	$5,714.88
5.3	**Mortar mix**								
	20,403	lbs							
	341	bags	60 lb	@	$3.91	bag	=	$1,333.31	
	or 10.2	tons		@		ton	=	$0.00	
								$1,333.31	$1,333.31
5.4	**Basement windows, doors and foundation vents**								
	2	basement windows	36" × 24"	@	$42.69	piece	=	$85.38	
	1	basement doors	30" × 80"	@	$328.00	piece	=	$328.00	
		foundation vents		@			=	$0.00	
								$413.38	$413.38
5.5	**Lintels, beams, column posts, anchor bolts, and reinforcing bars**								
	3	lintels	48"	@	$84.80	piece	=	$254.40	
		beams		@		piece	=	$0.00	
	2	column posts	8'0" × 3½"	@	$67.82	piece	=	$135.64	
	46	anchor bolts	½" × 18"	@	$1.76	piece	=	$80.96	
		reinforcing steel		@		piece	=	$0.00	
								$471.00	$471.00
5.6	**Waterproofing and drain**								
5.6.1	4,400	mortar mix, lbs							
	74	bags	60 lb	@	$3.91	bag	=	$289.34	
5.6.2	2	5-gallon cans bituminous coating		@	$21.70	can	=	$43.40	
5.6.3	11	10' pieces, 4" perforated plastic pipe		@	$4.40	piece	=	$48.40	
5.6.4	10	couplings		@	$6.47	piece	=	$64.70	
5.6.5	2	ells		@	$6.47	piece	=	$12.94	
5.6.6	8.0	tons stone		@	$17.21	ton	=	$137.68	
								$596.46	$596.46
5.7	**Miscellaneous material**				$111.28				$111.28
			Cost of material						$8,949.91
			Sales tax	@	7.75	%			$693.62
			Total cost of material						$9,643.53
5.8	**Masonry labor**								
	2,143	blocks	150 blocks per day using						
	228.59	manhours	B9 crew (bricklayer + helper)	@	$28.74	hour	=	$6,569.68	
5.9	**Other labor**								
5.9.1	12	manhours - stone spreading, BL		@	$26.64	hour	=	$319.68	
5.9.2	18	manhours - parging, BL		@	$26.64	hour	=	$479.52	
5.9.3	14	manhours - bituminous coating, BL		@	$26.64	hour	=	$372.96	
5.9.4	11	manhours - footing drain, BL		@	$26.64	hour	=	$293.04	
			Total cost of labor					$8,034.88	$8,034.88
	Cost of foundation (entered on line 5 of Form 100)								**$17,678.41**

Chapter 6
Floor Systems

MOST OF THE INFORMATION you need for floor construction is found in the wall section of the blueprints, which show sill plate size, the size and spacing of the floor joists, and the type and thickness of the subfloor. Some plans also show the girder size recommended.

Board Measure

Framing lumber is normally priced and sold by the thousand board feet. Board measure is the standard basis of measuring lumber, and the board foot is the unit of measurement. Framing lumber is usually abbreviated like this:

BF or bf	Board foot
MBF or Mbf	Thousand board feet
BM or bm	Board measure
Mbm	Thousand board measure

A board foot is 1 inch thick × 1 foot wide × 1 foot long (Figure 6-1), often noted as *one square foot, 1 inch thick*. The formula to calculate board measure is:

$$\frac{\text{Thickness (inches)} \times \text{width (inches)} \times \text{length (feet)}}{12}, \text{ also indicated as } \frac{T \times W \times L}{12}$$

Let's look at a few examples of how you would use this formula, or its variation, to calculate board feet.

How many board feet are in a piece of lumber 1″ × 12″ × 1′0″?

$$\frac{1'' \times 12'' \times 1'0''}{12} = 1 \text{ bf (Figure 6-1)}$$

How many board feet are in a piece of lumber 2″ × 4″ × 1′0″?

$$\frac{2'' \times 4'' \times 1'0''}{12} = 0.67 \text{ bf (Figure 6-1)}$$

How many board feet are in a piece of lumber 2″ × 10″ × 16′0″?

$$\frac{2'' \times 10'' \times 16'0''}{12} = 26.67 \text{ bf}$$

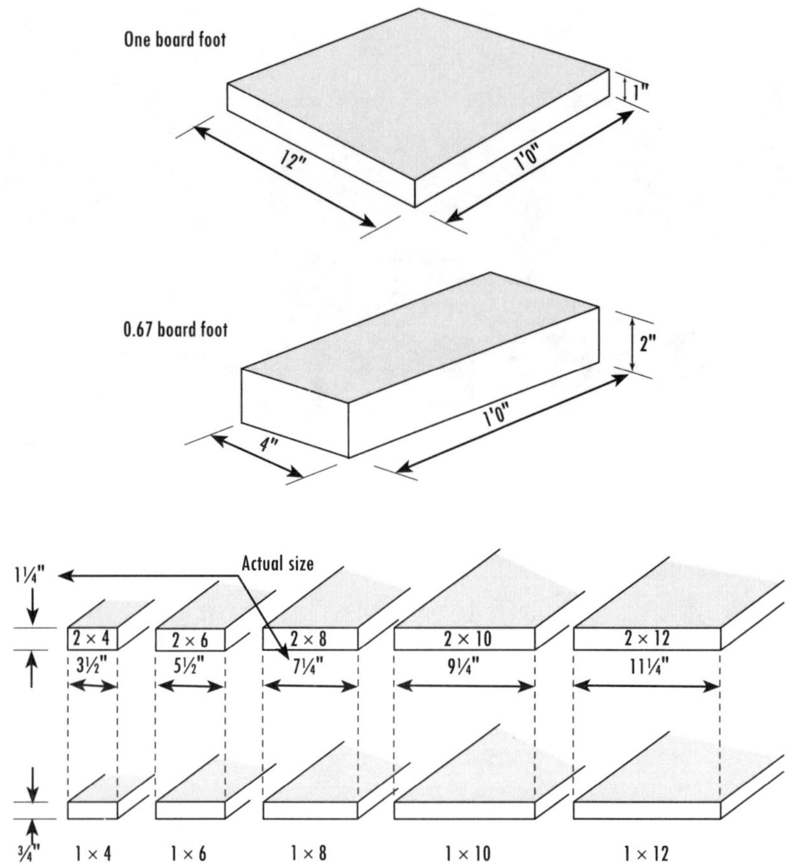

Figure 6-1
Nominal and actual lumber sizes

Converting Linear Feet to Board Measure

Lumber dealers often price sheathing by hundred linear feet. To convert linear feet to board feet, multiply the number of linear feet times its thickness times its width (in inches), and divide by 12.

How many board feet are in 1,250 linear feet of 1″ × 8″ sheathing?

$$\frac{1{,}250' \times 1'' \times 8''}{12} = 833.33 \text{ bf}$$

To convert board measure to linear feet, multiply the number of board feet by 12 and divide by its thickness times its width (in inches).

How many linear feet are in 833.33 board feet of 1″ × 8″ sheathing?

$$\frac{833.33 \times 12}{1'' \times 8''} = 1{,}250 \text{ lf}$$

Converting Piece Pricing to Board Measure Pricing

As I said earlier, the most common way to price lumber is by thousand board measure (Mbm), and the unit of measure is usually thousand board feet (MBF). But many lumber dealers price framing by the piece. For example: 2″ × 4″ × 8′0″ pieces at $4.59 each, or 2″ × 8″ × 12′0″ pieces at $13.20 each.

To convert these unit prices into the price per MBF:

1. Convert the framing piece into board feet.
2. Divide the unit cost by the number of board feet in each piece of framing.
3. Multiply the result by 1,000 to get the MBF cost.

If 2″ × 4″ × 8′0″s are $4.59 each, what's the price per MBF?

$$\frac{2'' \times 4'' \times 8'0''}{12} = 5.33\text{ bf}$$

$$\frac{\$4.59}{5.33\text{ bf}} = \$0.861\text{ per bf}$$

$$\$0.861 \times 1{,}000 = \$861.00\text{ per MBF}$$

If 2″ × 8″ × 12′0″s are $13.20 each, what's the cost per MBF?

$$\frac{2'' \times 8'' \times 12'0''}{12} = 16\text{ bf}$$

$$\frac{\$13.20}{16\text{ bf}} = \$0.825\text{ per bf}$$

$$\$0.825 \times 1{,}000 = \$825.00\text{ per MBF}$$

Calculating Lumber Pricing

You can also use lumber calculator tables to determine board feet for various sizes of lumber. Check online under "lumber calculator" for a quick reference. These tables can come in handy. However, most estimators keep a calculator on their desk. Using a pocket calculator and your supplier costs, you can often estimate your lumber costs just as fast – especially if you need a quick estimate in the field.

Let's try a few more examples. We'll calculate the costs for the following pieces of lumber, and then total them up to provide an estimated cost for lumber on a job.

1. 200 pieces of 2″ × 4″ × 8′0″ at $861.00 per MBF

$$\frac{200 \times 2'' \times 4'' \times 8'0''}{12 \times 1{,}000} = \frac{12{,}800}{12{,}000}$$

$$\frac{12{,}800}{12{,}000} = 1.067$$

$$1.067 \times \$861.00 = \$918.69$$

2. 65 pieces of 2″ × 6″ × 12′0″ at $791.00 per MBF

$$\frac{65 \times 2'' \times 6'' \times 12'0''}{12 \times 1{,}000} = \frac{9{,}360}{12{,}000}$$

$$\frac{9{,}360}{12{,}000} = 0.78$$

$$0.78 \times \$791.00 = \$616.98$$

3. 78 pieces of 2″ × 10″ × 16′0″ at $781.00 per MBF

$$\frac{78 \times 2'' \times 10'' \times 16'0''}{12 \times 1{,}000} = \frac{24{,}960}{12{,}000}$$

$$\frac{24{,}960}{12{,}000} = 2.08$$

$$2.08 \times \$781.00 = \$1{,}624.48$$

Now total the costs:

$$\$918.69 + \$616.98 + \$1{,}624.48 = \$3{,}160.15$$

The estimated lumber cost is $3,160.15.

Estimating the Sill Plate, Girder and Ledger or Joist Hangers

Now, let's break the floor system down into parts and estimate the lumber we need for our sample house.

Sill Plate

The sill plate attaches the floor system to the foundation using anchor bolts (refer back to Figure 5-17 in Chapter 5). The minimum thickness of the sill plate is 2 inches (nominal measurement), and the width can't be less than 1½ inches bearing for the ends of the joists. There should be no joints over window and door openings.

To estimate material for the sill plate:

- Check the sill plate dimensions on the wall section of the blueprints. In residential construction, it's usually 2″ × 6″.
- Total the lengths of the foundation walls where the sill plates are to be installed.
- Divide the total by the sill plate lumber length you're using to get the number of pieces required. For example, divide by 12 if you're using 12-foot sill plate lumber or by 14 for 14-foot sill plate lumber, etc.
- If you want your estimate expressed in board feet (bf), convert the number of pieces to board feet as explained at the beginning of this chapter.

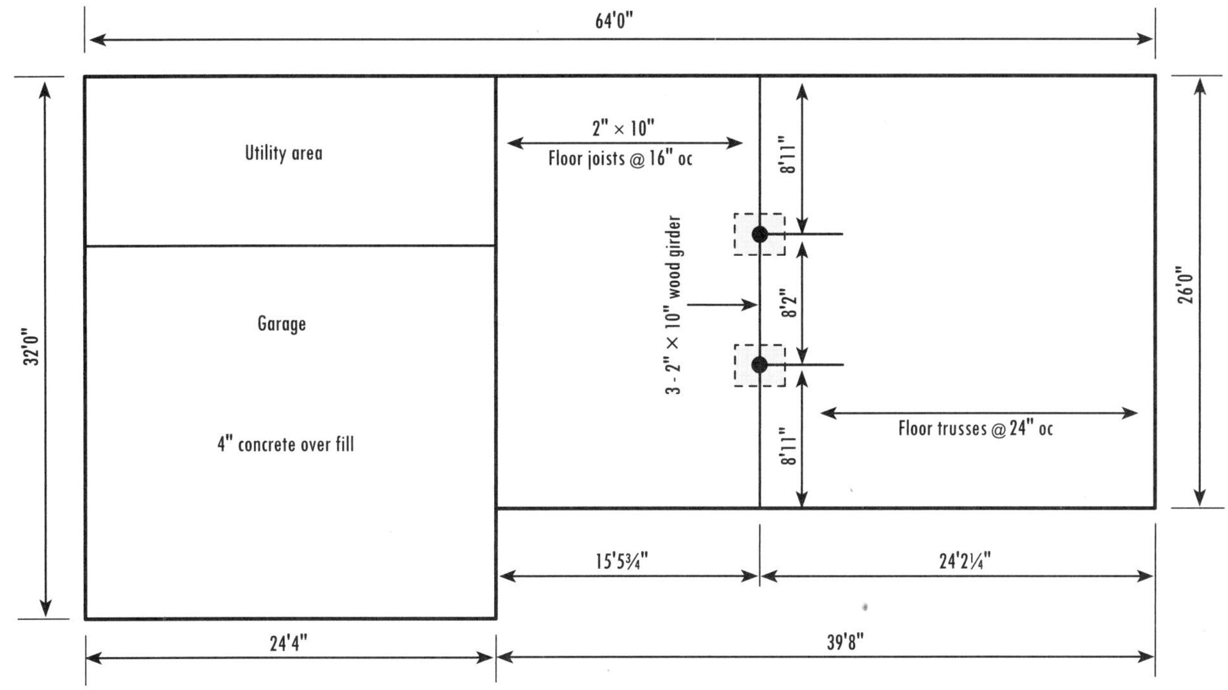

Figure 6-2
Foundation plan

Let's calculate how many pieces of 2″ × 6″ × 12′0″ sill plate lengths we'll need for the 2″ × 6″ foundation and basement sill plates in our house, shown in Figure 6-2. Since we know the dimension of the sill plate, our next step is to total the lengths needed.

Rear of house/utility area	64′0″	64.00′
Left side of garage/utility	32′0″	32.00′
Garage rear	24′4″	24.33′
Garage front	24′4″	24.33′
Garage offset	6′0″	6.00′
Basement, left wall	26′0″	26.00′
House front	39′8″	39.67′
Right side of house	26′0″	26.00′
Total:		242.33′

Next, divide the total needed by the length of the sill plate lumber to find the total number of pieces we'll need. Then add in 10 percent for waste:

$$\frac{242.33'}{12.00'} = 20.19 \text{ pieces}$$

Subtotal	20.19
Waste (20.19 × 0.10)	2.02
Total:	22.21 (rounded to 22) pieces of sill plate lumber

*Enter your estimate of the **number of pieces and size of sill plate** on line 6.1.1 of your Cost Estimate Worksheet. We'll use 22 pieces of 2″ × 6″ × 12′0″ in our Example Cost Estimate Worksheet.*

How many board feet of sill plate do we need?

$$\frac{22 \times 2'' \times 6'' \times 12'0''}{12} = 264 \text{ board feet of sill plate}$$

Basement Girder

The bearing on walls should be a minimum of 4 inches, with a minimum of 6 inches of solid masonry to support the girder. A ¼-inch bearing plate may be required where the girder rests on masonry block. All wood girder joints must fall over the column supports. Built-up girders made of three or more individual members should be nailed together from both sides, in this manner:

- Two 20d nails at the ends of each piece and each splice.
- Two rows of 20d nails between splices at the top and bottom of the girder 32 inches on center, with nails staggered.

The span of the girder must be in accordance with standard engineering practice, based on the allowable fiber stress of the lumber. The size of the girder shown on the blueprints will be designed in accordance with the maximum allowable span for the house shown. The type and size of the girder should never be downgraded from the size given in the blueprints (see Figure 6-2).

To estimate the material for girders:

- Check the blueprints for girder size and span lengths between the column posts or piers.
- From the total length of the girders, calculate the most practical lengths of material to use. You'll need at least 4 inches of girder bearing on each external wall, and all joints must be made over the column supports.

Figure 6-2 indicates that the cross-section of the girder is made of 3 pieces of 2″ × 10″, and the column posts are 8′11″ from each outside wall, with 8′2″ between the columns. Allow for at least 4 inches bearing on the external walls, and a 4-inch overlap at the column posts. Since it's not possible to span the entire 26 feet with a single length of lumber (the longest is usually 18 feet), we have two options:

1. Span the 26 feet with 3 short lengths of girder. The minimum length of material required would then be 10 feet. Nine pieces of 2″ × 10″ lumber would be needed to make 3 lengths, 3 pieces thick.
2. Span the 26 feet with one short and one long girder. The minimum length of the longer girder to bridge the rear and middle spans would need 3 pieces of 18-foot lumber and the remaining front span would require 3 pieces of 10-foot lumber.

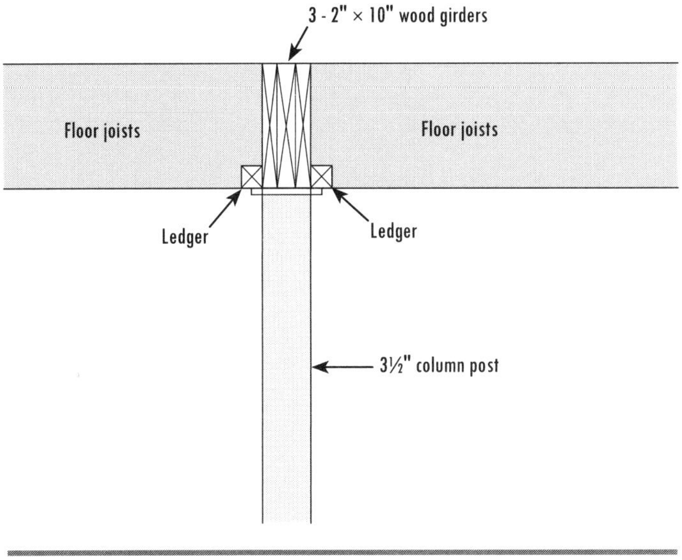

Figure 6-3
Floor joists framing into girder

There's some lumber waste regardless of the lengths you choose, but using fewer pieces means fewer splices, which reduces labor costs, saving money. So we'll use option two in our Example Cost Estimate Worksheet.

> *Enter your estimate of the* ***number of pieces and size of lumber*** *on line 6.1.2 and 6.1.3 (if necessary) of your Cost Estimate Worksheet. We'll use 3 pieces each of 2" × 10" × 18'0" and 2" × 10" × 10'0" lumber on our Example Cost Estimate Worksheet.*

Now we need to know how many 20d nails will be needed for our 26-foot one-splice girder. There are approximately thirty 20d nails per pound. The nails must be on 32-inch centers at the top and bottom of the girder, from both sides, which means four rows of nails. We also need two nails at each end, and two at each splice, also from both sides.

$$26' \times 12 = 312"$$

$$\frac{312"}{32} = 9.75 \text{ nails per row}$$

$$9.75 \times 4 \text{ (rows)} = 39 \text{ nails}$$

Nails needed	39
Add 4 nails for one end*	4
Add 4 nails for the splice	4
Subtotal	47
Add 10 percent for waste	5
Total nails:	52

*The other end was covered in the first calculation.

$$\frac{52}{30} = 1.73 \text{ pounds}$$

We'll round up to 2 pounds of 20d common nails.

> *Enter your estimate of the number of* ***pounds of 20d common nails*** *on line 6.1.6 of your Cost Estimate Worksheet. We'll use 2 pounds of 20d common nails in our Example Cost Estimate Worksheet.*

Ledger Strips

> **Note**: 2 × 4s can be ripped in half to make 2 × 2s.

Use steel joist hangers or wood ledger strips to frame floor joists into the side of a wood girder, as shown in Figure 6-3. If wood ledger strips are used, they must be at least 2 inches by 2 inches, and must be nailed to the girder with three 16d nails at each joist. You can use 2 x 4s ripped in half to make 2 × 2 wood ledger strips.

We'll need 48 feet of ledger strip (24 feet on each side of the girder). We can use two 2 × 4 × 12'0"s (ripped in half) to make our ledger strips.

> *Enter your estimate of the number and size of **pieces of ledger strip** on line 6.1.4 of your Cost Estimate Worksheet. We'll use 2 pieces of 2" × 4" × 12'0" ledger strip in our Example Cost Estimate Worksheet.*

Nail the ledger strips to the girder, 6 inches on center, using 16d nails. Let's calculate how many 16d nails we'll need for 48 feet of ledger strip at 6-inch centers.

$$48' \times 12 = 576"$$

$$\frac{576"}{6"\text{oc}} = 96 \quad \text{16d nails}$$

16d nails come in boxes of approximately 54 nails per pound, so we'll need 2 pounds to fasten our ledger strips to the girder.

> *Enter your estimate of the number of **pounds of 16d common nails** on line 6.1.7 of your Cost Estimate Worksheet. We'll use 2 pounds of 16d common nails in our Example Cost Estimate Worksheet.*

Floor Joists

The span of the joists is the clear span between the inner faces of supports. Joists can't exceed the allowable standard engineering design stresses. It's beyond the scope of this book and its intent to list maximum allowable span for every wood floor joist. The blueprints will show the size and spacing of floor joists for each span, which should never be downgraded.

Joists are required to have a minimum bearing of 1½ inches on the exterior wood sill plate. Install a continuous 2-inch header joist (band joist) to prevent lateral movement and provide subfloor nailing. Toenail each floor joist to the sill plate with three 8d nails, and use 8d nails spaced 16 inches on center to toenail the header to the sill plate. Nail the header to each floor joist with two 16d nails, minimum.

Floor joists that frame into the side of a wood girder or beam should be supported by a ledger strip (Figure 6-3), and toenailed to the girder with three 10d or 16d nails.

Floor joists framing over girders and bearing partitions must have a minimum lap of 4 inches (Figure 6-4) and be nailed together with three 16d nails. The joists should then be nailed to the girder or bearing partition with three 8d toenails.

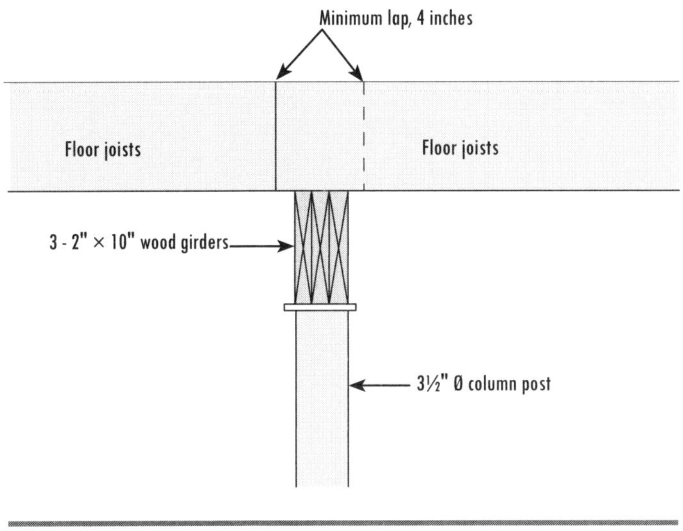

Figure 6-4
Floor joists resting on top of a girder

Double floor joists under all parallel partitions, and nail them with 10d or 16d nails, top and bottom, every 32 inches.

When you estimate floor joists from the blueprint information, remember to take-off the pieces needed for each section separately (Figures 6-5 and 6-6). If joists are 16 inches on center, multiply the linear feet (at right angles to the length of the joists in each section) by 0.75. Why do we use 0.75? Let's look at an example.

If the linear distance at right angles to the joists is 10 feet (120 inches) and the spacing between the joists is 16 inches on center, the number of joists needed is:

$$\frac{120''}{16''} = 7.5 \text{ joists}$$

For a 10-foot distance, with joist spacing 16 inches on center, you need 7.5 joists. So for each 1-foot length, you need 0.75 joists. That's our joist factor of 0.75.

What if the joist spacing is 24 inches on center? How would you calculate your joist factor? If the linear distance at right angles to the joists is 10 feet (120 inches) and the spacing between the joists is 24 inches on center, the number of joists needed is:

$$\frac{120''}{24''} = 5 \text{ joists}$$

For a 10-foot length with joist spacing 24 inches on center, you need 5 joists. So for each foot of length you need 0.5 joists. Our joist factor would be 0.50.

Joist Spacing	Joist Factor
16 inches on center	0.75
24 inches on center	0.50

The number of joists you need is based on the linear distance of the floor section at right angles to the joists and the joist spacing. Once you've calculated the number of joists you need using the joist factor, you must then add in one joist for the starter and one for each double joist. Don't deduct material for small openings — the material you cut out can be used for headers. Now let's estimate the joists we need for our sample house, beginning with the first floor.

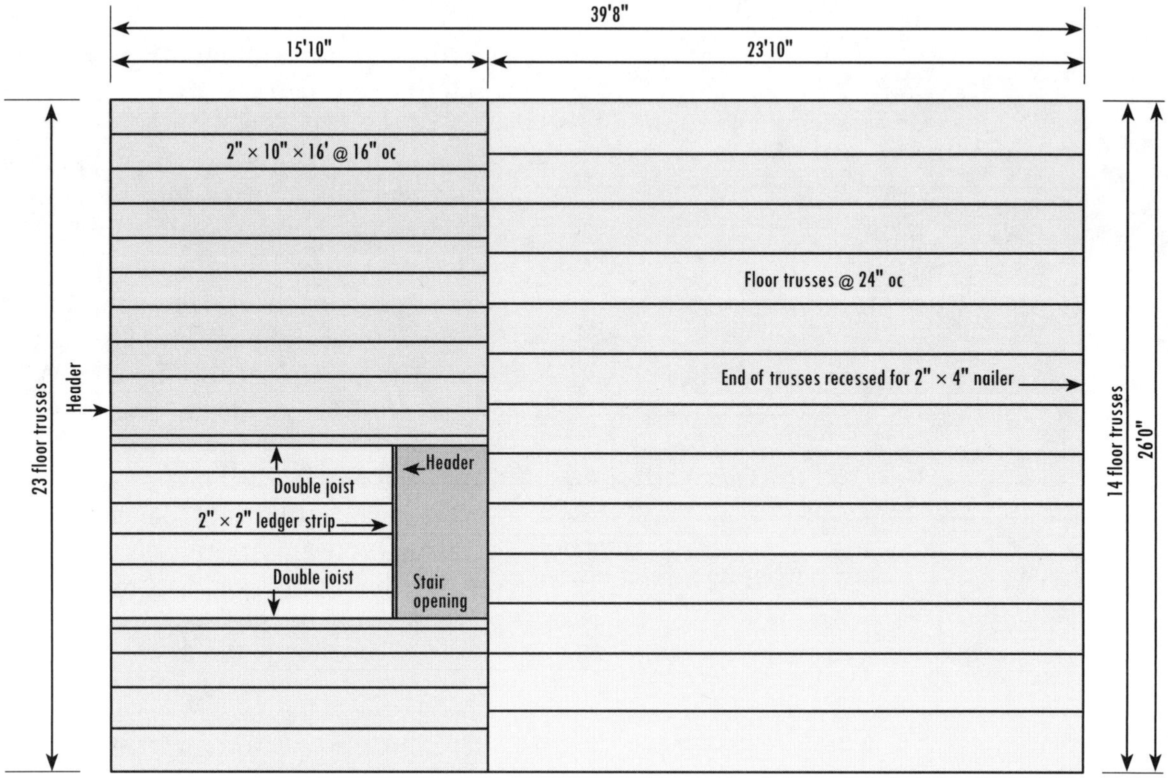

Figure 6-5
First floor joists

Look at Figure 6-5. How many joists do we need for the left side of the first floor (with the stair opening), and what length should they be? You'll need the following information to help you calculate.

- The joist spacing is 16 inches on center.
- The total length perpendicular to the joists is 26 feet.
- The span between supports is 15'10", requiring 16-foot joist lengths.

26' × 0.75* =	19.50
Starter joist	1.00
Double joists** (2)	2.00
Total:	22.50 (rounded to 23) joists

*0.75 joist factor for 16 inches oc.

**One extra joist is required for each double joist. Because only one joist was included in the calculation for that position, a second one is needed to make the single joist into a double joist.

We'll need 23 joists, 2" × 10" × 16'0" in size.

*Enter your estimate of the **number of joists** and the joist size, on line 6.2.1 of your Cost Estimate Worksheet. We'll use 23 2" × 10" × 16'0" joists in our Example Cost Estimate Worksheet.*

In Figure 6-5, there are two elevations on the first floor — the higher of the two is on the right side. The architect designed floor trusses to provide the two elevations. The floor trusses are spaced 24 inches on center. For the 23′10″ span, let's calculate how many floor trusses we'll need.

The total length at right angles to the trusses is 26 feet.

26′ × 0.50* =	13.00
Starter truss	1.00
Total:	14.00

* 0.50 is the joist (or truss) factor for 24-inch on center spacing.

We'll need 14 floor trusses, 23′10″ in length, for the 23′10″span.

> *Enter your estimate of the **number and length of floor trusses** on line 6.2.5 of your Cost Estimate Worksheet. We'll use 14 floor trusses, 23′10″ long, in our Example Cost Estimate Worksheet.*

Second Floor

Now we'll do our calculations for the second floor. The second floor sits over the entry and family room and extends to the left over the garage and utility area. The floor joists on this floor (Figure 6-6) are 16 inches on center. We'll start with the rear section:

40′ × 0.75* =	30.00
Starter joist	1.00
Double joists** (5)	5.00
Total:	36.00 joists

*Joist factor for 16-inch on center spacing.
**One extra joist is required to make a double joist.

We'll need 36 joists, 2″ × 10″ × 12′0″ in size.

> *Enter your estimate of the **number of 2″ × 10″ × 12′0″ floor joists** on line 6.2.2 of your Cost Estimate Worksheet. We'll use 36 joists, 2″ × 10″ × 12′0″ in size, in our Example Cost Estimate Worksheet.*

Let's calculate the joists for the front section of the second floor:

40′ × 0.75* =	30.00
Starter joist	1.00
Double joists** (3)	3.00
Total:	34.00 joists

*Joist factor for 16-inch on center spacing.
**One extra joist is required to make a double joist.

We'll need 34 joists, 2″ × 10″ × 16′0″ in size. Remember, you don't deduct for the stair opening. The cutout pieces can be used for headers.

> *Enter your estimate of the **number of 2″ × 10″ × 16′0″ floor joists** on line 6.2.3 of your Cost Estimate Worksheet. We'll use 34 joists, 2″ × 10″ × 16′0″ in our Example Cost Estimate Worksheet.*

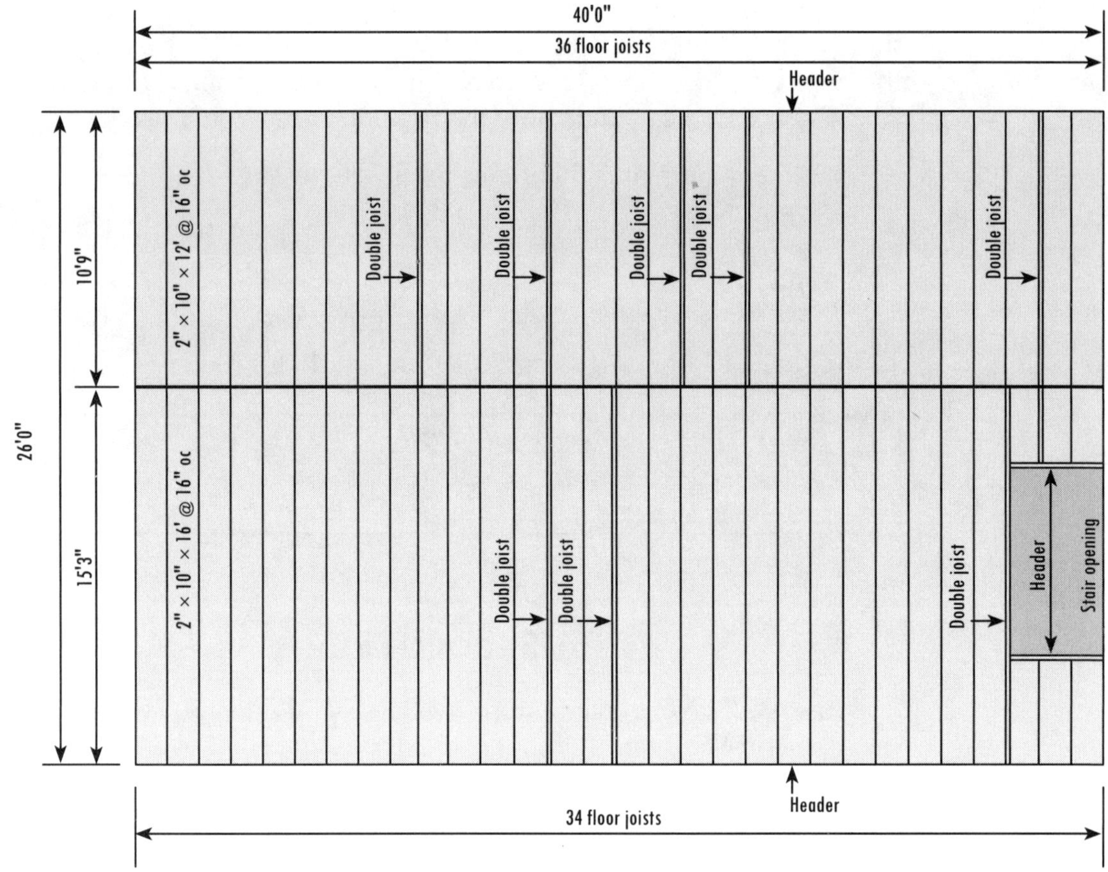

Figure 6-6
Second floor joists

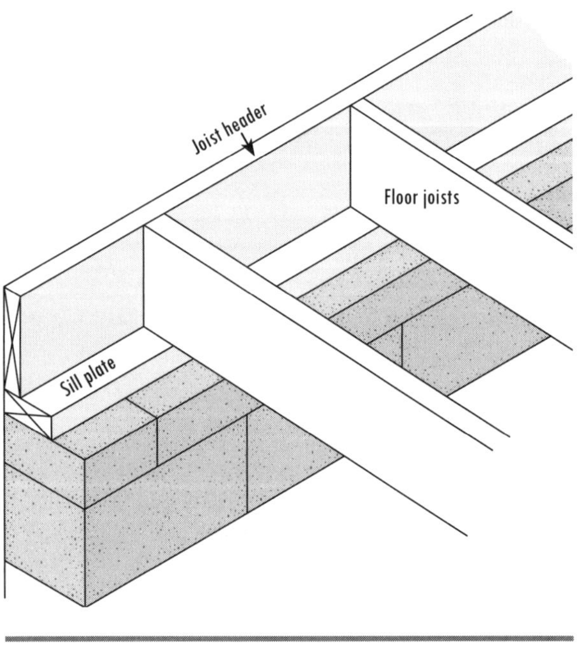

Figure 6-7
Joist header (band joist)

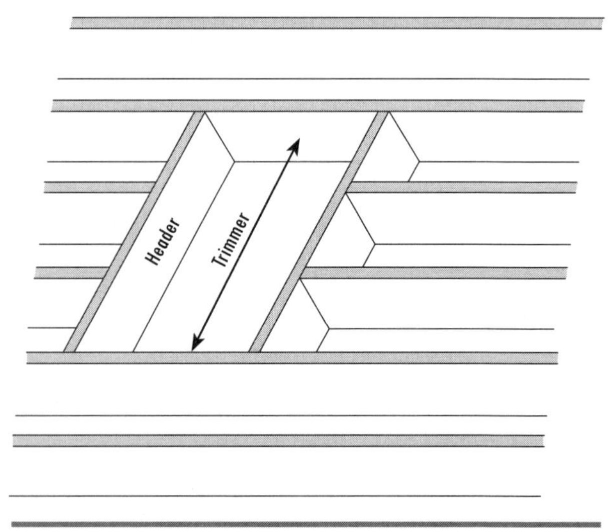

Figure 6-8
Joist header and trimmer

Headers

Joist headers (or band joists) are placed at the joists ends (Figure 6-7) to prevent lateral movement of the joists, and provide a nailing base for the subfloor.

To estimate the joist headers required, divide the total length of the headers by the length of the lumber to be used.

How many pieces of 2″ × 10″ × 16′0″ (the length we've chosen for the headers) are required for our sample house? First, add up the linear feet of joist headers indicated in the plans in Figures 6-5 and 6-6:

First floor*	26.00′
Second floor (40′ + 40′)	80.00′
Headers for stair openings**	0.00
Subtotal:	106.00′

*Only needed on the left side of the first floor; there are trusses on the other side.
**Cut-outs from the joists will be used for these headers.

$$\frac{106.00'}{16.00'} = 6.63 \text{ (rounded to 7)}$$

We'll need 7 pieces of 2″ × 10″ × 16′0″ for the joist headers.

> *Enter your estimate of the number of **2″ × 10″ × 16′0″** on line 6.2.6 of your Cost Estimate Worksheet. We'll use 7 lengths of 2″ × 10″ × 16′0″ headers in our Example Cost Estimate Worksheet.*

Trimmers

A joist adjacent to an opening where the header is framed is called a *trimmer* (Figure 6-8). Most trimmers are estimated as floor joists.

Bridging

Bridging should be placed between floor joists in any spans over 8 feet to stiffen the joists against twisting. Bridging also distributes the load from a single joist to adjacent joists. The types of bridging we'll discuss are *wood cross*, *metal-type*, and *solid* bridging.

To construct *wood cross* bridging (Figure 6-9), 1″ × 3″s or 1″ × 4″s are recommended. Attach with two 7d or 8d nails at each end. Cross bridging is based on joist length. Estimate the linear feet of bridging required by totaling the length of each section at right angles to the joists. Multiply by 2, if 2″ × 8″ or 2″ × 10″ joists are used. For 2″ × 12″ joists, multiply by 2.5. Cross bridging factors are shown at the top of the next page.

Joist length	Cross bridging factor
2″ × 8″	2.0
2″ × 10″	2.0
2″ × 12″	2.5

For *metal-type* bridging, estimate according to joist size and spacing, two pieces for each space between joists. For 16-inch on center joists, multiply the length of each section at right angles to the joists by 0.75. For 24 inches on center, multiply by 0.50.

For *solid* bridging, use the same size members as for the floor joists. Fit them tightly between joists.

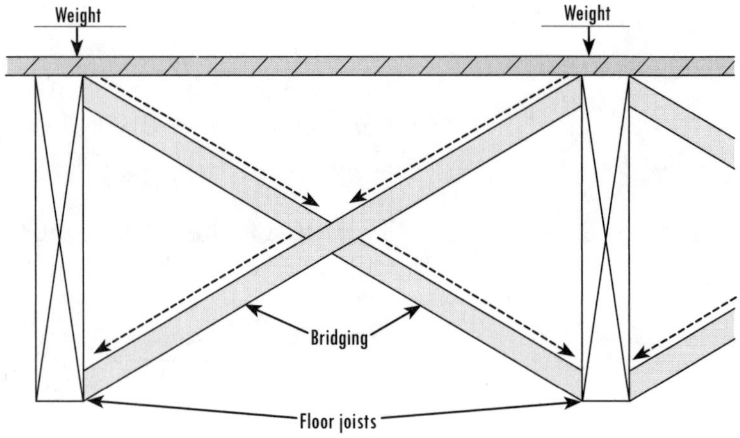

Figure 6-9
Bridging

Allow 3.5 pounds of 8d nails per 100 linear feet of bridging.

Estimating Wood Cross Bridging

We'll use wood cross bridging in our house. Look at Figures 6-5 and 6-6. Our joists are 2″ × 10″s and we'll be using 1″ × 4″ wood cross bridging. How many linear feet of bridging will we need?

First floor at right angles to joists (left)*	26.00
Second floor at right angles to joists (rear)	40.00
Second floor at right angles to joists (front)	36.00**
Subtotal:	102.00 linear feet

*No bridging is required between the floor trusses on the right side.
**Bridging stops at the stair opening.

Our cross bridging factor for 2″ × 10″ floor joists is 2.0.

$$102' \times 2.0 = 204.00 \text{ linear feet of bridging}$$

We need 204 linear feet of bridging, but we should round up our order to 210 linear feet. If we buy 2″ × 4″ × 12′0″s, we can rip them in half to make 1″ × 4″ cross bridging. So, how many pieces of 2″ × 4″ × 12′0″s will we need?

$$\frac{210 \text{ lf}}{12 \times 2} = 8.75 \text{ (rounded to 9) pieces of } 2 \times 4 \times 12$$

*Enter your estimate of the number of pieces of 2″ × 4″ × 12′0″ for wood cross **bridging** on line 6.2.9 of your Cost Estimate Worksheet. We'll use 9 pieces of 2″ × 4″ × 12′0″ for wood cross bridging in our Example Cost Estimate Worksheet.*

Nails

Estimating the quantity of nails for residential construction can feel like a hit-or-miss exercise. In addition to the quantity of nails that would normally be required following the recommended nailing schedule, there's often substantial waste, as well as extra nails needed for temporary scaffolds, and braces, etc. Few factor tables are 100 percent correct, but the following guidelines, allowing for 15 percent waste and extras, should provide a reasonably accurate means for estimating nails for floor joists, headers, and bridging.

Floor Joists

Allow 9 pounds of 16d common nails per thousand board feet for 2″ × 6″ or 2″ × 8″ joists. Allow 7 pounds of 16d common nails per thousand board feet for 2″ × 10″ or 2″ × 12″ joists. There are approximately 54 16d nails per pound.

Let's look at an example. How many pounds of 16d common nails will we need for the floor joists we just estimated? There were 23 joists, 2″ × 10″ × 16′0″ in size on the first floor, and 34 of that dimension on the second floor. That's a total of 57 2″ × 10″ × 16′0″ joists. There were also 36 2″ × 10″ × 12′0″ joists on the second floor. We need to calculate the total board feet for each length, then determine how many nails we'll need.

2″ × 10″ × 16′0″ joists:

$$\frac{57 \times 2'' \times 10'' \times 16'}{12 \times 1{,}000} = 1.52 \text{ MBF}$$

2″ × 10″ × 12′0″ joists:

$$\frac{36 \times 2'' \times 10'' \times 12'}{12 \times 1{,}000} = 0.72 \text{ MBF}$$

$$1.52 + 0.72 = 2.24 \text{ MBF}$$

2.24 MBF × 7 lbs per MBF = 15.68 (rounded to 16) lbs of 16d common nails

> *Enter your estimate of the number of **pounds of 16d common nails** on line 6.2.10 of your Cost Estimate Worksheet. We'll use 16 pounds of 16d common nails in our Example Cost Estimate Worksheet.*

Sill Plate

Use 8d nails to toenail the joist headers to the sill plate and the floor joists to the girder and sill plate. There are approximately 110 8d common nails per pound, and you should allow 2.5 pounds per thousand board feet. So, how many 8d nails will we need? We've already calculated the board feet for the 57 2″ × 10″ × 16′0″ joists and the 36 2″ × 10″ × 12′0″ joists. We need to do the same for the 7 2″ × 10″ × 16′0″ headers, and then add them into our total board feet.

2″ × 10″ × 16′0″ headers:

$$\frac{7 \times 2'' \times 10'' \times 16'}{12 \times 1{,}000} = 0.19\text{MBF}$$

$$2.24 + 0.19 = 2.43 \text{ MBF}$$

$$2.43 \text{ MBF} \times 2.5 \text{ lbs per MBF} = 6.08 \text{ (rounded to 6*) lbs of 8d common nails}$$

*Although we usually recommend rounding up, when it comes to .08 lb of nails, 6 lbs is adequate.

> *Enter your estimate of the number of* ***pounds of 8d common nails*** *on line 6.2.11 of your Cost Estimate Worksheet. We'll use 6 pounds of 8d common nails in our Example Cost Estimate Worksheet.*

Bridging

Fasten bridging with four 8d common nails per piece, and allow 3.5 pounds of nails per 100 linear feet. We have 210 linear feet of wood cross bridging. How many nails should we estimate?

$$\frac{210 \text{ lf} \times 3.5}{100} = 7.35 \text{(rounded to 7.5) lbs}$$

We'll need 7.5 pounds of 8d common nails to attach our bridging material.

> *Enter the number of* ***pounds of 8d common nails*** *on line 6.2.12 of your Cost Estimate Worksheet. We'll use 7.5 pounds of 8d common nails in our Example Cost Estimate Worksheet.*

Subfloor

Apply the subfloor, sometimes called the *floor deck* or *floor sheathing*, over the joists as a base for the finished floor. Subfloor material in residential construction is usually wood boards, plywood, or OSB (*oriented strand board*).

Wood Boards

The minimum thickness for wood boards is 3/8 inch, and the maximum width is 8 inches. Install the boards diagonal to, or at right angles to the floor joists (but not parallel). Each end cut should be parallel to and over

Floor Area Waste Factors		
Wood board nominal size	**Laid at right angles to joists**	**Laid diagonal to joists**
1″ × 6″ S4S	1.14	1.20
1″ × 8″ S4S	1.12	1.17
1″ × 6″ T&G	1.20	1.25
1″ × 8″ T&G	1.15	1.20

S4S - surfaced on 4 sides
T&G - tongue and groove

Figure 6-10
Waste factors

the center of a floor joist. Maximum joist spacing should be 16 inches on center. An exception is made for $^{25}/_{32}$-inch strip flooring when the subfloor is installed diagonally. In that case, joist spacing may be 24 inches on center. Nail the wood boards to joists at each bearing with 8d common or 7d threaded (anchor-down) nails. Use two nails in 6-inch boards and three in 8-inch boards.

Floor Area Waste Factor

Because there's always some waste, you need to use a floor area waste factor to figure the board foot area that'll be covered with subfloor. To find the effective floor area, multiply the actual floor area by the applicable factor from the table in Figure 6-10.

Let's calculate the number of board feet of subfloor we need for the second floor of our sample house. If we use 1″ × 8″ S4S boards laid diagonally (1.17 waste factor), how many board feet are needed for the 40-foot × 26-foot subfloor system? (Remember, one board foot is one square foot, for one-inch-thick boards.)

$$40' \times 26' = 1{,}040 \text{ sf}$$

$$1{,}040 \text{ sf} \times 1.17 = 1{,}216.80 \text{ (rounded to 1,220) bf}$$

Allowing 27 pounds of 8d common or 7d threaded (anchor-down) nails per 1,000 square feet (which includes 15 percent allowance for waste and extras), how many pounds of 7d threaded nails will we need to fasten the 1,040 square feet of subfloor.

$$\frac{27 \text{ lbs x } 1{,}040 \text{ sf}}{1{,}000} = 28.08 \text{ (rounded to 28) lbs of 7d threaded nails}$$

The waste factors we've used are guidelines. Remember, the best source of information comes from your own job records.

Plywood or OSB

Plywood or OSB are also used for subflooring. Install plywood sheets with the face grain at right angles to the joists, and stagger it so the end joints meet over different joists in adjacent panels. The only disadvantage to using plywood or OSB for subflooring is its deflection under loads at the edges.

Plywood must be continuous over two or more joists, with the face grain across supports. Nail the plywood to joists with 8d common or 6d threaded nails (see the recommended nail sizes in Figure 6-11). Space the nails 6 inches on center along all panel edges, and 10 inches on center along intermediate joists. Figure 6-12 shows the nailing detail for plywood or OSB subfloor.

To estimate plywood or OSB, divide the square feet of floor area by the square feet in one piece of plywood (typically, 4 × 8 = 32 square feet). Round your answer up to the next whole number. For example, how many pieces of 4 × 8 plywood or OSB sheets do you need for a 40′3″ × 32′0″ floor area?

$$40.25'\ (40'3'') \times 32.00'\ (32'0'') = 1{,}288 \text{ sf}$$

$$\frac{1{,}288 \text{ sf}}{32} = 40.25 \text{ (rounded to 41) pieces}$$

Deduct only large openings when you estimate plywood. A plywood layout drawn to scale helps in eliminating waste. If floor dimensions for both width and length are on a 4-foot module (24 feet, 48 feet, 64 feet, etc.), waste will be zero. If dimensions are on a 2-foot module (26 feet, 34 feet, 50 feet, 66 feet, etc.), waste will be 3 to 5 percent. If any of the dimensions vary from the module system, waste can be much higher. Uneven joist spacing can also add to waste.

Let's calculate how many pieces of 4 × 8 plywood we'll need for the first floor subfloor in our sample house, shown in

Nail Sizes for Plywood or OSB		
Plywood or OSB thickness (in.)	**Maximum span (in.)**	**Nail size**
½	16	6d threaded or 8d common
⅝	16	8d common
⅝, ¾	20	8d common
¾, ⅞	24	8d common

Figure 6-11
Nail sizes for subfloor

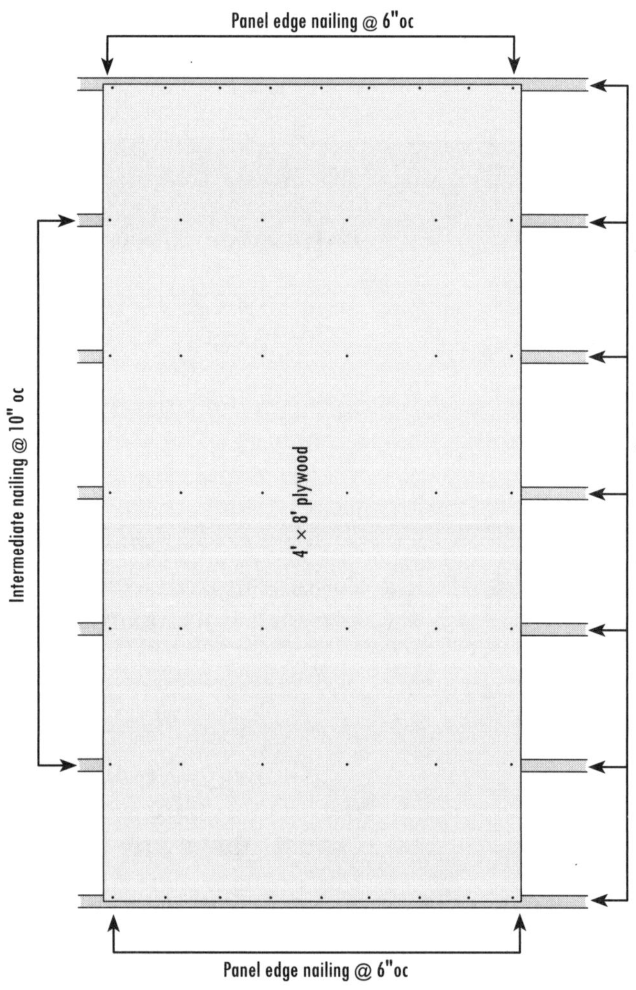

Figure 6-12
Plywood nailing detail

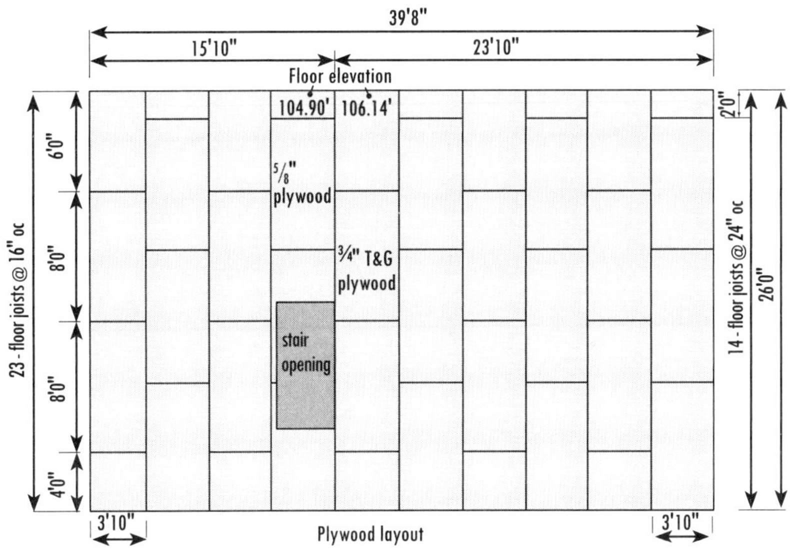

Figure 6-13
Plywood layout – first floor

Figure 6-13. The right side floor area is 23′10″ × 26′0″. The floor trusses are 24 inches on center, and the plans call for ¾-inch tongue and groove subfloor.

$$23.83' \ (23'10'') \times 26.00' = 619.58 \text{ sf}$$

$$\frac{619.58 \text{ sf}}{32} = 19.36 \text{ (rounded to 20) pieces}$$

We need 20 pieces of plywood for this section of the subfloor.

> *Enter your estimate of the number of* ***pieces of ¾-inch 4 × 8 T&G plywood*** *on line 6.3.2 of your Cost Estimate Worksheet. We'll use 20 pieces of ¾-inch × 4 × 8 T&G plywood in our Example Cost Estimate Worksheet.*

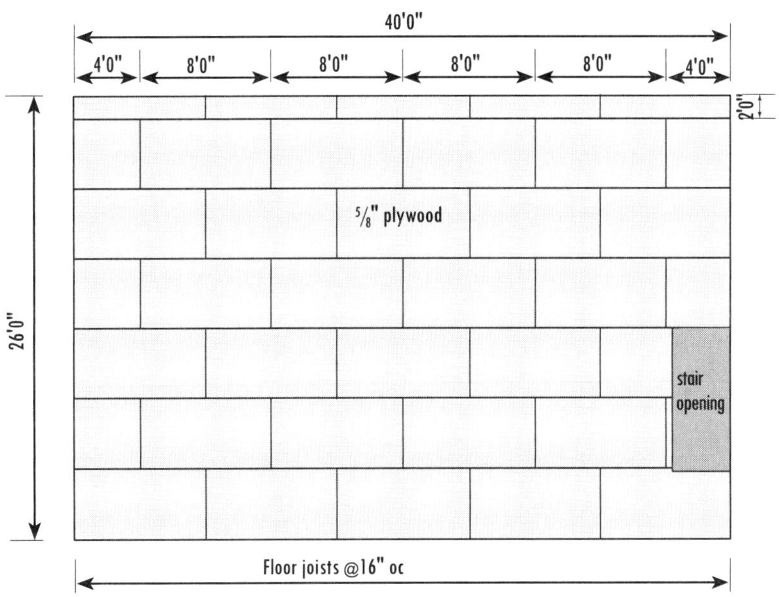

Figure 6-14
Plywood layout – second floor

The left side floor in Figure 6-13 is 15′10″ × 26′0″. Floor joists are 16 inches on center, and the plans call for a ⅝-inch C-D plywood subfloor.

$$15.83' \ (15'10'') \times 26.00' = 411.58 \text{ sf}$$

$$\frac{411.58 \text{ sf}}{32} = 12.86 \text{ (rounded to 13) pieces}$$

We need 13 pieces of plywood for this subfloor section, and a total of 33 pieces of plywood for the entire first floor.

> *Enter your estimate of the number of* ***pieces of ⅝-inch 4 × 8 C-D plywood*** *on line 6.3.3 of your Cost Estimate Worksheet. We'll use 13 pieces of ⅝-inch × 4 × 8 C-D plywood in our Example Cost Estimate Worksheet.*

Next we'll find how many pieces of 4 × 8 plywood we need for the second floor of our sample house, shown in Figure 6-14.

Joist spacing	Nail type	Pounds per 1,000 sf of subfloor*
16″	8d common	16
16″	6d threaded	8
20″	8d common	10
24″	8d common	12

*For 4 × 8 plywood subfloor (including 15 percent waste)

Figure 6-15
Joist spacing and nail types

The floor dimensions are 40 × 26 feet. Floor joists are 16 inches on center, and the plans call for ⅝-inch C-D plywood.

$$40.00' \times 26.00' = 1{,}040 \text{ sf}$$

$$\frac{1{,}040 \text{ sf}}{32} = 32.50 \text{ (rounded to 33) pieces}$$

We need 33 pieces of plywood for the second floor subfloor.

Enter your estimate of the number of ***pieces of ⅝-inch 4 × 8 C-D plywood*** *on line 6.3.4 of your Cost Estimate Worksheet. We'll use 33 pieces of ⅝-inch 4 × 8 C-D plywood in our Example Cost Estimate Worksheet.*

Nails

The nailing pattern in Figure 6-12 uses 48 nails per 4 × 8 board for 16-inch joist spacing. That's 48 nails per 32 square feet of plywood subfloor, or:

$$\frac{48 \times 1{,}000}{32} = 1{,}500 \text{ nails per } 1{,}000 \text{ sf of subfloor}$$

There are approximately 110 8d common nails per pound, so allow:

$$\frac{1{,}500}{110} = 13.64 \text{ lbs per } 1{,}000 \text{ sf of subfloor}$$

Add to that total 15 percent for waste:

Subtotal	13.64
Waste (13.64 × 0.15)	+ 2.05
Total:	15.69 (rounded to 16) pounds per 1,000 square feet

Figure 6-15 shows pounds of nails per 1,000 square feet of plywood subfloor, based on the joist spacing and nail type used.

Using the nail factors from the table in Figure 6-15, how many 8d common nails do we need for the subfloors of our sample house?

First floor, right side, we have 619.58 (rounded to 620) square feet, with joists 24 inches on center:

$$\frac{620 \text{ sf} \times 12 \text{ (nail factor)}}{1{,}000} = 7.44 \text{ lbs}$$

First floor, left side, we have 411.58 square feet (which we'll round up to 412 square feet), with joists 16 inches on center:

$$\frac{412 \text{ sf} \times 16 \text{ (nail factor)}}{1{,}000} = 6.59 \text{ lbs}$$

The second floor is all one level, 1,040 square feet, with joists 16 inches on center:

$$\frac{1{,}040 \text{ sf} \times 16 \text{ (nail factor)}}{1{,}000} = 16.64 \text{ lbs}$$

$$7.44 + 6.59 + 16.64 = 30.67 \text{ (rounded to 31) lbs of 8d common nails}$$

*Enter your estimate of the **number of pounds of 8d common nails** on line 6.3.5 of your Cost Estimate Worksheet. We'll use 31 pounds of 8d common nails in our Example Cost Estimate Worksheet.*

Determining Labor Costs for Floor Systems

Labor costs are estimates and can be difficult to calculate. Many labor factor tables are available to help you estimate manhours for each phase of residential construction, and they're good under average conditions. But, again, the most accurate estimates are based on your records from similar jobs. The use of proportions can save a lot of time in estimating, especially where there's a high waste factor (in masonry cement and nails, for example). We'll do a brief review of proportions to show how they can help you.

Proportions

The ratio of two numbers is the result (*quotient*) of one quantity divided by another of the same kind, or the relationship of one number to another. This relationship is indicated by the sign (:). For example, if 6:8 is the ratio of 6 to 8, and is equivalent of the fraction you get dividing the first number by the second, then the ratio of 6 to 8 is:

$$\frac{6}{8} = \frac{3}{4}$$

The same applies for any multiples of those numbers, such as:

$$\frac{12}{16} = \frac{6}{8} = \frac{3}{4}$$

When two ratios are equal it's called a *proportion*. So:

$$6 : 8 = 12 : 16$$

The first and last numbers in a proportion are called the *extremes*. The second and third numbers (middle numbers) are called the *means*. In this proportion, the extremes are **6** and **16**, and the means are *8* and **12**. The product (multiplication result) of the means is always equal to the product of the extremes.

Example: $6 : 8 = 12 : 16$

Product of means ($8 \times 12 = 96$) is equal to the **product of extremes** ($6 \times 16 = 96$).

So, if any three parts of the proportion are known, the fourth one can be calculated. The product of two known numbers divided by the third known number gives us the fourth (unknown) number. This unknown number is usually designated as (X).

$$6 : 8 = 12 : X$$

$$8 \times 12 = 96$$

therefore

$$6 \times X = 96$$

$$X = \frac{96}{6}$$

$$X = 16$$

The solution in a proportion is easier to understand when using numbers that represent the same thing in the ratio, like dollars to dollars, manhours to manhours, board feet to board feet, etc. But you can also use proportions to ratio things of different types, like apples and oranges. If 8 apples weigh the same amount as 6 oranges, we could say:

8 apples : 6 oranges = 16 apples : 12 oranges

How many oranges weigh the same as 24 apples? The answer you want to calculate is X, so the number of oranges in the second ratio is X. Your proportion looks like this:

8 apples : 6 oranges = 24 apples : X oranges

Product of means: $6 \times 24 = 144$

therefore

Product of the extremes: $8 \times X = 144$

$$X = \frac{144}{8}$$

X = 18 oranges

Let's see how you would apply this to estimating manhours on a job. If recent records show you need 161 manhours to install the floor system for the first floor of a residence with 2,224 square feet of living space, how many manhours should you estimate for the first floor of a house having

1,084 square feet of living space? Use X to represent the number of manhours to install the 1,084 square foot floor system. Set up the proportion like this:

$$161 \text{ manhours} : 2{,}224 \text{ sf} = X \text{ manhours} : 1{,}084 \text{ sf}$$

$$\text{Product of extremes: } 161 \times 1{,}084 = 174{,}524$$

therefore

$$\text{Product of means: } 2{,}224 \times X = 174{,}524$$

$$X = \frac{174{,}524}{2{,}224}$$

$$X = 78.47 \text{ (rounded to 79) manhours}$$

Let's try a few more examples using different types of materials in the proportion. If you need 64 bags of masonry cement to lay 2,139 masonry blocks, how many bags of cement (X) should you estimate to lay 1,784 blocks?

$$64 \text{ bags} : 2{,}139 \text{ blocks} = X \text{ bags} : 1{,}784 \text{ blocks}$$

$$\text{Product of extremes: } 64 \times 1{,}784 = 114{,}176$$

therefore

$$\text{Product of means: } 2{,}139 \times X = 114{,}176$$

$$X = \frac{114{,}176}{2{,}139}$$

$$X = 53.38 \text{ (rounded to 54) bags}$$

If you used 29 pounds of 7d threaded nails to install 1,176 square feet of 1″ × 8″ S4S subfloor, how many pounds of nails (X) should you estimate per 1,000 square feet for a job using the same materials?

$$29 \text{ lbs of nails} : 1{,}176 \text{ sf} = X \text{ lbs} : 1{,}000 \text{ sf}$$

$$\text{Product of extremes: } 29 \times 1{,}000 = 29{,}000$$

therefore

$$\text{Product of means: } 1{,}176 \times X = 29{,}000$$

$$X = \frac{29{,}000}{1{,}176}$$

$$X = 24.66 \text{ (rounded to 25) lbs}$$

Labor Estimate

In one of the examples we just covered, we used payroll records from a recent job to estimate manhours for a house using the same materials. These records showed that it took 161 manhours to install 2,224 square feet of subfloor on the first floor. The house shown in Figure 6-13 has a first floor area of 1,032 square feet (rounded up from 1031.42 square feet). Using the same materials, what would the estimated manhours (X) be to install this subfloor system?

$$161 \text{ manhours} : 2{,}224 \text{ sf} = X \text{ manhours} : 1{,}032 \text{ sf}$$

$$\text{Product of extremes: } 161 \times 1{,}032 = 166{,}152$$

therefore

$$\text{Product of means: } 2{,}224 \times X = 166{,}152$$

$$X = \frac{166{,}152}{2{,}224}$$

$$X = 74.71 \text{ (rounded to 75) manhours}$$

Records show that it took 118 manhours to lay a 1,360 square feet second-story subfloor. Estimate the manhours (X) to install the same type of subfloor system in the second floor of the house shown in Figure 6-14, with 1,040 square feet.

$$118 \text{ manhours} : 1{,}360 \text{ sf} = X \text{ manhours} : 1{,}040 \text{ sf}$$

$$118 \times 1{,}040 = 1{,}360 \times X$$

$$122{,}720 = 1{,}360 \times X$$

$$X = \frac{122{,}720}{1{,}360}$$

$$X = 90.24 \text{ (rounded to 91) manhours}$$

The total manhours estimated to install the first- and second-floor subfloor systems for our sample house is:

First floor	75
Second floor	91
Total:	166 manhours

We'll use one carpenter and one laborer (craft code B1 in the *National Construction Estimator*) to lay the floor system.

> *Enter your estimate of the total number of **manhours** for floor system labor on line 6.4.1 of your Cost Estimate Worksheet. We'll use 166 hours in our Example Cost Estimate Worksheet.*

COST ESTIMATE WORKSHEET FOR FLOOR SYSTEMS

#	Qty	Item	Size	MBF		Cost	Per		Amount	Subtotal
6.1		**Sill plate, girder, ledger or joist hangers**								
6.1.1	22	pieces, sill plate	2 " × 6 " × 12 ft	0.264	@	$718.00	MBF	=	$189.55	
6.1.2	3	pieces, girder	2 " × 10 " × 18 ft	0.090	@	$725.00	MBF	=	$65.25	
6.1.3	3	pieces, girder	2 " × 10 " × 10 ft	0.050	@	$725.00	MBF	=	$36.25	
6.1.4	2	pieces, ledger	2 " × 4 " × 12 ft	0.016	@	$697.00	MBF	=	$11.15	
6.1.5		pieces, joist hangers	" × " × ft		@		each	=	$0.00	
6.1.6	2	lbs, nails	20d common		@	$1.84	lb	=	$3.68	
6.1.7	2	lbs, nails	16d common		@	$2.04	lb	=	$4.08	
									$309.96	$309.96
6.2		**Floor joists, headers, trimmers and bridging**								
6.2.1	23	pieces, floor joists	2 " × 10 " × 16 ft	0.613	@	$630.00	MBF	=	$386.19	
6.2.2	36	pieces, floor joists	2 " × 10 " × 12 ft	0.720	@	$630.00	MBF	=	$453.60	
6.2.3	34	pieces, floor joists	2 " × 10 " × 16 ft	0.907	@	$630.00	MBF	=	$571.41	
6.2.4		pieces, floor joists	" × " × ft	0.000	@		MBF	=	$0.00	
6.2.5	14	pieces, floor trusses	24' 2¼" long		@	$90.86	truss	=	$1,272.04	
6.2.6	7	pieces, headers	2 " × 10 " × 16 ft	0.187	@	$630.00	MBF	=	$117.81	
6.2.7		pieces, headers	" × " × ft	0.000	@		MBF	=	$0.00	
6.2.8		pieces, trimmers	" × " × ft	0.000	@		MBF	=	$0.00	
6.2.9	9	pieces, bridging	2 " × 4 " × 12 ft	0.072	@	$689.00	MBF	=	$49.61	
6.2.10	16	lbs, 16d common nails			@	$2.04	lb	=	$32.64	
6.2.11	6	lbs, 8d common nails			@	$2.04	lb	=	$12.24	
6.2.12	7.5	lbs, 8d common nails			@	$2.04	lb	=	$15.30	
									$2,910.84	$2,910.84
6.3		**Subfloors**								
6.3.1		board ft, wood boards			@			=	0.00	
6.3.2	20	pieces, T&G plywood	¾ " × 4 ft × 8 ft		@	$19.95	piece	=	$399.00	
6.3.3	13	pieces, C-D plywood	⅝ " × 4 ft × 8 ft		@	$18.95	piece	=	$246.35	
6.3.4	33	pieces, C-D plywood	⅝ " × 4 ft × 8 ft		@	$18.95	piece	=	$625.35	
6.3.5	31	lbs, 8d common nails			@	$2.04	lb	=	$63.24	
6.3.6					@			=	$0.00	
									$1,333.94	$1,333.94
			Cost of material							$4,554.74
			Sales tax		@	7.75	%			$352.99
			Total cost of material							$4,907.73
6.4		**Labor costs**								
6.4.1	166	manhours - B1 crew (carpenter plus laborer)			@	$29.37	hour	=	$4,875.42	
6.4.2		manhours			@		hour		$0.00	
			Total cost of labor						4,875.42	$4,875.42
		Cost of floor system (entered on line 6 of Form 100)								**$9,783.15**

Chapter 7
Superstructure

THE SUPERSTRUCTURE OF A house, as defined in this chapter, consists of the following:

- Framing for the exterior walls
- Framing for the interior walls
- Sheathing for the exterior walls
- Framing for the roof system
- Sheathing for the roof
- Stringers for the stairs from the first to the second floor
- Framing and sheathing for porches
- Labor and other materials necessary to assemble the superstructure

Exterior and Interior Walls

We'll estimate the materials and labor required for these systems in our sample house, beginning with the exterior and interior walls. You'll find the wall system dimensions for the first floor shown in Figure 7-1, and the second floor in Figure 7-2.

Sole and Top Plates

The bottom or *sole* plate is a horizontal member, resting on the subfloor, to which studs are attached (Figures 7-3 and 7-4). Normally, sole plates in residential construction are made of 2 × 4s. They should be nailed to the floor joists with 16d common nails or 3½-inch spiral-thread nails, spaced not more than 16 inches on center.

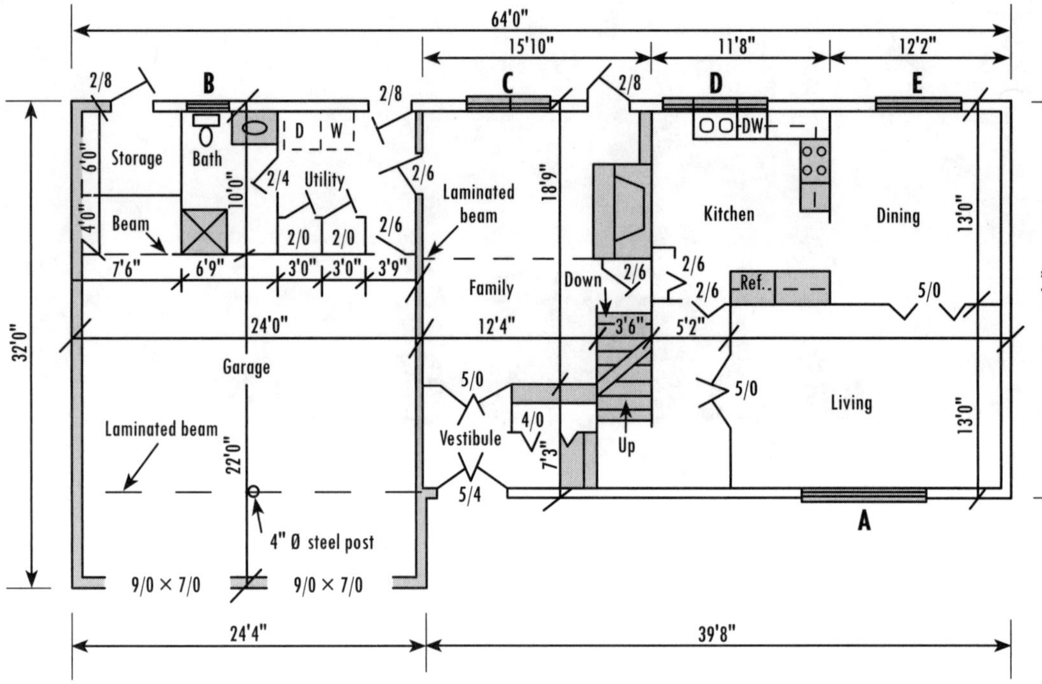

Figure 7-1
First floor

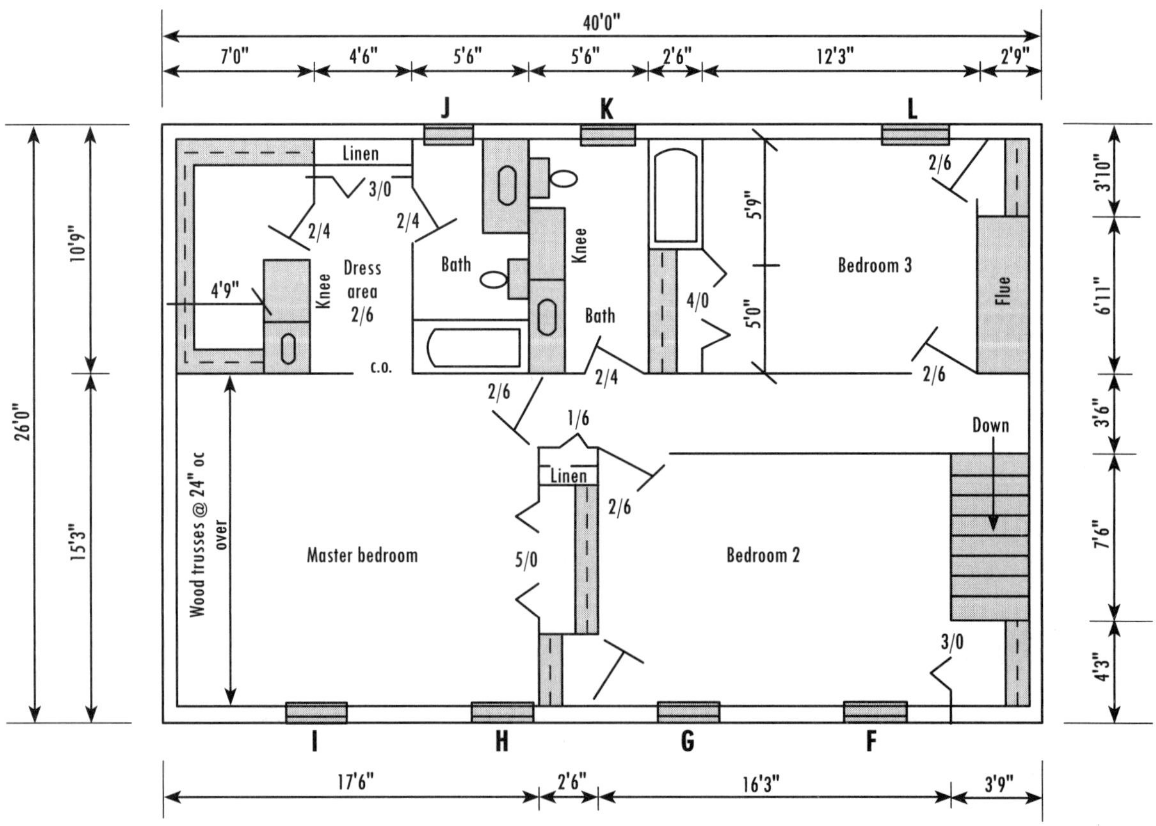

Figure 7-2
Second floor

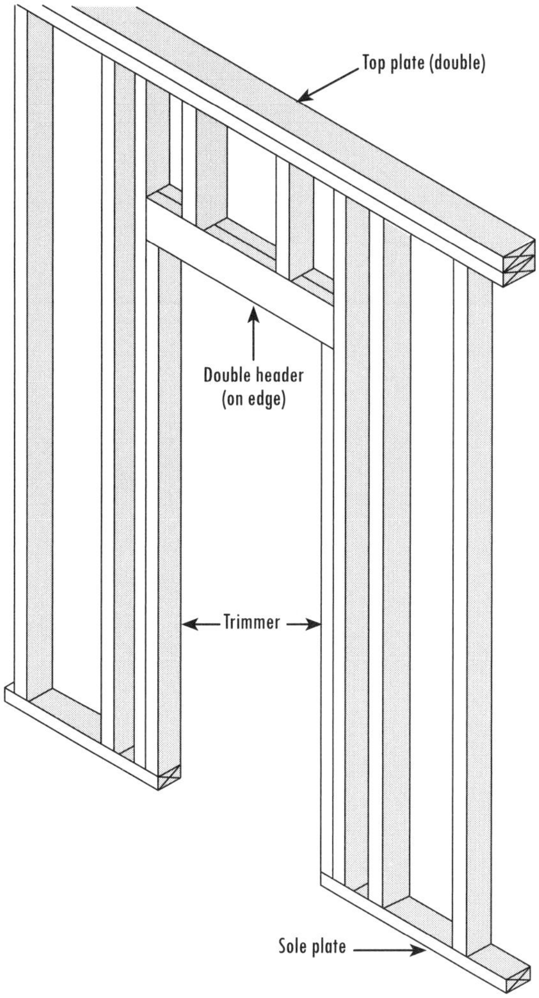

Figure 7-3
Bearing wall with sole plate and double top plate

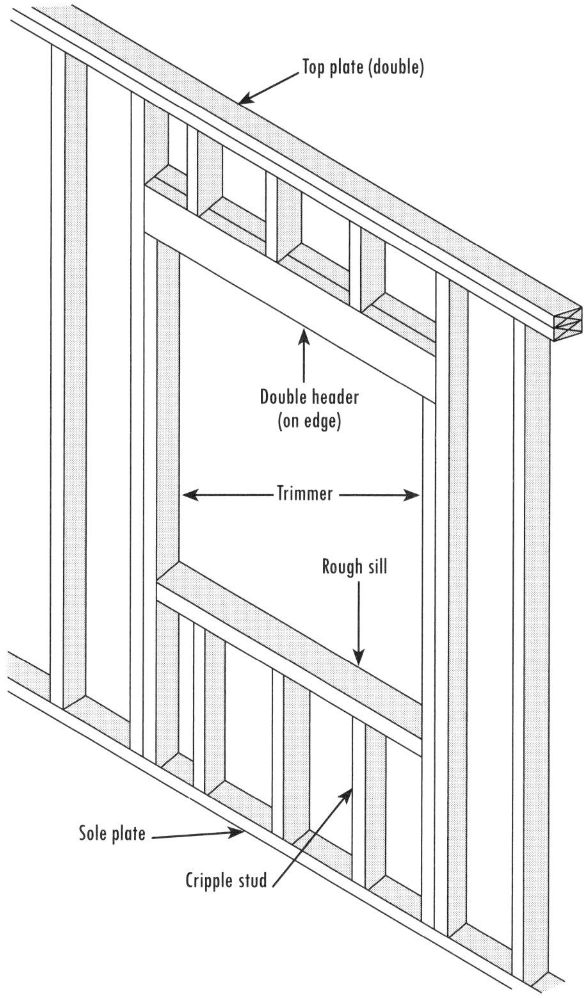

Figure 7-4
Bearing wall with double top plate and opening

The top plates must be doubled for all bearing walls (Figures 7-3 and 7-4), but may be single for nonbearing walls, as shown in Figure 7-5. However, many builders use double top plates for both bearing and nonbearing walls.

When using doubled top plates, the ends of the top plates on lower members must meet over studs. Joints in the upper members of top plates must be at least 24 inches from any joints in the lower members. Use two 16d common nails at each stud for the lower top plate member, and nail the upper plate to the lower plate with 16d common or 3½-inch spiral-thread nails, 16 inches on center.

Calculate exterior and interior walls separately when estimating sole and top plates. Some walls vary from the normal 2 × 4 dimensions. For example, a wall may be 6 inches wide to accommodate the plumbing for a bathroom, while a closet partition may be only 2 inches wide, to gain additional space. These differences can change an estimate considerably.

Door openings aren't normally deducted from an estimate unless they exceed 8 feet in width, as is the case for a garage door. The wood cut from smaller openings can be used in other locations during the construction of the house.

If double top plates are to be used on all walls, multiply the total linear feet of the exterior and interior walls by the number of plates used to get the total linear feet of top and sole plate lumber required. In this estimate, the number will be 3: one sole plate and 2 top plate members. If 2 × 4 × 12s are used, divide the total linear feet of the plates by 12 to determine the number required. If 2 × 4 × 14s are used, divide by 14, and so on.

Let's calculate how many 2 × 4 × 12s will be required for the sole and top plates for our house. We'll use the dimensions given in Figures 7-1 and 7-2 to determine the linear feet of interior wall required. There are no walls that deviate from the normal 2 × 4 walls, and we'll use double top plates on both bearing and nonbearing walls.

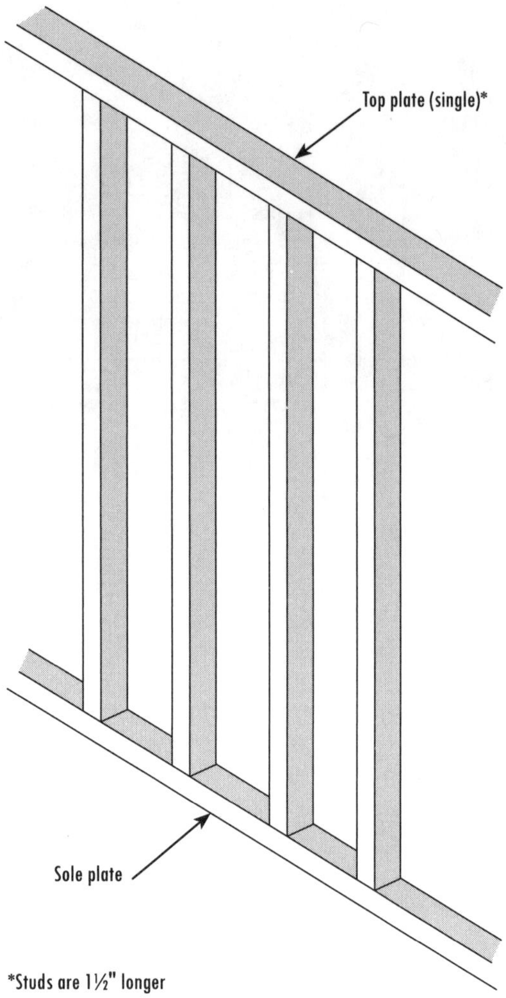

Figure 7-5
Nonbearing wall

First floor (from Figure 7-1)

Exterior walls	192 lf
Interior walls	213 lf
Subtotal	405 lf

Second floor (from Figure 7-2)

Exterior walls	132 lf
Interior walls	185 lf
Subtotal	317 lf

Total (405 lf + 317 lf) = 722 lf

We need to multiply our linear feet by 3 to include 1 sole plate and 2 top plate members, then subtract the linear feet of sole plate not needed for the garage door opening:

722 lf × 3 =	2,166 lf
	− 18 lf
Total:	2,148 lf

We'll use 2 × 4 × 12′ lumber for the sole and top plates.

$$\frac{2{,}148\ \text{lf}}{12'\ \text{per piece}} = 179\ \text{(rounded to 180) pieces, or 1,440 bf}$$

We need a total of 180 pieces of sole and top plate lumber.

Enter your estimate of the number of pieces of 2 × 4 × 12 ***sole and top plates*** *on line 7.1.1 of your Cost Estimate Worksheet. We'll use 180 pieces of 2 × 4 × 12s in our Example Cost Estimate Worksheet.*

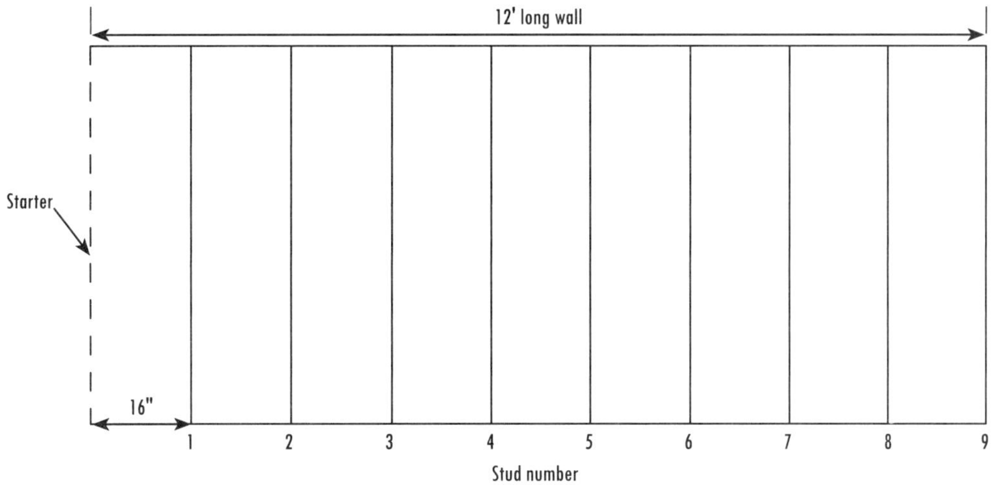

Figure 7-6
Number of studs required

Studs

Studs are the vertical framing members in a wall to which sheathing material is fastened. Studs are the "bones" of a house. The wall section of the structure's blueprints will give the size, spacing, and wall height for studs. They must be continuous lengths with no splicing and be strong enough to safely support the design loads. Each stud should be nailed to the sole plate with four 8d or two 16d common nails.

Studs in Plain Walls

There are two important points to remember when estimating the number of studs — always add one starter stud, as shown in Figure 7-6, and use the correct *stud factor*.

It's fairly easy to estimate the number of studs in a solid wall, one without any window or door openings or intersecting walls. For instance, how many studs, spaced at 16 inches on center, are required to frame a 12-foot-long wall?

12′0″ × 0.75 =	9
Add for starter stud	+ 1
Total:	10 studs

Why did we multiply by 0.75? To get the quantity we need based on 16-inch centers. Let's look at an example.

A 10-foot wall is 120 inches long (10′ × 12″ = 120″). If the studs are placed at 16-inch centers, the number of studs required will be:

$$\frac{120''}{16''\text{oc}} = 7.5 \text{ studs}$$

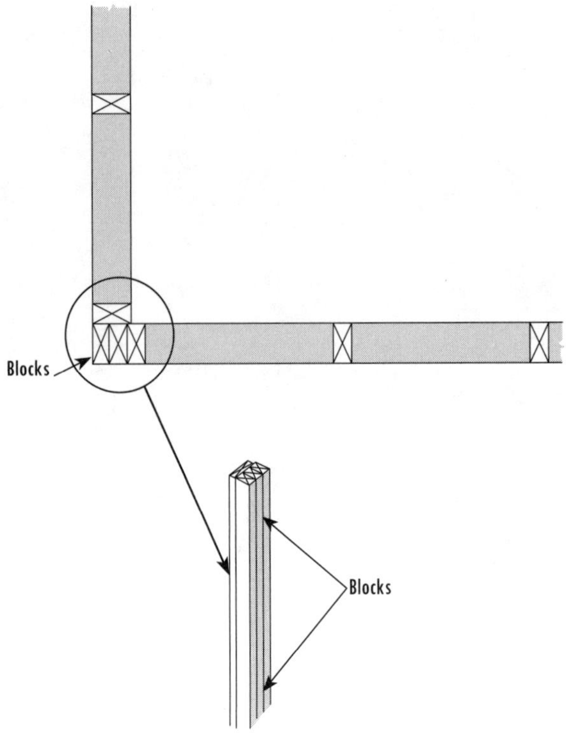

Figure 7-7
Corner post

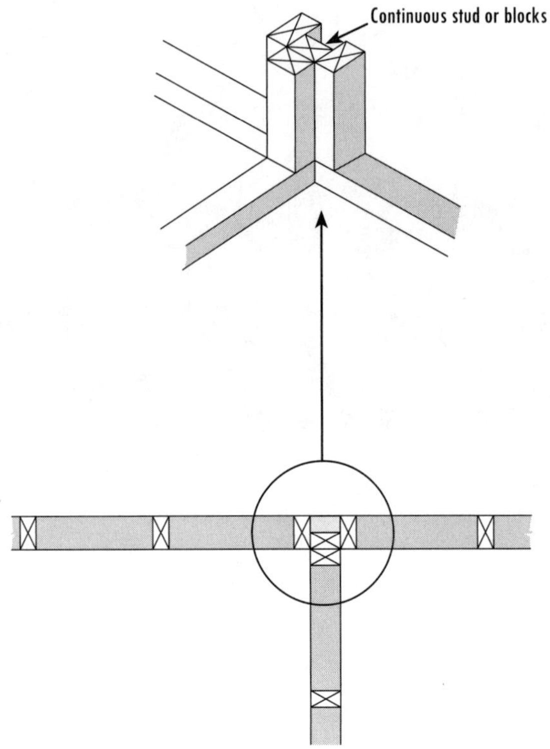

Figure 7-8
Wall intersection

So if you need 7.5 studs for a 10-foot wall, a 1-foot wall will need 0.75 stud, or a stud factor of 0.75 for 16-inch on center studs.

$$\frac{7.5}{10} = 0.75 \text{ stud (stud factor)}$$

You can calculate the stud factor this way for any stud spacing. Here are stud factors for the most frequently used spacing:

Stud spacing	Stud factor
12-inch centers	1.00
16-inch centers	0.75
20-inch centers	0.60
24-inch centers	0.50

Studs in Walls with Window and Door Openings

When you're building an entire house, some of the walls get a little more complicated, and estimating the number of studs is also more complicated. You don't have just plain walls any more. There are window and door openings with trimmer studs and headers, corner posts, T-posts at intersecting walls, and blocking for partitions. All these require additional studs (see Figures 7-3, 7-4, 7-7, and 7-8).

You can estimate the number of studs you need by allowing one stud for each linear foot of exterior and interior wall. Some estimators make an additional allowance of two studs for each corner, and some arbitrarily allow 50 to 100 additional studs, depending on the size of the house. Always allow more studs than the number calculated. Your previous records can help you determine the additional allowance. Don't forget to account for double walls when they're used for plumbing or sound reduction.

Note: Studs can be ordered precut to 7′8⅝″ (92⅝″). Adding 4½″ for the sole plate and two top plates (i.e. 3 × 1½″ actual thickness) will give us a ceiling height of 8′1⅝″ from the subfloor to the ceiling joists. There's very little difference between the cost of precut studs and regular 2 × 4 × 8s when large quantities are ordered, but there's a significant savings in labor when you use precut studs.

We'll use the one stud per linear foot approach for our sample house. We've already calculated the linear feet of interior and exterior walls for our sole plates and top plates. The number of studs required for both exterior and interior walls will be equal to the total linear feet.

First floor exterior walls	192 lf
First floor interior walls	213 lf
Second floor exterior walls	132 lf
Second floor interior walls	185 lf
Total:	722 lf

Allowing one stud per linear foot of sole and top plates, we'll need 722 studs.

Studs in Gables

If you use rafters, rather than roof trusses, for a gable roof, you must also estimate studs for the gable ends. Use the following steps:

1. For each gable, multiply the width of the gable by the stud factor to determine the number of studs needed. Add one to your total for the starter stud.
2. Determine the length of the longest stud from the blueprint elevation, and divide by 2 to get the average stud length.
3. Multiply the number of studs in each gable by the average stud length to find the total linear feet of gable studs required.
4. Divide the total linear feet by 8 and round up to get the number of 2 × 4 × 8 studs you need for each gable.
5. If all gables are the same size, multiply the number of studs in each gable by the number of gables to get the total number of gable studs required. If gables vary in size, calculate each separately, then add them together to get the number of studs you'll need.

Let's do a sample problem using the information from Figure 7-9. How many studs will be required for the three roof gables plus the gables on the 6-foot offset for the garage area? The width of each gable is 26 feet, and the studs will be spaced 16 inches on center. The stud factor is 0.75.

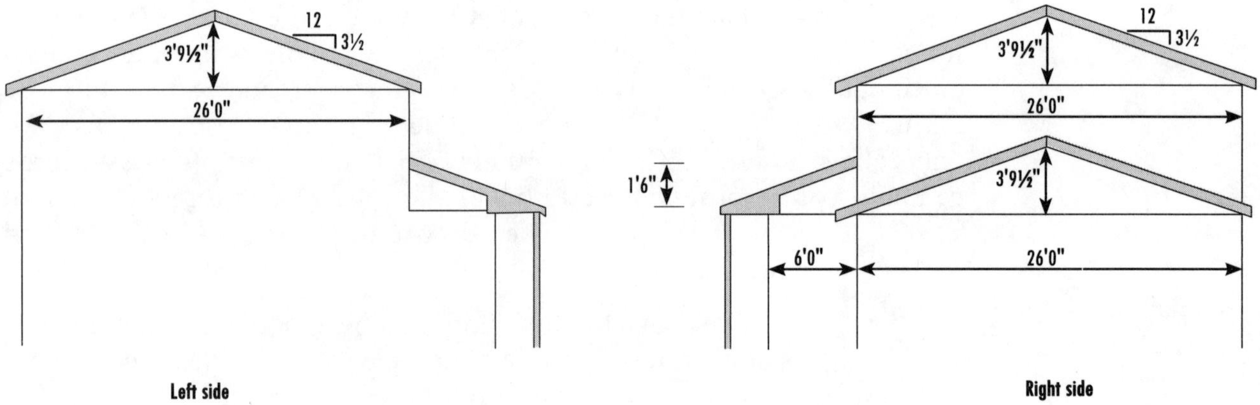

Figure 7-9
Gable ends

1. Multiply the gable width by the stud factor to find the number of studs needed:
   26′ × 0.75 = 19.50 (rounded to 20) studs,
   add 1 for the starter stud = 21 studs per gable.
2. The longest gable stud is 3′9½″ (3.79′). Divide by 2 to get the average gable stud length:
   $$\frac{3.79'}{2} = 1.895' \text{ (rounded to } 1.90'\text{) average stud length}$$
3. Multiply the number of studs per gable by the average length to get the linear feet per gable.
   21 studs × 1.90′ = 39.90 linear feet per gable.

Multiply by the number of gables to get the total linear feet of studs required.

39.90 lf × 3 gables	119.70 lf
Estimate for the 2 offset gables	11.00 lf
Total:	130.70 lf

Divide the total linear feet by 8 to determine how many 8-foot studs are required for the 3 gables and 2 offset gables.

$$\frac{130.70 \text{ lf}}{8'} = 16.34 \text{ (rounded to } 17\text{) studs}$$

Now, let's total the number of 2 × 4 × 8 studs required for the exterior, interior, and gable walls in Figures 7-1, 7-2, and 7-9:

Exterior and interior walls	722
Gables	17
Subtotal	739 studs

Based on the waste allowance of 50-100 studs mentioned previously, our cost records show that an allowance of 61 extra studs should satisfy our waste requirement.

Door Width	Header Size (on edge)	Header Length	Estimate for Each Double Header
2'0"	2 × 4s	2'5"	1 - 2 × 4 × 8'
2'4"	2 × 4s	2'9"	1 - 2 × 4 × 8'
2'6"	2 × 4s	2'11"	1 - 2 × 4 × 8'
2'8"	2 × 6s	3'1"	1 - 2 × 6 × 8'
3'0"	2 × 6s	3'5"	1 - 2 × 6 × 8'
4'0"	2 × 6s	4'5"	1 - 2 × 6 × 10'
5'0"	2 × 6s	5'5"	1 - 2 × 6 × 12'
6'0"	2 × 8s	6'5"	1 - 2 × 6 × 14'

Figure 7-10
Headers for interior door openings

Subtotal	739
Allowance for extras	61
Total:	800 studs

We'll need 800 2 × 4 × 8 studs (4,267 board feet).

*Enter your estimate of the number and length of **studs** required on line 7.1.2 of your Cost Estimate Worksheet. We'll enter 800 pieces of 2 × 4 × 8s in our Example Cost Estimate Worksheet.*

Headers

Headers support the weight over wall openings (Figures 7-3 and 7-4). Header size varies depending on the span of the opening. Both the size and the length of each header can be found on the blueprints. Multiply the header length by 2 if you're installing a double header.

Interior Wall Headers

When estimating a header for an interior door, the length of the header is the door width plus 5 inches. This is calculated as follows:

Thickness of door jambs (2 at ¾")	1½"
Allowance for clearance	½"
Bearing on 2 studs (2 at 1½")	3"
Total:	5"

So the header length for a 2'6" door opening would be 2'11". Because the header is doubled, we'll need 5'10" (2 × 2'11"). One 2 × 4 × 8 should be estimated for this door opening. Estimates for other door widths are given in Figure 7-10.

Exterior Wall Headers

Header Size (on edge)	Maximum Width of Rough Stud Opening	Header Length
2 × 4s	3'0"	Rough stud opening width plus bearing on two studs, i.e. 3"
2 × 6s	6'0"	
2 × 8s	8'0"	
2 × 10s	*10'0"	Rough stud opening width plus bearing on four studs, i.e. 6"
2 × 12s	*12'0"	

*Triple studs at jamb opening; headers to bear on two 2 × 4s on each end

Figure 7-11
Headers for exterior wall openings

Estimating the header lengths for openings in exterior walls requires extra time and care. When windows are twin or triple, an additional allowance must be made for the rough stud opening for each mullion.

A window with a sash opening 2'4" wide may require a rough stud opening of 2'6½" for a single unit, 5'½" for a twin unit, and 7'6½" for a triple unit. The thickness of outside door jambs is normally 1¼ inches or 15/16-inch. Some doors are double (like the front door in Figure 7-1) and some have side lights. When you're estimating headers for exterior openings, add 3 inches to any rough stud opening up to 8 feet. Add 6 inches for openings over 8 feet wide (Figure 7-11).

Let's look at an example. A 2/4 × 4/2 twin window with a rough stud opening width of 5'1" will require two 2 × 6s for the header. The bearing on two studs is 3 inches (2 × 1½ inches), plus the rough stud opening width of 5'1", which equals 5'4". This will be the length of each header. The double header, then, will be 10'8" (2 × 5'4"). You would estimate one 2 × 6 × 12 for this window opening's header.

Manufacturers of windows and sliding doors provide specifications for rough stud openings for their products. These specs are available where the products are sold, so refer to a copy of these specifications before making your estimate.

Figure 7-12 is a form for estimating the headers for exterior walls, and Figure 7-13 is a form for estimating headers for interior door openings. These forms can be found on the accompanying CD-ROM, in a sleeve at the back of this book.

The consolidated list of material for the headers for the exterior and interior walls in our house, in board feet, shown in Figures 7-1 and 7-2, is:

2	2 × 12 × 12	48 bf
4	2 × 10 × 10	67 bf
2	2 × 8 × 8	21 bf
7	2 × 6 × 12	84 bf
2	2 × 6 × 10	20 bf
2	2 × 6 × 8	16 bf
9	2 × 4 × 12	72 bf
3	2 × 4 × 10	20 bf
6	2 × 4 × 8	32 bf
Total:		380 bf

*Enter your estimate of the number and length of **headers** required on line 7.1.3 of your Cost Estimate Worksheet. We'll use the figures just listed in our Example Cost Estimate Worksheet.*

Headers for Exterior Openings					
Unit	Rough stud opening	Add for bearing	Header length	Header size (on edge)	Estimate for each double header

Figure 7-12
Form for estimating headers for exterior openings

Headers for Interior Wall Openings					
Location	Door width	Add for allowance	Header length	Header size (on edge)	Estimate for each double header

Figure 7-13
Form for estimating headers for interior wall openings

Temporary Wall Braces

After the exterior and interior walls are in place, you'll need temporary wall braces to keep them plumb and aligned until the upper top plate and ceiling joists or trusses are installed. These temporary wall braces are normally 2 × 4 × 12s. When the top plate and joists are in place, these braces can be reused for blocking or as nailers, drop ceilings, or lookouts. Estimating the number of 2 × 4 × 12s for temporary wall bracing isn't an exact science; allowing 20 lengths or 160 board feet of 2 × 4 × 12s for the entire house (the braces from the first floor can be reused on the second floor) is a reasonable estimate for most jobs. As always, the actual numbers you've used in similar jobs serve you better as ballpark figures.

Enter your estimate of the number of ***temporary wall braces*** *required on line 7.1.4 of your Cost Estimate Worksheet. We'll use 20 lengths of 2 × 4 × 12s in our Example Cost Estimate Worksheet.*

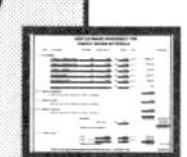

Corner Bracing

Corner bracing adds rigidity to the structure by protecting against lateral forces (such as wind) pushing against the walls. Any of the following materials may be used for corner bracing; the thicknesses listed are minimum requirements.

- Wood sheathing installed at 45-degree angles, in opposite directions from each corner.
- 1 × 4 or wider boards let into either the inner or outer face of studs, sole plate, and top plate located near each corner, and set at approximately 45-degree angles.
- $5/16$-inch-thick 4 × 8 plywood or OSB sheathing applied vertically for studs at 16 inches on center.
- $3/8$-inch-thick 4 × 8 plywood or OSB sheathing applied vertically for studs at 24 inches on center.
- $7/16$-inch-thick 4 × 8 sheathing-grade fiberboard applied vertically for studs at 16 inches on center.

- ½-inch-thick 4 × 8 exterior-grade gypsum board applied vertically for studs at 16 inches on center.
- 3/8-inch-thick 4 × 8 particleboard applied vertically for studs at 16 or 24 inches on center.

Generally, ½-inch-thick 4 × 8 plywood sheathing is used for corner bracing, and ½-inch-thick 4 × 8 OSB sheathing is installed vertically on the balance of the walls. When estimating corner bracing, use one of these two methods:

1. If let-in bracing is used, estimate two 1 × 4 × 12s for each corner.
2. If sheet material is used, estimate two 4 × 8 pieces for each corner, one on each side of the corner.

The corner bracing for our house will be made of ½-inch-thick 4 × 8 plywood. There are six corners on the first floor, and four on the second floor, for a total of 10 corners. Each corner requires two panels, for a total of 20 panels.

> *Enter your estimate of the number of panels or let-in **corner bracing** required on line 7.1.5 of your Cost Estimate Worksheet. We'll use 20 pieces of ½-inch-thick 4 × 8 C-D plywood sheathing in our Example Cost Estimate Worksheet.*

Wall Sheathing

Wall sheathing is installed to strengthen and add rigidity to exterior walls. It also provides additional insulation for the house. Special polystyrene foam boards with reflective vapor retarders increase the "R factor" for greater energy savings, and are now being used by many builders. (We'll discuss the R factor later when we cover *Insulation.*)

The following materials may be used for wall sheathing:

1. Wood board sheathing (T & G, square-edge or shiplapped); minimum thickness ¾-inch; maximum width 12 inches. Corner bracing is required unless boards are installed diagonally.
2. Exterior grade plywood or OSB sheathing; minimum thickness 5/16 inch for studs at 16 inches on center; 3/8 inch for studs at 24 inches on center. Plywood sheathing, installed vertically, is acceptable for corner bracing. Nail plywood or OSB with 6d nails spaced 6 inches on center along edges, and 12 inches on center along intermediate studs.
3. Sheathing grade fiberboard; minimum thickness ½ inch. Corner bracing is required. Nail sheathing to studs at each bearing with 1½-inch roofing nails (1¾-inch roofing nails for 25/30-inch sheathing), using the same spacing as for plywood.

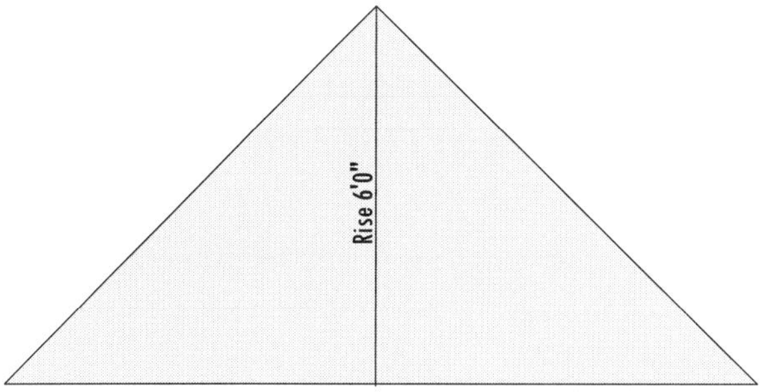

Figure 7-14
Gross area of gable

4. Exterior grade gypsum sheathing; minimum thickness ½ inch. Corner bracing is required. Nail sheathing to studs at each bearing with 1½-inch roofing nails with 3/8- to 7/16-inch heads, 4 inches on center at edges and 8 inches at intermediate supports.

5. Polystyrene foam board, cement board, cement/steel board, fiber cement board and foil-backed board sheathing. Follow the manufacturer's recommendation for installation and nail schedule.

Sheet or panel material has virtually replaced wood boards for wall sheathing in most new houses. When estimating wall sheathing, check the wall section of the blueprints for the type and thickness of the material required. Some estimators deduct for large openings, such as garage doors and picture windows. However, the cost of material you save by deducting for these openings can be lost in the extra labor required to cut them. There's less labor involved in sheathing the walls solid and cutting out openings for windows and doors, than in cutting and fitting small pieces of sheathing around them. The cutouts from these openings can be used as fillers around band joists, and to help offset waste for the gable sheathing.

The first step in estimating wall sheathing is to calculate the total linear feet of the perimeter of the exterior walls on each floor. Then multiply that figure by the wall height to get the gross wall area, in square feet.

The second step is to calculate the gross area in square feet of the gables. Do this by multiplying the rise from the top plate to the ridge (found in the elevation section on your blueprints) by the width of the gable, then dividing by 2.

Let's calculate how much sheathing will be required for the gable area shown in Figure 7-14. It's 24 feet wide, with a 6-foot rise from the top plate to the ridge.

$$\frac{6' \text{ Rise} \times 24' \text{ Width}}{2} = 72.00 \text{ sf}$$

The gable area will require 72 square feet of sheathing.

Calculate each gable separately if they're different sizes, then add those numbers for the total material needed for the gable areas.

The third step is to add the wall and gable areas to determine the combined total gross area of wall sheathing. If deductions for openings are made, subtract the openings from the gross area to get the net area in square feet to be sheathed.

- If 4 × 8 sheets or panels are used, divide the total area (in square feet) by 32 (4 × 8 = 32 square feet) and round up to get the necessary number of sheets.

- ➢ If plywood corners are used for bracing, and the balance of the walls will be another type of panel, subtract the number of plywood bracing sheets from the total panels required, to get the amount of sheathing to estimate.

Refer to Figures 7-1 and 7-2 to calculate how much wall sheathing we need to estimate for the first and second floors of our sample house. The dimensions for the three gables are in Figure 7-9. We'll use ½-inch-thick 4 × 8 plywood on all corners for bracing, and ½-inch-thick 4 × 8 OSB sheathing for the balance of the house.

In Figure 7-1, the perimeter of the first floor is 192 feet. In Figure 7-2, you can see that the perimeter of the second floor is 132 feet. The height of each floor is 9 feet (8-foot ceiling height plus 1 foot for the floor joists).

First floor	192′ × 9′ = 1,728 sf
Second floor	132′ × 9′ = 1,188 sf

Each of the three gables in Figure 7-9 is 26 feet wide with a rise of 3.79 feet (3′9½″).

$$\frac{3.79' \text{ Rise} \times 26.00' \text{ Width}}{2} = 49.27 \text{ sf}$$

49.27 sf × 3 gables = 147.81 sf

The 6-foot extension of the garage at the left front corner of the house has a shed-type roof with two gables. The area of these two gables is 9 square feet.

The total gable area is:

3 main gables	147.81 sf
2 shed gables	9.00 sf
Total:	156.81 (rounded to 157) sf

Sheathing gables produces more waste, but the cutouts left from the picture window and door openings offset this waste, so no additional allowance needs to be made.

The total gross area for wall sheathing, in square feet, is:

First floor	1,728 sf
Second floor	1,188 sf
Gables	157 sf
Subtotal	3,073 sf
Minus garage door opening	–126 sf
Total:	2,947 sf

$$\frac{2{,}947 \text{ sf}}{32 \text{ sf per panel}} = 92.09 \text{ (rounded to 93) panels}$$

We need 93 panels less the 20 plywood corner panels that we previously estimated (6 corners on the first floor, and 4 corners on the second floor, for a total of 10 corners, each corner requiring 2 plywood panels, one on either side of the corner):

$$\begin{array}{ll} & 93 \\ & \underline{-20} \\ \textbf{Total:} & 73 \text{ OSB panels} \end{array}$$

> *Enter your estimate of the number of panels required for* ***wall sheathing*** *on line 7.1.6 of your Cost Estimate Worksheet. We'll use 73 pieces of ½-inch-thick 4 x 8 OSB sheathing in our Example Cost Estimate Worksheet.*

Nails

Next, estimate the quantity of nails needed to secure the studs, plates and headers, and the plywood and OSB sheathing.

Nails for Studs, Plates and Headers

The nail factor per 1,000 board feet of framing for studs, plates, and headers is about twice that of the floor joists. That's because there's a high percentage of waste, and many nails are used to temporarily brace walls and build scaffolds. You must make allowance for this waste. A reasonable guideline to use in the absence of previous records is 22 pounds of 16d common nails per 1,000 board feet. Using records from previous jobs, you can get a more accurate nail factor. You can set up a proportion to estimate for similar new jobs.

If you used 129 pounds of 16d common nails on a previous job for the studs, plates and headers totaling 5,980 board feet of lumber, what will the nail factor (X) per 1,000 board feet be for a similar job? Remember, the product of means is equal to the product of extremes:

$$129 \text{ lbs} : 5{,}980 \text{ bf} = X \text{ lbs} : 1{,}000 \text{ bf}$$

$$5{,}980 \times X = 129 \times 1{,}000$$

$$X = \frac{129 \times 1{,}000}{5{,}980} = 21.57 \text{ (rounded to 22) lbs}$$

The new nail factor (X) is 22 pounds per 1,000 board feet of studs, plates, and headers.

Using this nail factor, how many pounds of 16d common nails will be required for the board feet of sole and top plates, studs, headers, and temporary braces we have previously estimated for the house shown in Figures 7-1, 7-2, and 7-9?

Sole and top plates	180 lengths	2 × 4 × 12′0″	1,440 bf
Studs	800 lengths	2 × 4 × 8′0″	4,267 bf
Headers, various			380 bf
Temporary braces	20 lengths	2 × 4 × 12′0″	160 bf
Total:			6,247 bf

Set up the proportion:

$$22 \text{ lbs} : 1{,}000 \text{ bf} = X \text{ lbs} : 6{,}247 \text{ bf}$$

$$1{,}000 \times X = 22 \times 6{,}247$$

$$X = \frac{22 \times 6{,}247}{1{,}000} = 137.43 \text{ (rounded to 138) lbs}$$

We need 138 pounds of 16d common nails for the 6,247 board feet of material.

> *Enter your estimate of the number of pounds of 16d common **nails** on line 7.1.7.1 of your Cost Estimate Worksheet. We'll use 138 pounds of 16d common nails in our Example Cost Estimate Worksheet.*

Nails for Corner Bracing with Plywood Panels

We've estimated 20 panels of ½-inch-thick 4 × 8 plywood sheathing for the corner bracing, for a total of 640 square feet (20 × 32 square feet). 6d common nails will be used for the plywood. The nail factor is 11 pounds of 6d common nails per 1,000 square feet.

How many pounds of 6d common nails (X) should we estimate for the 640 square feet of plywood corner bracing?

Set up the proportion:

$$11 \text{ lbs} : 1{,}000 \text{ sf} = X \text{ lbs} : 640 \text{ sf}$$

$$1{,}000 \times X = 11 \times 640$$

$$X = \frac{11 \times 640}{1{,}000} = 7.04 \text{ (rounded to 7) lbs}$$

We need 7 pounds of 6d common nails for the plywood corner bracing.

> *Enter your estimate of the number of pounds of 6d common **nails** on line 7.1.7.2 of your Cost Estimate Worksheet. We'll use 7 pounds of 6d common nails in our Example Cost Estimate Worksheet.*

Nails for OSB Sheathing

We've estimated 73 panels of ½-inch-thick 4 × 8 OSB sheathing, for a total of 2,336 square feet (73 × 32 square feet). If we use 1½-inch roofing nails, how many pounds will we need to install the sheathing? The nail factor is 10 pounds of 1½-inch roofing nails per 1,000 square feet.

Nominal Size (inches)	Spacing (inches oc)	Select Structural	Dense Construction	Construction	Standard
	12"	9'6"	-	8'2"	6'4"
2 x 4	16"	8'6"	-	7'2"	5'6"
	24"	7'6"	-	5'10"	4'6"
	12"	14'4"	14'4"	14'4"	14'4"
2 x 6	**16"**	**13'0"**	**13'0"**	**13'0"**	**12'10"**
	24"	11'4"	11'4"	11'4"	10'6"
	12"	18'4"	18'4"	18'4"	18'4"
2 x 8	*16"*	*17'0"*	*17'0"*	*17'0"*	*17'0"*
	24"	15'4"	15'4"	15'4"	14'4"
	12"	21'0"	21'10"	21'10"	21'10"
2 x 10	16"	20'4"	20'4"	20'4"	20'4"
	24"	18'4"	18'4"	18'4"	18'0"

Figure 7-15
Calculating ceiling joist spans

Set up the proportion:

$$10 \text{ lbs} : 1{,}000 \text{ sf} = X \text{ lbs} : 2{,}336 \text{ sf}$$

$$1{,}000 \times X = 10 \times 2{,}336$$

$$X = \frac{10 \times 2{,}336}{1{,}000} = 23.36 \text{ (rounded to 24) lbs}$$

We need 24 pounds of 1½-inch roofing nails to install the sheathing.

Enter your estimate of the number of pounds of 1½-inch roofing ***nails*** *on line 7.1.7.3 of your Cost Estimate Worksheet. We'll use 24 pounds of 1½-inch roofing nails in our Example Cost Estimate Worksheet.*

Ceiling Joists

Ceiling joists are designed to support the ceiling, not support floor loads. If you intend to use the attic for more than limited storage, floor joists are required. Preassembled roof trusses can replace the ceiling joists. The blueprint's floor plan shows the size, spacing, and the direction the ceiling joists will run. The size and spacing shown on the plans is usually calculated by the designer of the house, but you can also use the information in Figure 7-15 as a guide for ceiling joist spans. Determine the length of the joists from the floor plans by the dimensions of the rooms, and direction the joists run. The direction the joists run is normally the shorter span.

Ceiling joists seldom span the entire width of a building as trusses do, and the direction they run varies, as shown in Figures 7-16 and 7-17. When estimating these joists, calculate each section separately.

Headers and trimmers are required for any openings in the ceiling, such as chimneys and access doors to the attic. Frame these openings as you would openings in floor joists (refer back to Chapter 6, Figure 6-8). Estimate these headers and trimmers the same as for floor joists.

Although the layout is different, estimating ceiling joists is the same for both hip and gable roofs. The run of regular joists must stop short of the outside wall to permit the hip and jack rafters to clear them. Short ceiling joists are installed perpendicular to the regular joists to permit the rafters to reach the outside wall plate (Figure 7-18).

To estimate ceiling joists, multiply the length (in feet) of the wall perpendicular to the joists by the appropriate joist factor. Use the same joist factor (based on joist spacing) that you use for estimating floor joists. You'll use 0.75 if the joist spacing is 16 inches on center, adding one joist for the starter. If the joist spacing is 24 inches on center, multiply by 0.50 and add one joist for the starter.

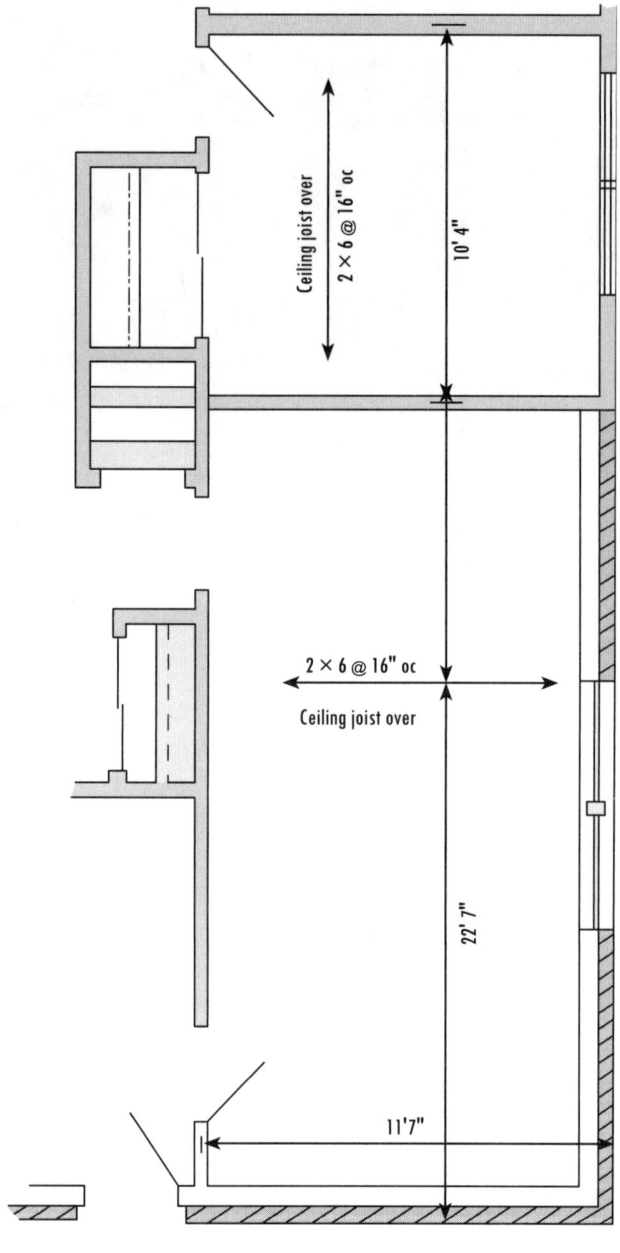

Figure 7-16
Ceiling joists running in two directions

Estimating First Floor Ceiling Joists

We'll use ceiling joists for the first and second floors of our sample house. All joists will be 16 inches on center.

For the first floor, ceiling joists are only required for the living room, kitchen, and dining room. The second floor extends over the remainder of the first floor, including the family room, vestibule, garage, utility room, etc. The ceiling in these areas will need floor joists to support the second floor rooms.

Living Area

How many, and what size joists should be estimated for the living area ceiling shown in Figure 7-1?

The ceiling joists in the living room area will span the shorter dimension, 13 feet. Figure 7-15 (**bolded numbers**) shows that 2 × 6 joists set at 16

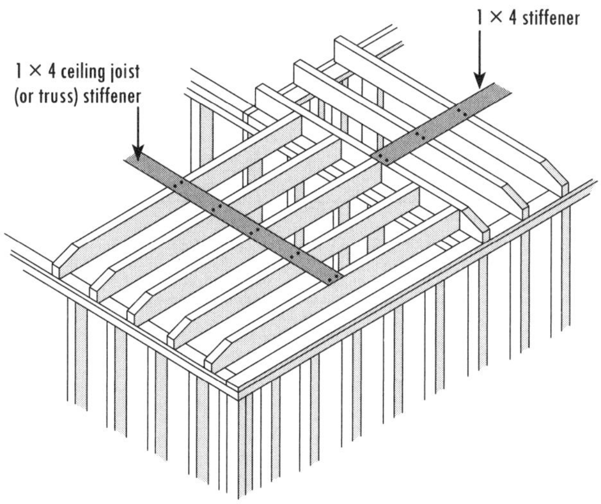

Figure 7-17
Ceiling joists running in two directions

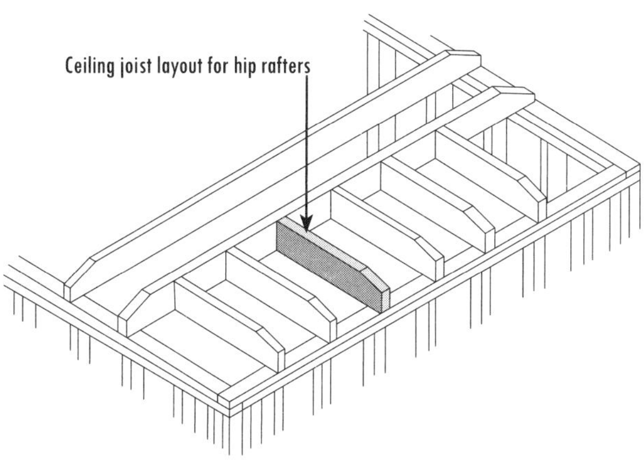

Figure 7-18
Ceiling joists for hip roof

inches on center can be used for a 13-foot (13′0″) span using select structural, dense construction, or construction (but not standard) grades of lumber. The length of the wall perpendicular to the joists is 23′10″ (11′8″ + 12′2″). The number of ceiling joists required for the living room is:

23.83' (23'10") × 0.75 = 17.87 (plus one for the starter)

We'll need 18.87 (rounded to 19) 2 × 6 joists.

> *Enter your estimate of the number and size of **ceiling joists** on line 7.2.1.1 of your Cost Estimate Worksheet. We'll use 19 2 × 6 × 14 joists in our Example Cost Estimate Worksheet.*

Kitchen and Dining Areas

The ceiling joists in the kitchen and dining area will also span 13 feet. Again, Figure 7-15 (**bolded numbers**) shows that 2 × 6 joists set at 16 inches on center can be used for a 13-foot (13′0″) span, using select structural, dense construction, or construction (but not standard) grades of lumber. The length of the wall perpendicular to the joists is the same as the living area, 23′10″ (11′8″ + 12′2″). The number of ceiling joists required for the kitchen and dining area is:

23.83' (23'10") × 0.75 = 17.87 (plus one for the starter)

We'll need 18.87 (rounded to 19) 2 × 6 joists.

> *Enter your estimate of the number and size of **ceiling joist** pieces on line 7.2.1.2 of your Cost Estimate Worksheet. We'll use 19 2 × 6 × 14 joists in our Example Cost Estimate Worksheet.*

Estimating Second Floor Ceiling Joists

The span of the front section of the second floor where the master bedroom and bedroom 2 are located is 15′3″(see Figure 7-2). How many, and what size joists should be estimated for the ceilings in these rooms?

Figure 7-15 (*italicized numbers*) shows that 2 × 8 joists set at 16 inches on center are required for a 15′3″ span. You can use any of the four grades of lumber. The length of the wall perpendicular to the joists is 40 feet. Let's calculate the number of ceiling joists required for the master bedroom and bedroom 2.

40' × 0.75 = 30 (plus one for the starter)

We'll need to order 31 2 × 8 ceiling joists.

Enter your estimate of the number and size of ***ceiling joist*** *pieces on line 7.2.1.3 of your Cost Estimate Worksheet. We'll use 31 2 × 8 × 16 joists in our Example Cost Estimate Worksheet.*

The span of the back section of the second floor (bedroom 3, the master bathroom, walk-in closet and dressing area, and the second bathroom) is 10'9". How many, and what size, joists should we estimate for the ceiling of the back section of the second floor in Figure 7-2?

Figure 7-15 (**bolded numbers**) shows that we can use 2 × 6 joists set 16 inches on center. The length of the wall perpendicular to the joists is 40 feet here as well. The number of ceiling joists required for the back section of the second floor is the same as for the front.

$$40' \times 0.75 = 30 \text{ (plus one for the starter)}$$

We'll need to order 31 2 × 6 ceiling joists for this section, too.

Enter your estimate of the number and size of ***ceiling joists*** *on line 7.2.1.4 of your Cost Estimate Worksheet. We'll use 31 2 × 6 × 12 joists in our Example Cost Estimate Worksheet.*

The total board feet of ceiling joists required for the first and second floors shown in Figures 7-1 and 7-2 is:

First floor			
Living room	19	2 × 6 × 14	266
Kitchen and dining room	19	2 × 6 × 14	266
Second floor			
Front	31	2 × 8 × 16	661
Back	31	2 × 6 × 12	372
Total:			1,565 board feet

Note: These calculations are based on the use of ceiling joists and rafters. Roof trusses could be used as an alternative. We'll discuss how to estimate for them later in this chapter.

Ceiling Backing

Ceiling backing provides nailing support for gypsum board, gypsum lath, or ceiling panels. It's used on wall top plates, parallel to ceiling joists or trusses, as shown in Figure 7-19. This backing should be continuous. Many builders use 2 × 6 material on a 2 × 4 wall because the thickness of the framing provides a firm base on which to nail, and the width is sufficient to support the finished ceiling.

When estimating ceiling backing, add the linear feet of all walls running parallel to the ceiling joists or trusses. You can find this information on the floor plans of the blueprints. Divide the total linear feet by the length of the material to be used. For example, if 2 × 6 × 12s are used, divide the total by 12. Round up to get the number of pieces required.

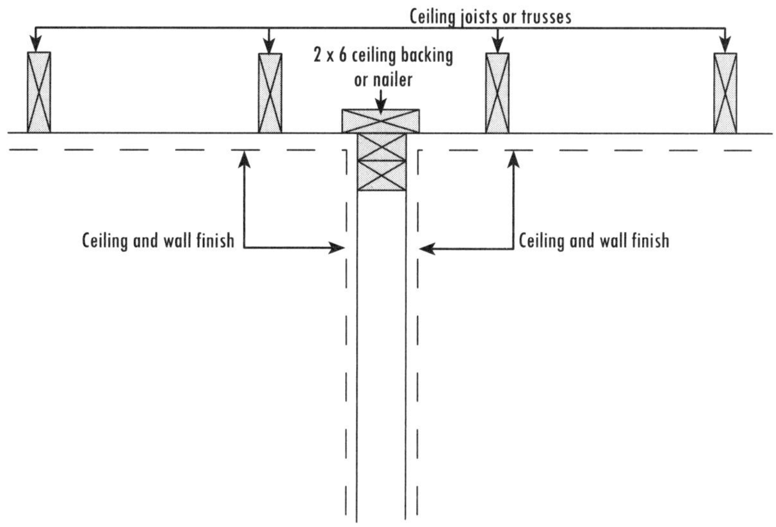

Figure 7-19
Wall parallel to ceiling joists

The total linear feet of the walls running parallel to the joists or trusses for the first and second floors in Figures 7-1 and 7-2 is:

First floor	87
Second floor	105
Total:	192 linear feet

Note: You may be able to use some of the waste material for ceiling backing.

So, how many pieces of 2 × 6 × 12 ceiling backing will we need for 192 linear feet of wall?

$$\frac{192 \text{ lf}}{12' \text{ per piece}} = 16 \text{ pieces of } 2 \times 6 \times 12$$

Enter your estimate of the number of pieces of 2 × 6 × 12 ***ceiling backing*** *on line 7.2.2 of your Cost Estimate Worksheet. We'll use 16 pieces of 2 × 6 × 12 in our Example Cost Estimate Worksheet.*

Ceiling Joist Stiffeners

Ceiling joist *stiffeners*, as the name implies, stiffen ceiling joists or trusses and hold them in alignment (Figure 7-17). 1 × 4 material is generally used, nailed to each ceiling joist or truss. Use two 7d or 8d nails at the approximate center of the span in each section. Estimate material by adding the linear feet of each section. Material typically comes in 12-foot lengths.

Let's calculate how many pieces of 1 × 4 × 12 will be required for ceiling joist stiffeners for the first and second floors of our house.

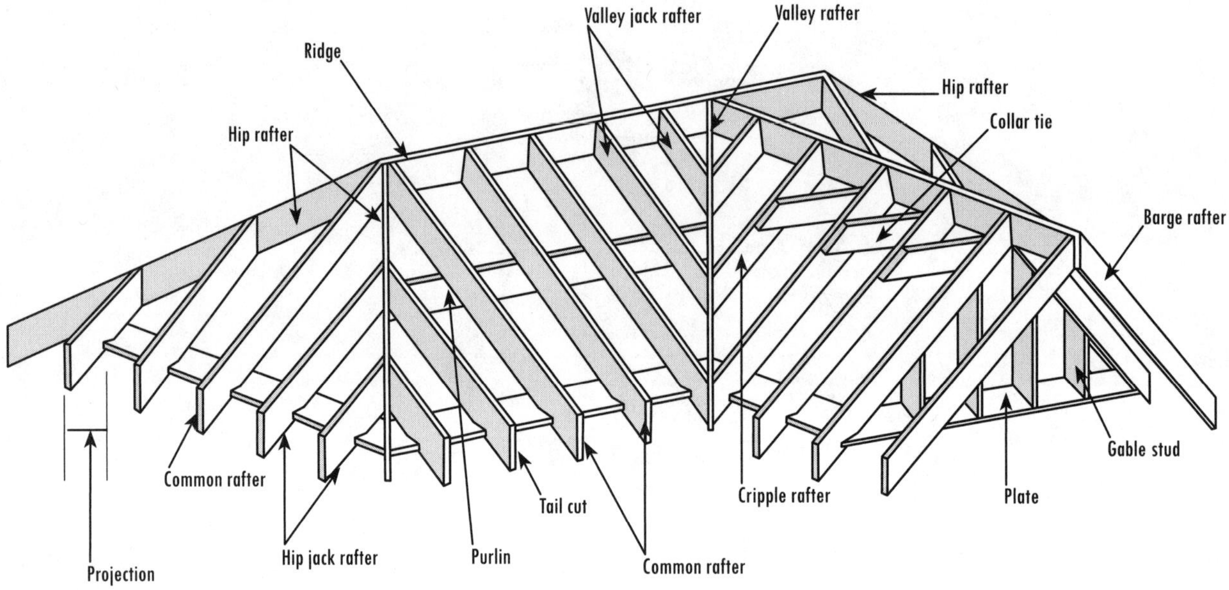

Common Rafters: Run from the ridge to the plate
Hip Rafters: Run from the ridge to the plate at a 45-degree angle to the common rafters at the outside corners
Valley Rafters: Run from the ridge to the plate at a 45-degree angle to the common rafters at the inside corners
Jack Rafters: Run from hip or valley rafters to the plate or ridge
Cripple Rafters: Run between hip and valley rafters
Ridge: The hightest horizontal member in the roof system

Figure 7-20
Different types of rafters

First, total the number of linear feet perpendicular to the ceiling joists in the front and back sections of the first and second floors.

First floor (Figure 7-1)	
Front section (11′8″ + 12′2″)	23′10″
Back section (11′8″ + 12′2″)	23′10″
Subtotal	47′8″
Second floor (Figure 7-2)	
Front section	40′0″
Back section	40′0″
Subtotal	80′0″
Total:	127′8″ or 127.66 linear feet

Now, divide the total by 12 (for 12-foot material), and round up to the nearest whole number.

$$\frac{127.66 \text{ lf}}{12' \text{ per piece}} = 10.64 \text{ (rounded to 11) pieces (44 bf)}$$

We'll need 11 pieces of joist stiffeners.

*Enter your estimate of the number of pieces of 1 × 4 × 12 lumber to be used as **joist stiffeners** on line 7.2.3 of your Cost Estimate Worksheet. We'll use 11 pieces of 1 × 4 × 12 in our Example Cost Estimate Worksheet.*

Nominal Size (inches)	Spacing (inches oc)	Select Structural	Dense Construction	Construction	Standard
	12"	11'6"	-	9'6"	7'4"
2 × 4	16"	10'6"	-	8'4"	6'4"
	24"	9'2"	-	6'10"	5'2"
	12"	16'10"	16'10"	16'10"	16'1"
2 × 6	16"	15'8"	15'8"	15'8"	15'0"
	24"	13'10"	13'10"	13'6"	12'2"
	12"	21'2"	21'2"	21'2"	21'2"
2 × 8	16"	19'10"	19'10"	19'10"	19'10"
	24"	17'10"	17'10"	17'10"	16'8"

Figure 7-21
Maximum clear spans

Rafters

Rafters are the supporting members of the roof system. The size and spacing of the rafters must be in accordance with recognized engineering practices. There are different types of rafters; the most common are shown in Figure 7-20.

Estimating Common Rafter Length

Common rafters run from the ridge to the plate (see Figure 7-20). Common rafter *maximum clear spans* (the actual length of members between inner faces of support, measured along the slope) are shown in Figure 7-21. You'll usually find the size and spacing of common rafters on the wall section of your blueprints.

Figure 7-22 gives illustrations and definitions of terms you'll use to calculate common rafter lengths.

The roof *pitch factor* is the number of inches of roof rise (N) per one foot of roof run (12 inches), and is normally designated on the elevation section of the blueprints, as shown in Figure 7-23. This is also called the *slope* of the roof; pitch and slope mean the same thing to most builders.

The example in Figure 7-24 tells you that the roof rises 6 inches for every 12 inches of run.

The common rafter length per foot run (CR pitch factor) is calculated using the following formula:

$$\text{Common Rafter Run} = \sqrt{\text{Rise}^2 + \text{Run}^2}$$

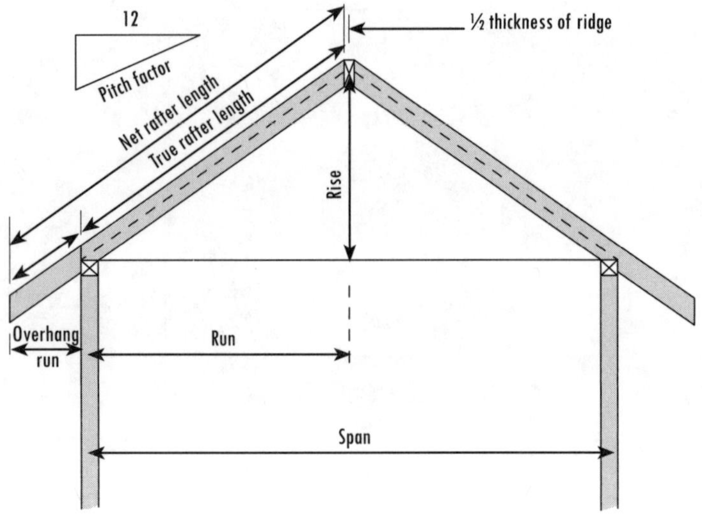

Span: The width

Run: One-half the span

Overhang: The distance from the outside of the wall to the tail cut of the rafter

Rise: The vertical distance from the plate line to the measuring line vertex (the 2 sloping dashed lines)

True Rafter Length: The distance between the face of the ridge and the outer face of the wall framing

Net Rafter Length: The distance between the face of the ridge and the tail cut; the total length of the rafter

Pitch Factor: The factor multiplied by the Run to get the True Rafter Length

Vertex: The top or summit; the point where two lines intersect

Figure 7-22
Rafter spans and definitions

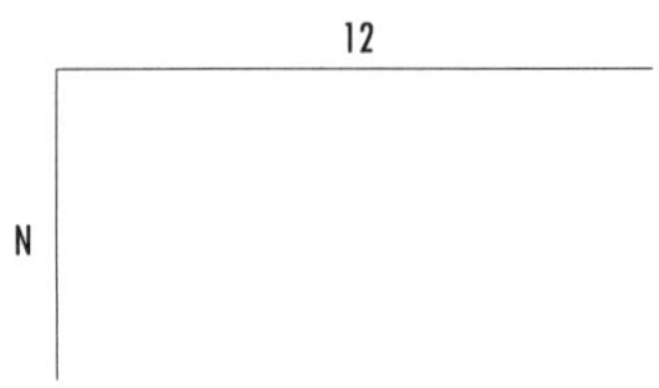

Figure 7-23
Roof pitch factor

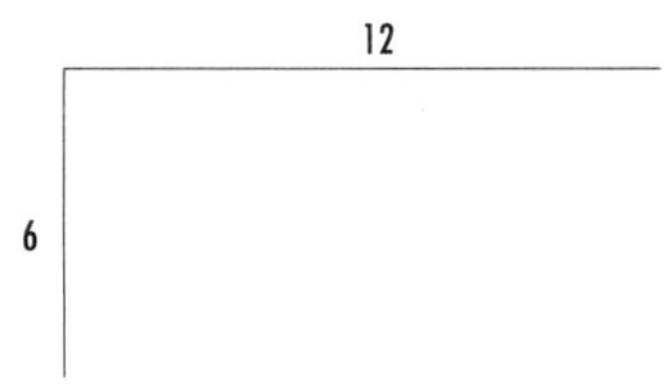

Figure 7-24
Roof pitch as indicated on blueprint

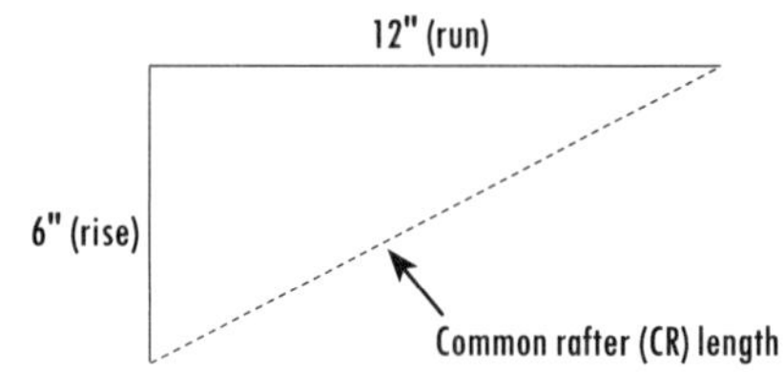

Figure 7-25
Common rafter length

Let's calculate the common rafter length per foot run, using the information in Figure 7-25:

$$\text{Common Rafter (CR) Length} = \sqrt{6^2 + 12^2} = 13.42''$$

The common rafter (CR) pitch factor for a 6/12 roof pitch is 13.42.

Now, calculate the common rafter length for a 15-foot run with a CR pitch factor of 13.42:

$$\text{CR Pitch Factor} = \frac{\text{Common Rafter(CR) Length (in)}}{\text{Run (ft)}}$$

$$\text{CR Length (in)} = \text{Run (ft)} \times \text{CR Pitch Factor}$$

$$\text{CR Length (in)} = 15' \times 13.42$$

$$\text{CR Length (in)} = 201.30''$$

For length in feet, divide by 12.

Cost Estimate Worksheet for Common Rafters

1	**Rafter length from plate to center of ridge:**				
	______ ft × ______			=	0.00 inches
	Run / CR Pitch Factor				
2	**Rafter length from plate to edge of ridge:**				
	0.00	minus	______ inches (½ thickness of ridge)	=	0.00 inches True rafter length
3	**Rafter overhang:**				
	______ ft × 0.00			=	0.00 inches
	Overhang / CR Pitch Factor				Rafter overhang
4	**Net rafter length:**				
	0.00 True rafter length	plus	0.00 Rafter overhang	=	0.00 inches Net rafter length
	0.00 in	divided by 12	= 0.00 ft	=	______ feet/inches

Figure 7-26
Cost estimate worksheet for common rafters

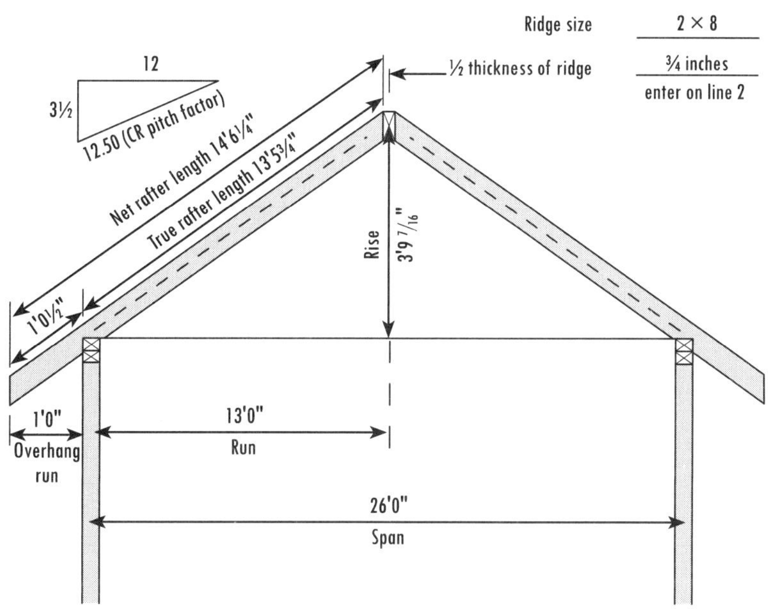

Figure 7-27
Example of common rafters

Figure 7-26 shows a sample worksheet that you can use for calculating common rafter lengths using CR factors for different roof pitches. This worksheet shows how the computations are made, and helps you compute the common rafter lengths for making the top (*plumb*) cut, the *bird's mouth* cut, and the *tail* of rafters. There's a blank worksheet on the accompanying CD-ROM at the back of this book.

Figure 7-27 is a working model of Figure 7-22, and Figure 7-28 shows the completed Figure 7-26 worksheet for that example. Note the CR pitch factor for this roof is shown as 12.50 (top left corner of Figure 7-27).

Cost Estimate Worksheet for Common Rafters

1 **Rafter length from plate to center of ridge:**
13.00 ft × 12.50 = 162.50 inches
Run CR Pitch Factor

2 **Rafter length from plate to center of ridge:**
162.50 minus 0.75 inches = 161.75 inches
(½ thickness of ridge) True rafter length

3 **Rafter overhang:**
1.00 ft × 12.50 = 12.50 inches
Overhang CR Pitch Factor

4 **Net rafter length:**
161.75 plus 12.50 = 174.25 inches
True rafter length Rafter overhang

174.25 in divided by 12 = 14.52 ft = 14' 6¼"

Figure 7-28
Sample worksheet for common rafters

Fractional Parts of an Inch

Whole Inches	0	1/8	1/4	3/8	1/2	5/8	3/4	7/8
0	0.00	0.01	0.02	0.03	0.04	0.05	0.06	0.07
1	0.08	0.09	0.10	0.11	0.13	0.14	0.15	0.16
2	0.17	0.18	0.19	0.20	0.21	0.22	0.23	0.24
3	0.25	0.26	0.27	0.28	0.29	0.30	0.31	0.32
4	0.33	0.34	0.35	0.36	0.38	0.39	0.40	0.41
5	0.42	0.43	0.44	0.45	0.46	0.47	0.48	0.49
6	0.50	0.51	0.52	0.53	0.54	0.55	0.56	0.57
7	0.58	0.59	0.60	0.61	0.62	0.64	0.65	0.66
8	0.67	0.68	0.69	0.70	0.71	0.72	0.73	0.74
9	0.75	0.76	0.77	0.78	0.79	0.80	0.81	0.82
10	0.83	0.84	0.85	0.86	0.88	0.89	0.90	0.91
11	0.92	0.93	0.94	0.95	0.96	0.97	0.98	0.99

Figure 7-29
Decimal to fraction conversion table

Figure 7-29 shows decimals (hundredths) of a foot and their equivalents in inches and fractions of an inch. Use this table to convert the calculation in Figure 7-28 (14.52 feet) to feet, inches, and fractions of an inch. In this case, the decimal to be converted is .52. Locate .52 on the decimal side of Figure 7-29, in the 6-inch row. Now, look up to the column head and you'll see ¼. That tells you that .52 of a foot equals 6¼ inches. So 14.52 feet calculated out in Figure 7-28 is equivalent to 14' 6¼".

Figure 7-30 shows the lengths per foot run (or *pitch factors*) for common rafters, hip and valley rafters, and jack rafters. Some estimators scale the rafter lengths directly from the blueprints; however, that particular method isn't recommended. Consider the following scenario. You scale 13'10" as

Pitch of Roof		Common Rafters: Length per foot of run in inches	Hip and Valley Rafters: Length per foot of run on common rafters in inches	Jack Rafters: Length of shortest jack on 16 inch centers	Jack Rafters: Length of shortest jack on 24 inch centers
Rise	Run				
	12				
1/12		12.04			
1½/12		12.09			
2/12		12.17			
2½/12		12.26			
3/12		12.37	17.23	16.49	24.74
3½/12		12.50	17.33	16.66	25.00
4/12		12.65	17.44	16.86	25.30
5/12		13.00	17.69	17.33	26.00
6/12		13.42	18.00	17.88	26.83
7/12		13.89	18.36	18.52	27.79
8/12		14.42	18.76	19.22	28.85
9/12		15.00	19.21	19.99	30.00
10/12		15.62	19.70	20.82	31.24
11/12		16.28	20.22	21.70	32.56
12/12		16.97	20.79	22.62	33.94

Figure 7-30
Factors per foot run

Roof Pitch	CR Pitch Factor	Divided By	CR Pitch Ratio
1/12	12.042	12	1.004
1½/12	12.093	12	1.008
2/12	12.166	12	1.014
2½/12	12.258	12	1.022
3/12	12.369	12	1.031
3½/12	12.500	12	1.042
4/12	12.649	12	1.054
5/12	13.000	12	1.083
6/12	13.416	12	1.118
7/12	13.892	12	1.158
8/12	14.422	12	1.202
9/12	15.000	12	1.250
10/12	15.620	12	1.302
11/12	16.279	12	1.357
12/12	16.971	12	1.414

Figure 7-31
Rafter length to factors per foot run

the rafter length from the blueprints, when the net rafter length is actually 14′2″. Consequently, you order the rafters in lengths of 14′0″, when they should be lengths of 16′0″. Big problem!

The table in Figure 7-31 includes common rafter pitch ratios to help you estimate rafter lengths. This is especially useful when you're ordering material. Use these CR pitch ratios as follows: Select the CR pitch ratio for the roof pitch you're planning for the house. Then, multiply this CR pitch ratio by the run of the house plus the overhang, in feet.

Let's look at an example. Figure the net length required for the common rafters in Figure 7-27. The roof pitch is 3½/12. Looking at Figure 7-31, we find that the CR pitch ratio for a 3½/12 roof pitch is 1.042. The run of the house plus the roof overhang is 14 feet (13'0" + 1'0").

14.00' × 1.042 = 14.59' (14'7⅛")

Refer to conversion chart in Figure 7-29 to convert from decimal feet to inches. The net rafter length in feet and inches is 14'7⅛".

The precise net rafter length (calculated on the sample worksheet in Figure 7-28) for Figure 7-27 is 14'6¼". You can see that the CR pitch ratios in Figure 7-30 can be used to get quick, reasonable estimates for most common rafter lengths. These rafters should be ordered in 16-foot lengths, in either case.

Estimating Common Rafter Quantity

To estimate the number of common rafters you need, multiply the linear feet of the eave (including the gable overhang) perpendicular to the rafters by 0.75 for 16-inch on-center rafter spacing, and add one for the starter. Multiply by 0.50 (and add one for the starter) for 24-inch on-center spacing.

Using 16-inch on-center spacing, how many common rafters would you need for the two-story and the single-story roof systems of our sample house?

We'll calculate the number of common rafters as follows: The roof overhang is 1 foot at each gable end. From Figure 7-2 (the second floor, two-story part of the house), the length of the wall perpendicular to the rafters is 40 feet. The linear feet of the eave is 40 feet plus the 1-foot overhang on each of the two gable ends, for a total of 42 feet (42.00').

42.00' × 0.75 = 31.50 (rounded to 32) rafters

32.00 + 1 starter = 33 each for the back and front sides of the house

33.00 × 2 (front + back) = 66 rafters for the two-story part of the house

The length of the eave for the single-story part of our house is 24 feet. That's the total building length, 64 feet, minus the 40-foot length of the two-story section. The eave will be 24 feet plus 1 foot (there's only one gable end for the single-story part of the house), or 25 feet (25.00').

25.00' × 0.75 = 18.75 (rounded to 19) rafters

19 + 1 starter = 20 rafters for each section (back and front)

20 × 2 (front + back) = 40 rafters for the single-story section

Note: The rafter that would rest on the gable can be omitted, and the material used for the overhang lookouts, if the lookouts are spaced 24 inches on center. As long as the length of each lookout doesn't exceed 2 feet, no extra material needs to be estimated for the gable end overhang.

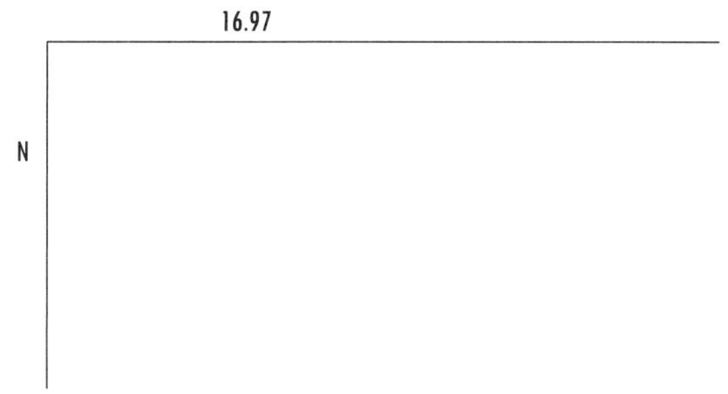

Figure 7-32
Roof pitch for hip and valley rafters

The sample worksheet in Figure 7-28 has a net rafter length of 14′6¼″. The span tables in Figure 7-21 show that 2 × 6 rafters on 16-inch centers can be made of any of the four grades of lumber. Estimate the material in 16-foot lengths, as follows:

Two-story section	2 × 6 × 16	66
Single-story section	2 × 6 × 16	40
Total:		106

We'll need 106 pieces (1,696 board feet) of 2 × 6 × 16 material.

> *Enter your estimate of the number and size of **rafters** on line 7.2.4 of your Cost Estimate Worksheet. We'll use 106 pieces of 2 × 6 × 16s in our Example Cost Estimate Worksheet.*

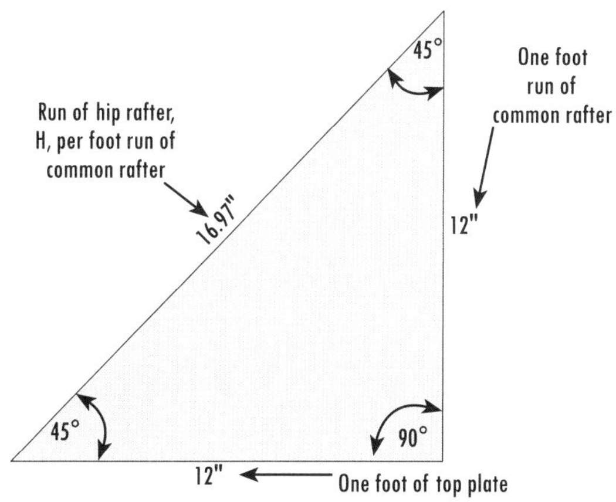

Figure 7-33
Hip rafter to common rafter run

Estimating the Length of Hip and Valley Rafters

Our sample house has a simple gable roof, with no hip or valley rafters. However, you'll come across those kinds of rafters in other houses, so we'll discuss how to estimate them here.

Hip rafters run from the roof ridge to the top plate at a 45-degree angle to the common rafters at the outside corners of the house. Valley rafters run from the ridge to the top plate at a 45-degree angle to the common rafters at the inside corners of the roof (see Figure 7-20).

Just as the roof pitch for common rafters is designated in inches of rise (N) per foot run, or N/12, the pitch for hip and valley rafters is designated in inches of rise (N) per 16.97 inches of run, or N/16.97 (Figure 7-32). The number 16.97 is to the hip and valley rafter as 12 is to the common rafter. Where do we get the number 16.97? Let's look at an example.

In a right triangle, the sides opposite 45-degree angles are equal. Because the length of the hip and valley rafter is calculated per foot run of the common rafter, use the following formula to calculate the run of the hip (or valley) rafter, which we'll call H:

$$H = \sqrt{12^2 + 12^2} = 16.97$$

This calculation is illustrated in Figure 7-33. You can find this same triangle within the larger illustration of triangle (BCD) in Figure 7-34. The run of the hip (or valley) rafter, H, is part of side BD (the hypotenuse) of that triangle.

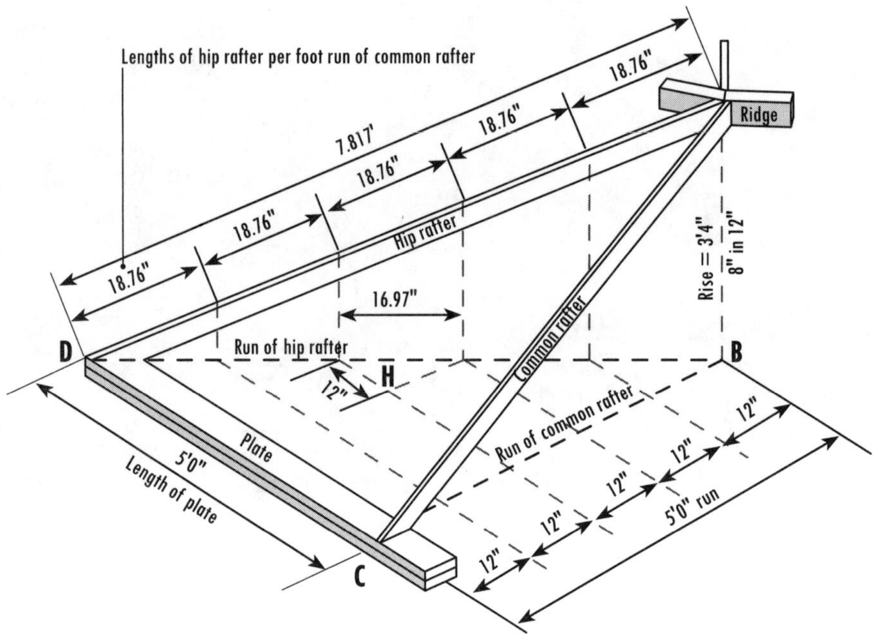

Figure 7-34
Run of hip rafter per run of common rafter

The roof pitch (shown on the right side of the illustration) for the common rafters in Figure 7-34 is 8/12 (Figure 7-35). So, the length of the hip rafter per foot run of the common rafter would be 8/16.97 (Figure 7-36). We can use the following formula to find the pitch factor (length per foot of run on common rafters in inches) for calculating the length of the hip (or valley) rafters (HR).

$$\text{Hip Rafter (HR) Length} = \sqrt{8^2 + 16.97^2} = 18.76$$

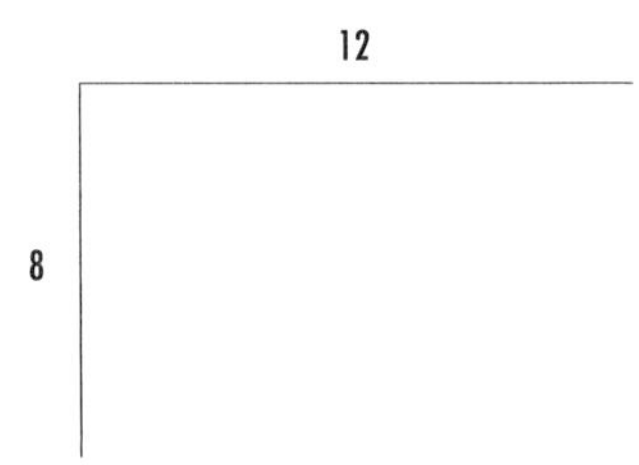

Figure 7-35
Roof pitch factors from Figure 7-34

The HR pitch factor for an 8/12 roof pitch is 18.76. Looking back at Figure 7-30, you'll see that's the same pitch factor provided in the table.

Now let's calculate the hip rafter length for a common rafter run of 5 feet (5.00') and an HR pitch factor of 18.76. Here's our formula:

$$\text{HR Pitch Factor} = \frac{\text{Hip Rafter Length (in)}}{\text{Common Rafter Run (ft)}}$$

$$\text{Common Rafter Run (ft)} \times \text{HR Pitch Factor} = \text{Hip Rafter Length (in)}$$

$$5.00' \times 18.76 = 93.80''$$

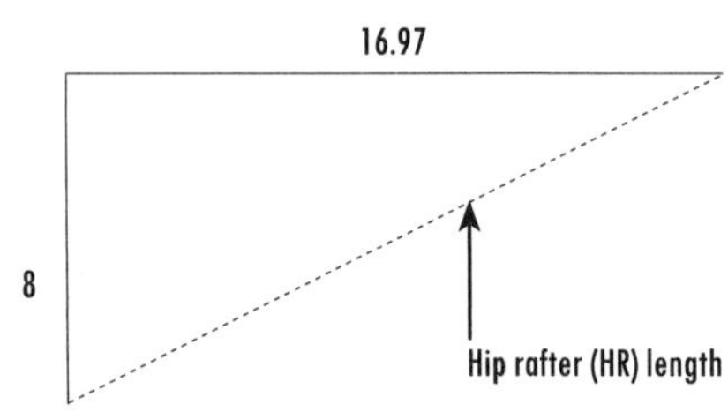

Figure 7-36
Hip rafter length

Divide by 12 to get length in feet:

$$\frac{93.80''}{12} = 7.82' \text{ or } 7'9\,{}^{7}/_{8}''*$$

*This is the true hip rafter length. You'll need to add the overhang to get the net hip rafter length.

Roof Pitch	HR Pitch Factor	Divided By	HR Pitch Ratio
1/12	-	-	
1½/12	-	-	
2/12	-	-	
2½/12	-	-	
3/12	17.234	12	1.436
3½/12	17.328	12	1.444
4/12	17.436	12	1.453
5/12	17.692	12	1.474
6/12	18.000	12	1.500
7/12	18.358	12	1.530
8/12	18.762	12	1.564
9/12	19.209	12	1.601
10/12	19.698	12	1.642
11/12	20.224	12	1.685
12/12	20.785	12	1.732

Figure 7-37
Ratio of hip and valley rafter length to run of common rafter

Figure 7-30 shows the length of the hip (and valley) rafters in inches per foot run of the common rafter (HR pitch factor) for different roof pitches.

To estimate the length of hip (and valley) rafters, refer to Figure 7-37, which shows the *ratio* of hip (and valley) rafter lengths to the run of common rafters for different roof pitches.

To calculate the length of hip and valley rafters using the ratios in Figure 7-37, select the HR pitch ratio for the roof pitch of the house and multiply the HR pitch ratio by the run of the common rafter.

Let's calculate the hip rafter length for a house when the run of the common rafter is 5 feet (5.00′) and the roof pitch is 8/12. In Figure 7-37, we see that the HR pitch ratio for an 8/12 roof pitch is 1.564.

Hip Rafter Length = Common Rafter Run × HR pitch ratio

5.00' × 1.564 = 7.82' or 7'9⅞" *

*This is true hip rafter length only.

If there's an overhang, add that to the run of the common rafter, then multiply by the HR pitch ratio to get the net hip (or valley) rafter length. For example, if there's an overhang of 1 foot (1.00′) for the common rafters in the problem we just did, you'd make the following adjustment to your calculations:

Net Hip Rafter Length = (Common Rafter Run + Overhang) × HR Pitch Ratio

(5.00' + 1.00') × 1.564 = 9.38' or 9'4½"

Estimating Hip and Valley Rafter Quantity

To estimate the number of hip and valley rafters, check the elevation section of the blueprints for the number shown on the roof plan.

Estimating Jack Rafters

Any rafter that doesn't extend from the top plate to the ridge is called a *jack rafter* (refer back to Figure 7-20). Jack rafters are classified as:

- *Hip jack*: running from the hip rafter to the top plate
- *Valley jack*: running from the valley rafter to the ridge
- *Cripple jack*: running from the hip rafter to the valley

Figure 7-30 shows factors for calculating jack rafters based on various roof pitches, as well as the length of the shortest jack rafter at 16 inches on center, and the shortest jack rafter at 24 inches on center. Let's look at how these lengths are calculated.

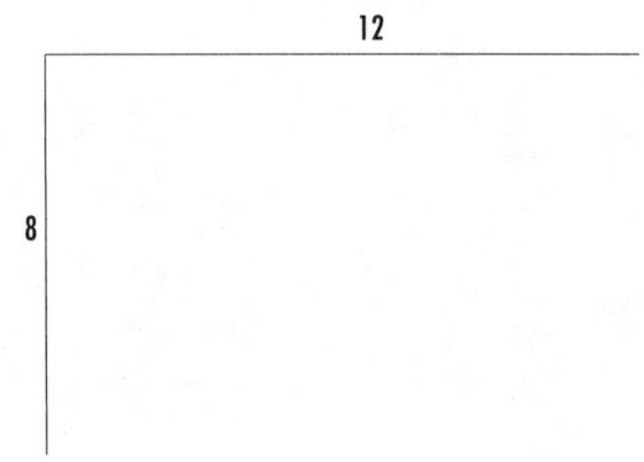

Figure 7-38
8/12 roof pitch

Jack Rafters Spaced 16 Inches on Center

Let's assume we have an 8/12 roof pitch, as shown in Figure 7-38.

First, calculate the length of hip (or valley) rafter (H in Figure 7-39) per 16-inch run of common rafter at 16 inches on center.

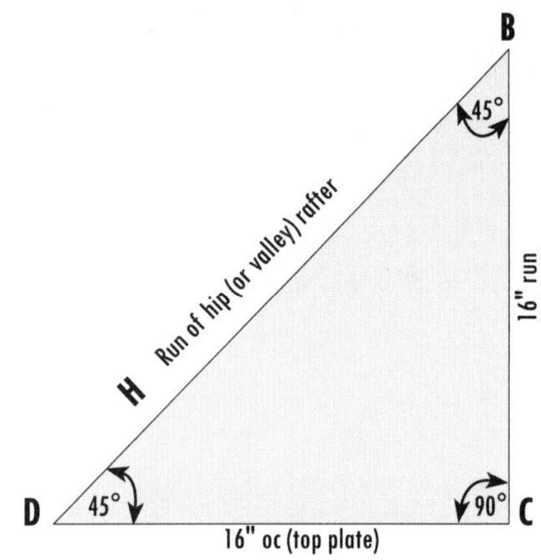

Figure 7-39
Run of hip or valley rafter

The HR pitch factor for hip (and valley) rafters for a roof pitch of 8/12 is 18.76 (Figure 7-30).

Length of Hip (or Valley) Rafter = Run of Common Rafter (in feet) × HR Pitch Factor

$$1.333' (16") \times 18.76 = 25.01"$$

The length of hip (or valley) rafter per 16-inch run of common rafter is 25.01 inches.

Locate the jack rafter (XY) at the bottom left corner of Figure 7-40. It forms part of the right triangle XYZ, along with the 16 inches of top plate, YZ, and the length of hip rafter, XZ. This is shown in more detail in Figure 7-41.

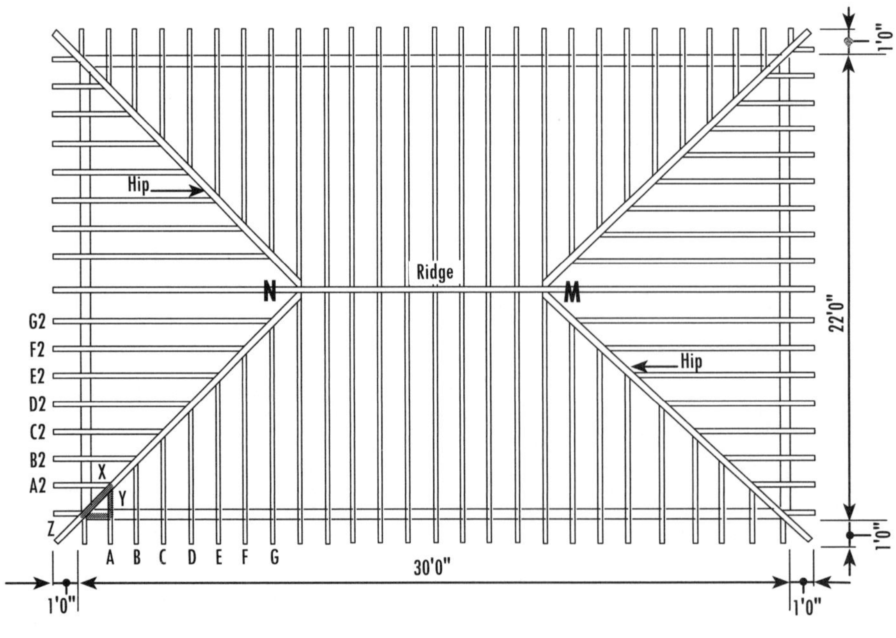

Figure 7-40
Framing detail for a hip roof

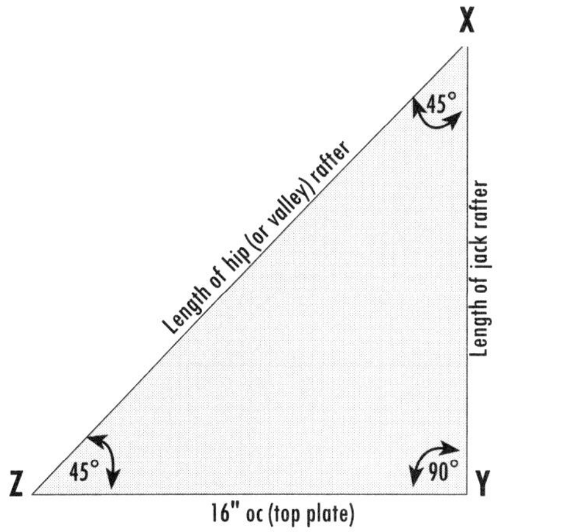

Figure 7-41
Length of hip or valley rafter, 16" oc

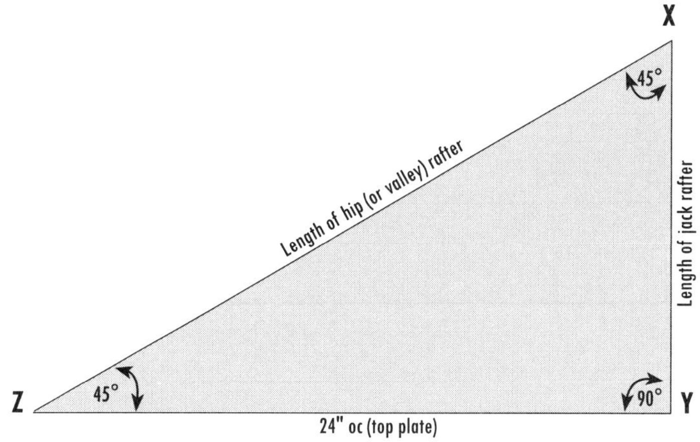

Figure 7-42
Length of hip or valley rafter, 24" oc

Remember the Pythagorean Theorem, $a^2 + b^2 = c^2$? This is a right triangle, so the length of the jack rafter, XY in this case, can be solved in the following manner:

$$(\text{Length of Hip Rafter, XZ})^2 = (\text{Length of Jack Rafter, XY})^2 + (16\text{" of Top Plate, YZ})^2$$

or

$$(\text{Length of Jack Rafter, XY})^2 = (\text{Length of Hip Rafter, XZ})^2 - (16\text{" of Top Plate, YZ})^2$$

$$\text{Length of Jack Rafter, XY} = \sqrt{(\text{Length of Hip Rafter, XZ})^2 - (16\text{" of Top Plate, YZ})^2}$$

$$\text{Length of Jack Rafter, XY} = \sqrt{25.01^2 - 16^2} = 19.22\text{"}$$

The length of the shortest jack rafter, XY, spaced 16 inches on center for a roof pitch of 8/12 is 19.22 inches. You'll find this same length provided in the table in Figure 7-30.

Jack Rafters Spaced 24 Inches on Center

If the jack rafters are spaced 24 inches on center (Figure 7-42), use the following calculations. The HR pitch factor for a roof pitch of 8/12 is, again, 18.76.

$$\text{Length of Hip (or Valley) Rafter} = \text{Run of Common Rafter (in feet)} \times \text{HR Pitch Factor}$$

$$2.00' (24\text{"}) \times 18.76 = 37.52\text{"}$$

$$(\text{Length of Hip Rafter, XZ})^2 = (\text{Length of Jack Rafter, XY})^2 + (24\text{" of Top Plate, YZ})^2$$

$$(\text{Length of Jack Rafter, XY})^2 = (\text{Length of Hip Rafter, XZ})^2 - (24\text{" of Top Plate, YZ})^2$$

$$\text{Length of Jack Rafter, XY} = \sqrt{(\text{Length of Hip Rafter, XZ})^2 - (24\text{" of Top Plate, YZ})^2}$$

$$\text{Length of Jack Rafter, XY} = \sqrt{37.52^2 - 24^2} = 28.84$$

At 24 inches on center, length of the shortest jack rafter (side XY in Figure 7-40) for a roof pitch of 8/12 is 28.84 inches. Again, you can verify this with the table in Figure 7-30. Once you've established the length of the shortest jack rafter, A or A2 in Figure 7-40, the lengths of the others are

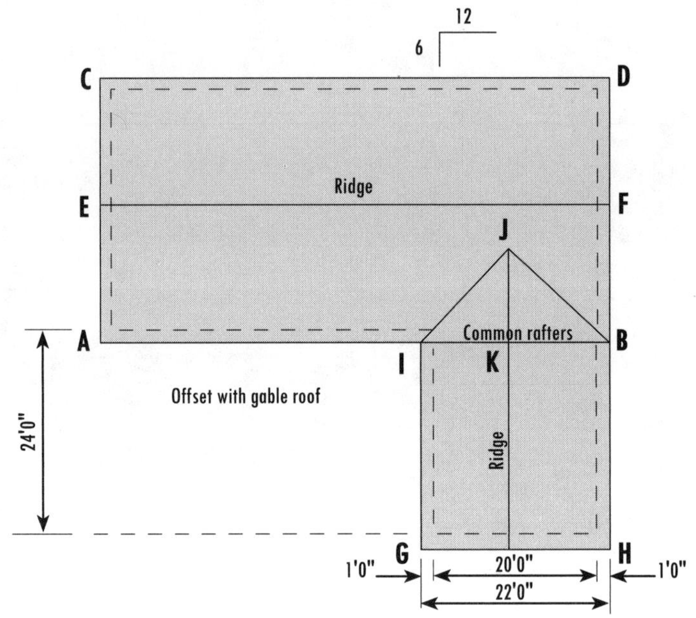

Figure 7-43
Gable roof offset

very easy to calculate. The second jack rafter, B or B2, will be twice that of the shortest jack rafter; the third jack rafter, C or C2, will be three times the shortest jack rafter, and so on. This is true for both 16 and 24 inches on center spacing.

When estimating jack rafters, it isn't necessary to calculate each length separately. From Figure 7-40, you can see that jack rafters A (or A2) and G (or G2) can both be cut from one common rafter. The same is true with B and F or C and E jack rafters. Two D jack rafters can always be cut from one common rafter. Count the number of jack rafters required, and divide by two to get the equivalent number of common rafters needed. Add this to the number of common rafters already estimated, then add two common rafters per hip to allow for waste.

The Ridge

The ridge is the highest horizontal member in the roof system, and is the spine to which rafters are attached (Figure 7-20). The ridge should be one size wider than the rafters. For example, if the rafters are 2 × 6, the ridge should be 2 × 8.

The length of the ridge on a straight gable roof is the length of the building plus the overhangs. On intersecting gable roofs, the ridge length is the intersection plus the extension of the ridge to the adjacent roof (Figure 7-43). Normally, this ridge extension is equal to the total length of the common

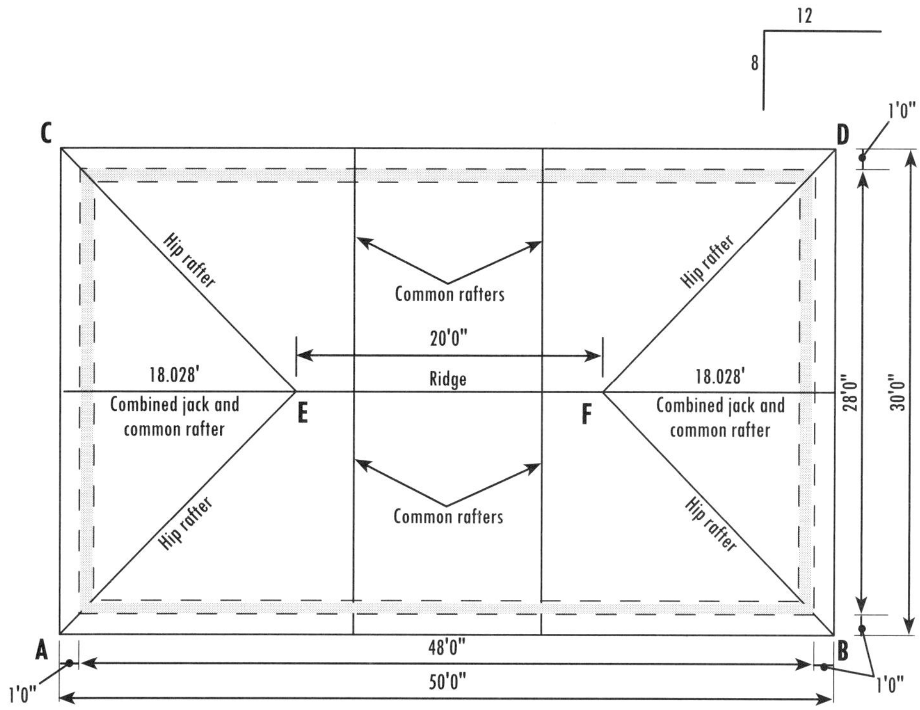

Figure 7-44
Hip roof area

rafter on the building extension. For example, if the building extension (or offset) is 16 feet and the length of the common rafter on the extension is 10 feet, the length of the ridge for the extension should be 26 feet (16 + 10).

The length of the ridge on a full hip roof is the length of the building minus its width (Figure 7-44). On intersecting hip roofs, the length of the ridge is the length of the intersection plus the overhang.

Some estimators feel safe scaling the ridge length from plans drawn to scale directly from the blueprints. This is generally accurate enough to estimate the material.

Let's calculate the ridge lengths for our sample house from the plans shown in Figures 7-1, 7-2 and 7-9.

Single-story section (right side)	25′0″
Two-story section (left side)	42′0″
Total:	67′0″

These two ridges (25′0″ and 42′0″) will be made of 5 pieces of 2 × 8 × 14 lumber (93 board feet).

*Enter your estimate of the number and size of **ridge** lumber on line 7.2.5 of your Cost Estimate Worksheet. We'll use 5 pieces of 2 × 8 × 14s in our Example Cost Estimate Worksheet.*

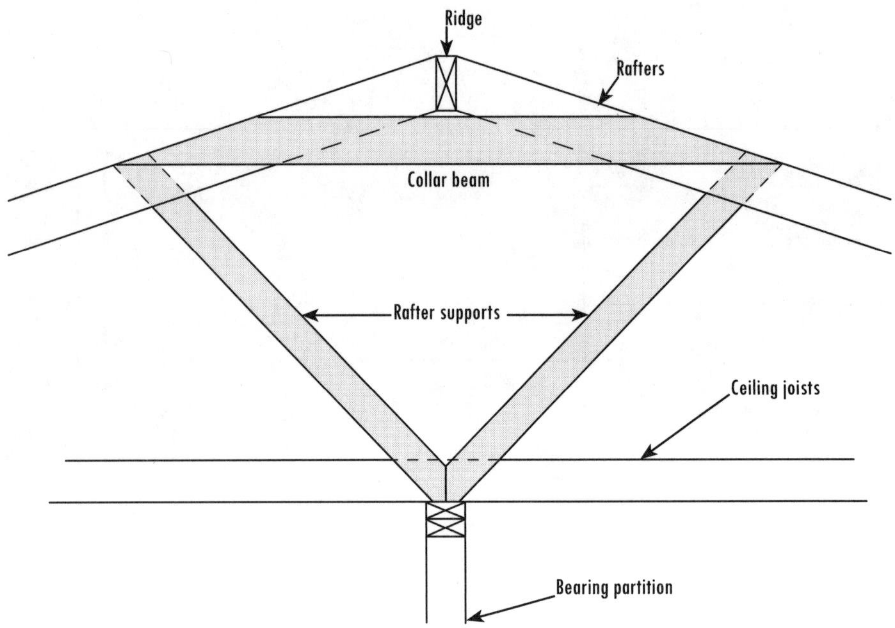

Figure 7-45
Collar beam and rafter supports

Collar Beams

Collar beams are horizontal framing members that tie rafters together to prevent roof thrust (Figure 7-45). They're located in the upper third of the attic space, below the ridge. The maximum spacing of collar beams is 4 feet on center.

If collar beams are spaced at 4 feet on center, divide the length of the house by 4 to determine the number of collar beams required (don't include the overhang).

The number of collar beams and material for our sample house should be estimated as follows:

Single-story section (right side)	24'0"
Two-story section (left side)	40'0"
Total length:	64'0"

$$\frac{64'0''}{4'\text{oc}} = 16 \text{ collar beams}$$

Normally, 2 × 4 material, 8 feet in length, will make up one collar beam. We'll need 16 pieces of 2 × 4 × 8, which equals 85 board feet.

Enter your estimate of the number and size of ***collar beams*** *on line 7.2.6 of your Cost Estimate Worksheet. We'll use 16 pieces of 2 × 4 × 8 material in our Example Cost Estimate Worksheet.*

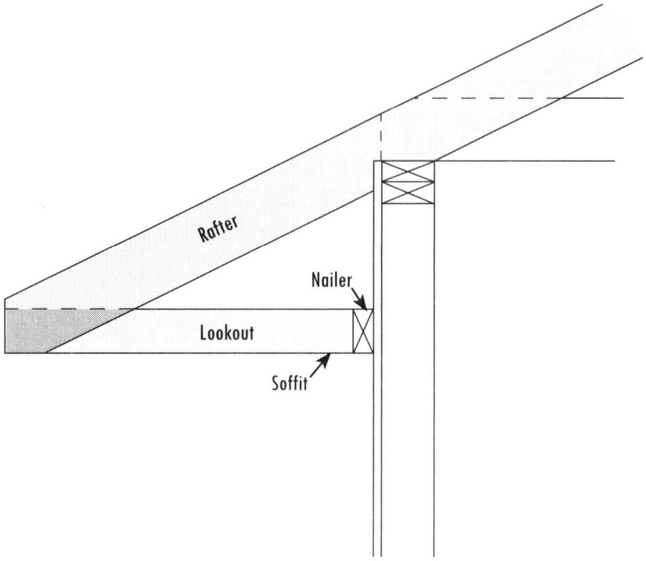

Figure 7-46
Framing for roof overhang soffit

Rafter Supports

Rafter supports, or *rafter braces*, prevent the roof from sagging under the weight of the dead and live loads on the rafters. Use 2 × 4s with the same spacing as the rafters for maximum support. Braces should run from the rafters to bearing partitions (Figure 7-45). The length of the rafter supports varies with the pitch of the roof. The roof pitch for our sample house is 3½/12, so we'll use 6-foot rafter supports. This is another length that you can generally estimate directly from the plans when they're drawn to scale.

Earlier in the chapter, we calculated that we would need 106 common rafters. Excluding the three roof overhangs where there's no partition to rest on (6 common rafters), we'll need a total of 100 rafter supports. If we use 2 × 4s in 12-foot lengths, we can calculate the amount of material to order as follows:

$$\frac{100 \times 6}{12} = 50 \text{ pieces of } 2 \times 4 \times 12\text{s or } 400 \text{ board feet}$$

Enter your estimate of the number and size of ***rafter supports*** *on line 7.2.7 of your Cost Estimate Worksheet. We'll use 50 pieces of 2 × 4 × 12s in our Example Cost Estimate Worksheet.*

Lookouts and Nailers

Lookouts are short framing members nailed to the sides of rafters that extend to the wall of the house (Figure 7-46). They provide the soffit for the roof overhang. There should be one lookout per rafter, including the gable overhang. You'll find their required lengths on the wall section of the blueprints. Normally, lookouts are cut from 2 × 4s and attached to the wall using a 2 × 4 nailer. The length per nailer is the length of the eave.

Let's calculate how much 2 × 4 lookout and nailer material we'll need for our sample house.

As previously figured, we need 106 rafters for the first and second stories. That means we'll also need 106 lookouts. The wall section of the blueprints tells us that the overhang will be 1 foot, so each lookout will be 1 foot long.

The eave length for each side of the house (see Figures 7-1 and 7-2) is:

First story, 23'10" (rounded to 24') plus a 1' overhang at one end	25'0"
Second story, 40' plus a 1' overhang at each end	42'0"
Total:	67'0"

$$67'0'' \times 2 = 134 \text{ lf of eaves, front and back}$$

Since 1 linear foot of nailer is required for every foot of eave, we need 134 linear feet of nailers.

Lookouts (106 × 1')	106 lf
Nailers	134 lf
Total:	240 lf

If we cut the lookouts and nailers from 2 × 4 × 12s, we'll need:

$$\frac{240 \text{ lf}}{12' \text{ per piece}} = 20 \text{ pieces of } 2 \times 4 \times 12\text{s } (160 \text{ bf})$$

Enter your estimate of the number of ***2 × 4 × 12 pieces*** *needed on line 7.2.8 of your Cost Estimate Worksheet. We'll use 20 pieces of 2 × 4 × 12 material in our Example Cost Estimate Worksheet.*

Roof Sheathing

Plywood and OSB are the most common materials used for roof sheathing in residential construction. For rafter spacing of either 16 or 24 inches on center, you can use ½-inch material. Use ⅝-inch material in regions where heavy snow loads are common. With 6d threaded or 8d common nails, nail the sheathing at each bearing, 6 inches on center along all edges, and 12 inches on center along intermediate members.

You'll need to calculate the roof area in order to estimate roof sheathing. Most roofs in residential construction have one or more of the shapes shown in Figure 7-47.

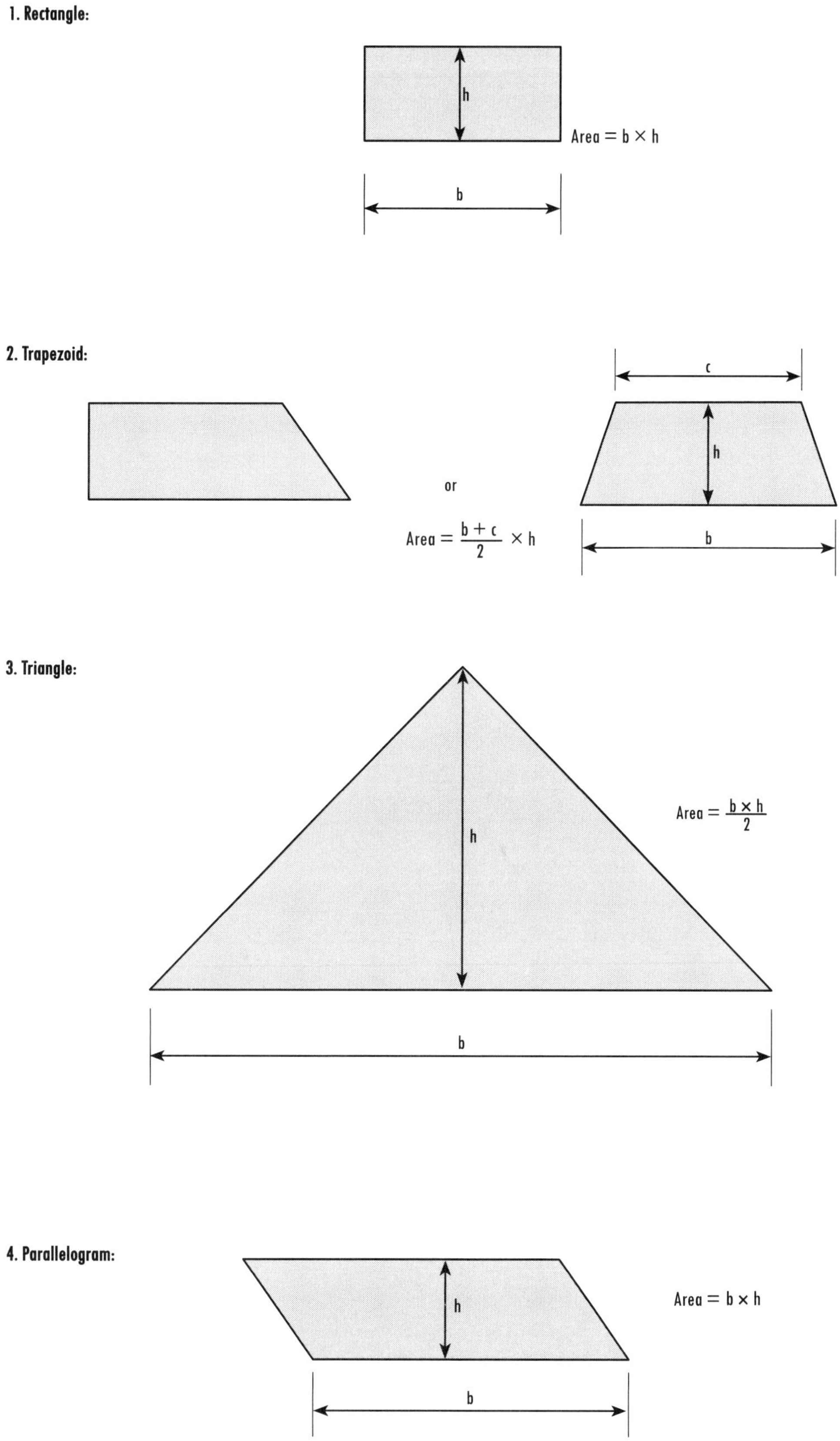

Figure 7-47
Typical roof shapes

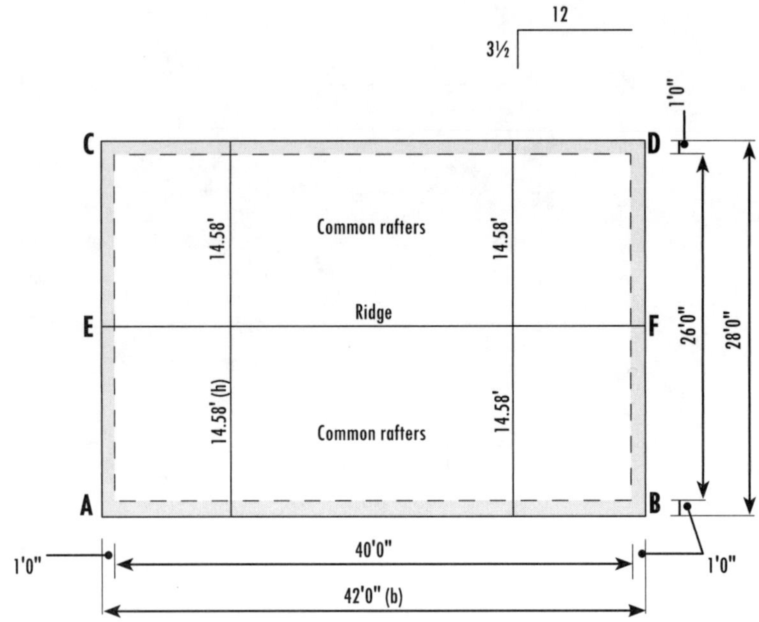

Figure 7-48
Gable roof area: two-story section

Plain Gable Roof

Each side of a plain gable roof, as shown in Figure 7-48, is a rectangle. The formula to find its area is:

$$\text{Area} = b \times h$$

If b (base) is the length of the eave AB, and h (height) is the rafter length to the center of the ridge EF, the area for roof section ABFE in Figure 7-48 is calculated as follows:

$$\text{Area} = b \times h$$
$$\text{Area} = 42.00' \times 14.58'$$
$$\text{Area} = 612.36 \text{ sf}$$

The roof section CDFE is the same size as ABFE, so the total roof area of the gable roof in Figure 7-48 is:

$$2 \times 612.36 = 1{,}224.72 \text{ sf}$$

Hip Roof

Turn back and look at Figure 7-44. A plain hip roof like the one in this drawing has two trapezoidal sections (AEFB and CEFD) and two triangular sections (AEC and BFD). The formula to find the area of a trapezoid is:

$$\text{Area} = \frac{(b + c)}{2} \times h$$

In our formula:

b is the length of the eave, AB or CD.
c is the length of the ridge, EF.
h is the length of the rafter to the center of the ridge.

The calculation for the area of AEFB in Figure 7-44 is:

$$\text{Area} = \frac{(AB + EF)}{2} \times h$$

$$\text{Area} = \frac{50.00' + 20.00'}{2} \times 18.028' = 630.98 \text{ sf}$$

The area for section AEFB is 630.98 square feet. Since both trapezoidal roof sections are equal, multiply the area of AEFB by 2 to get the area of the two trapezoidal roof sections:

$$630.98 \times 2 = 1{,}261.96 \text{ sf}$$

The formula for finding the area of a triangle is:

$$\text{Area} = \frac{b \times h}{2}$$

The length of the eave AC or BD is b, and h is the rafter length (the combined jack and common rafter lengths).

The triangular roof area, AEC, in Figure 7-44 is:

$$\text{Area} = \frac{AC \times \text{rafter length}}{2}$$

$$\text{Area} = \frac{30.00' \times 18.028'}{2} = 270.42 \text{ sf}$$

The triangular roof sections are equal, so multiply the area of AEC by 2 to get the combined area of AEC and BFD.

$$270.42 \times 2 = 540.84 \text{ sf}$$

Now we need to add the sections together to get the total area for the hip roof in Figure 7-44.

Two trapezoidal sections	1,261.96 sf
Two triangular sections	540.84 sf
Total roof area:	1,802.80 sf

Plain Hip Roof — Alternate Method

If the hip roof in Figure 7-44 was calculated as a simple rectangle, the calculations for the "bottom" half of the roof would be:

$$\text{Area} = AB \times \text{length of rafter}$$

$$50.00' \times 18.028 = 901.40 \text{ sf}$$

Since the "top" half of the roof is the same size as the bottom, multiply by 2 to get the area of the roof.

$$2 \times 901.40 = 1{,}802.80 \text{ sf}$$

Notice that this is exactly the same area as the previous calculations for the trapezoidal and triangular sections. So, as you can see, the roof area for a plain hip roof can be calculated as a simple rectangle.

Gable Roof Offset

Now look back to Figure 7-43. It shows a roof offset or intersection with a gable roof. To estimate the roof area of this offset, follow these guidelines:

- Compute the area of the main roof, section ABFE, as a complete rectangle. The area for this section will include the triangular section, IJB, in the offset. It doesn't need to be calculated separately.
- Multiply the eave length of the offset, IG, by the common rafter length in the offset to get the area of one section. Multiply the area of that section by 2 to get the total rectangular offset roof area.

Find the common rafter length for the offset:

$$11' \times 13.416 \text{ factor (for 6/12 roof pitch)} = 147.62" \text{ or } 12.30'$$

Now calculate the area:

$$\text{Area} = \text{IG} \times \text{rafter length}$$
$$24.00' \times 12.30' = 295.20 \text{ sf}$$
Multiply by 2 to get the total offset roof area:
$$2 \times 295.20 \text{ sf} = 590.40 \text{ sf}$$

Hip Roof Offset

Figure 7-49 shows an offset or intersection with a hip roof. The roof pitch is 6/12. To estimate the roof area of this offset, follow these guidelines:

- Compute the area of the main roof of section, ABFE, as solid. That will include the triangular area, IJB, of the offset. Use the method previously discussed in Plain Hip Roof — Alternative Method; this provides an accurate estimate.
- The remainder of the hip roof offset is two trapezoidal sections and one triangular section. But remember (from Plain Hip Roof — Alternative Method), you don't have to calculate these sections separately. Treat them as two

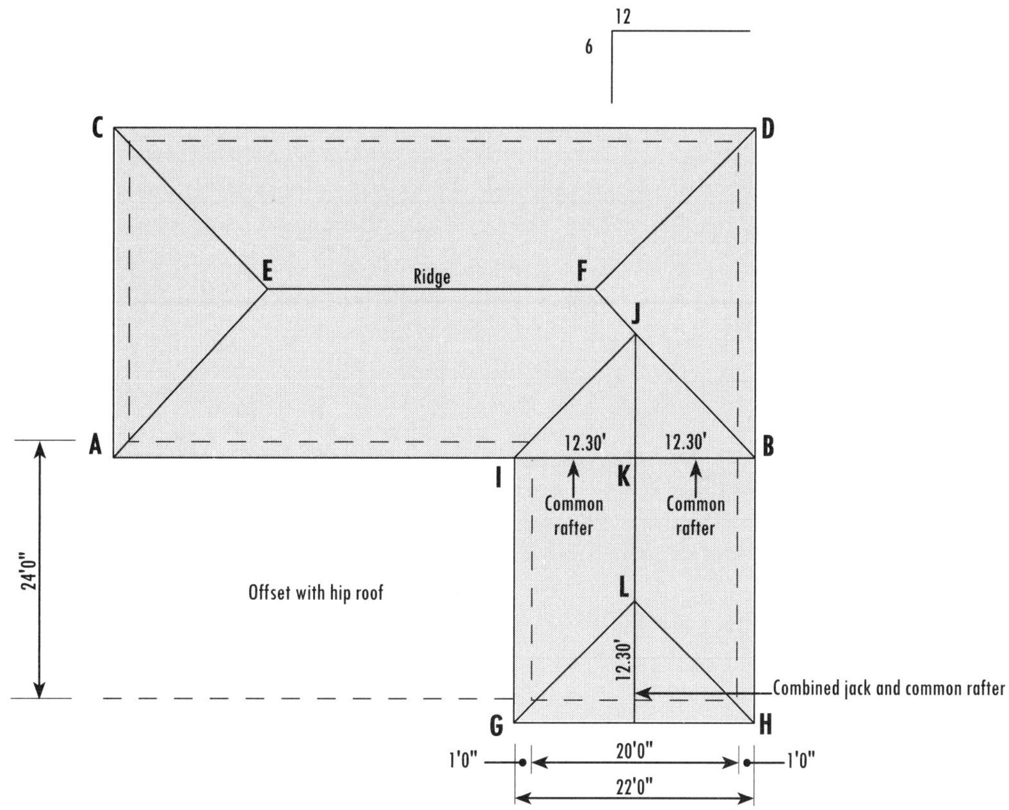

Figure 7-49
Hip roof offset

rectangular sections, multiply the eave length by the rafter length to get their areas. Then multiply by 2 to get the total area for the offset.

Find the common rafter length for the offset:

11' × 13.416 factor (for 6/12 roof pitch) = 147.62" or 12.30'

Now calculate the area:

Area = IG × rafter length

24'0" × 12.30' = 295.20 sf

Multiply by 2 to get the total offset roof area:

2 × 295.20 = 590.40 sf

Compare this to the area for the gable offset we just figured. You can see that the roof area of the offset is the same, whether it's a plain gable or a hip roof.

Again, add the offset area of the main roof to get the total area of a roof with a hip offset.

Alternate Method for Calculating Roof Areas

Another method of estimating roof area is to multiply the square feet of the floor area, including the roof overhang, by the CR pitch ratio shown in the table in Figure 7-31.

Let's look at an example using Figure 7-48. The floor area, including the roof overhang, is 1,176 square feet (42 feet × 28 feet). The roof pitch is 3½/12. The table in Figure 7-31 gives us a CR pitch ratio of 1.042 for a 3½/12 roof pitch.

(Floor Area + Overhang) × CR Pitch Ratio = Roof Area

1,176 sf × 1.042 = 1,225.39 sf

Our previous calculation for the area of the plain gable roof in Figure 7-48 was 1,224.72 square feet. Though these two calculations are slightly different (0.67 square foot), the result for both is sufficiently accurate for estimating purposes.

Here's another example of this method using the hip roof in Figure 7-44. The floor area, including the roof overhang, is 1,500 square feet (50 feet × 30 feet). The roof pitch is 8/12. The CR pitch ratio for an 8/12 roof pitch from Figure 7-31 is 1.202.

(Floor Area + Overhang) × CR Pitch Ratio = Roof Area

1,500 sf × 1.202 = 1,803.00 sf

Our previous calculation for the plain hip roof area in Figure 7-44 was 1,802.80 square feet (a difference of 0.20 square foot). Again, both calculations are sufficiently accurate for estimating purposes.

Note: If you're using ratios to calculate roof areas, be especially careful with offsets or intersections. Any break in the roof line can alter your calculations.

Now let's calculate the roof area for our sample house, using the dimensions shown in Figures 7-1, 7-2, and 7-9. The roof area consists of two plain gable roof sections. We'll use the Alternate Method for Calculating Roof Areas.

For the two-story section, the floor area, including 1-foot roof overhang all around, is:

(40'0" + 2'0" overhang) × (26'0" + 2'0" overhang) = 1,176 sf

The roof pitch is 3½/12. In Figure 7-31, the CR pitch ratio for a 3½/12 roof is 1.042.

(Floor Area + Overhang) × CR Pitch Ratio = Roof Area

1,176 sf × 1.042 = 1,225.39 sf

The dimensions for the single-story section are 24 feet (rounded from 23.83 feet) by 26 feet. There's a 1-foot roof overhang at the front, back and the right side only (the roof abuts the second story on the left). The square footage for the first story is:

(24'0" + 1'0" overhang) × (26'0" + 2'0" overhang) = 700 sf

The roof pitch is 3½/12. From Figure 7-31, the CR pitch ratio for a 3½/12 roof is 1.042.

$$\text{(Floor area + overhangs)} \times \text{CR pitch ratio} = \text{Roof area}$$

$$700.00 \text{ sf} \times 1.042 = 729.40 \text{ sf}$$

Two-story roof section area	1,225.39 sf
Single-story roof section area	729.40 sf
Total roof area:	1,954.79 sf

Calculating the Roof Sheathing

Now that we know the square feet of roof area for our sample house, we can figure how much sheathing to order. We'll use 4 × 8 OSB, ½-inch thick. Each sheet of 4 × 8 material is 32 square feet, so the number of pieces we need to order is:

$$\frac{1{,}954.79 \text{ sf}}{32 \text{ sf per piece}} = 61.01 \text{ (rounded to 62) pieces}$$

> *Enter your estimate of the number of ½-inch-thick 4 × 8 OSB sheets for **roof sheathing** on line 7.2.9 of your Cost Estimate Worksheet. We'll use 62 pieces of ½-inch-thick 4 × 8 OSB sheets in our Example Cost Estimate Worksheet.*

Nails

The quantity of nails for the ceiling, roof framing, and sheathing is calculated using the following allowances:

- Framing allowance
  - 8d common nails, allow 2 pounds per 1,000 board feet
  - 16d common nails, allow 10 pounds per 1,000 board feet
- Plywood (or OSB) allowance
  - 6d threaded nails, allow 12 pounds per 1,000 square feet

Framing Nails

First, total the number of board feet of framing lumber in the roof system. We entered material estimates previously in the following sections in our Cost Estimate Worksheet, which calculated thousand board feet (MBF):

7.2.1.1 Joists, 19 pieces	0.266 MBF
7.2.1.2 Joists, 19 pieces	0.266 MBF
7.2.1.3 Joists, 31 pieces	0.661 MBF
7.2.1.4 Joists, 31 pieces	0.372 MBF
Subtotal:	1.565 MBF

7.2.2 Ceiling backing, 16 pieces	0.192 MBF
7.2.3 Ceiling joist stiffeners, 11 pieces	0.044 MBF
7.2.4 Common rafters, 106 pieces	1.696 MBF
7.2.5 Ridge, 5 pieces	0.093 MBF
7.2.6 Collar beams, 16 pieces	0.085 MBF
7.2.7 Rafter supports, 50 pieces	0.400 MBF
7.2.8 Lookouts and nailers, 20 pieces	0.160 MBF
Subtotal:	2.670 MBF
Total:	4.235 MBF

Now let's figure what quantity of each nail size we need to estimate for the roof system framing of our sample house.

8d common nails (at 2 pounds per MBF):
4.235 MBF × 2 lbs = 8.47 (rounded to 9) lbs
16d common nails (at 10 pounds per MBF):
4.235 MBF × 10 lbs = 42.35 (rounded to 43) lbs

*Enter your estimate of the quantity and sizes of framing **nails** in sections 7.2.10.1 and 7.2.10.2 of your Cost Estimate Worksheet. We'll use 9 pounds of 8d common nails and 43 pounds of 16d common nails in our Example Cost Estimate Worksheet.*

Roof Sheathing Nails

We've calculated 1,954.79 square feet of OSB for the roof sheathing, and entered a total of 62 pieces of ½-inch-thick 4 × 8 OSB in our Example Cost Estimate Worksheet.

Let's calculate the quantity of nails we need to attach the roof sheathing, using 6d threaded nails. We need 12 pounds of nails per 1,000 square feet of sheathing.

$$\frac{1{,}954.79\text{ sf} \times 12\text{ lbs}}{1{,}000\text{ sf}} = 23.46\text{ (rounded to 24) lbs}$$

*Enter your estimate of the quantity and sizes of roof sheathing **nails** in section 7.2.10.3 of your Cost Estimate Worksheet. We'll use 24 pounds of 6d threaded nails in our Example Cost Estimate Worksheet.*

Trusses

As an alternative to assembling the ceiling and roof framing on-site, many contractors purchase factory-assembled roof trusses. Roof trusses are made of individual structural members formed into one unit, designed to span large distances. No support from interior partitions is necessary. Trusses must be designed by a qualified engineer or architect in accordance with standard engineering practice.

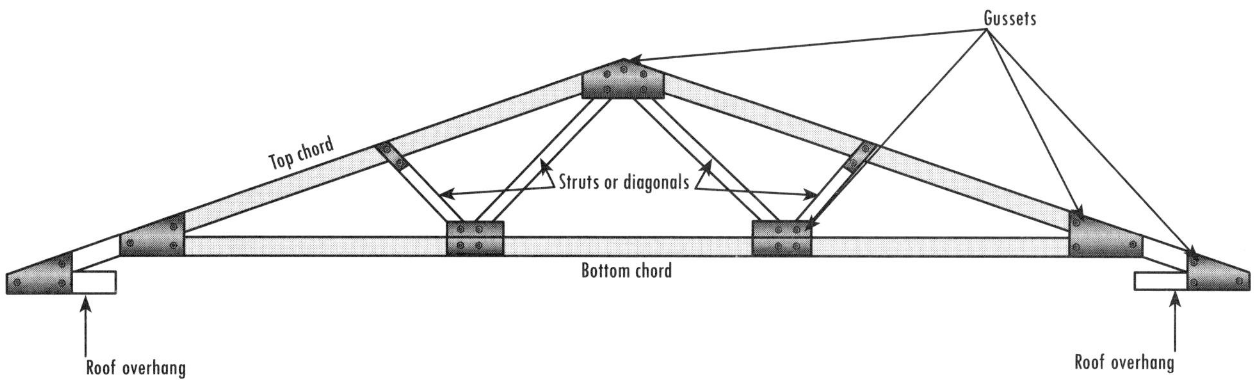

Figure 7-50
Truss with roof overhang

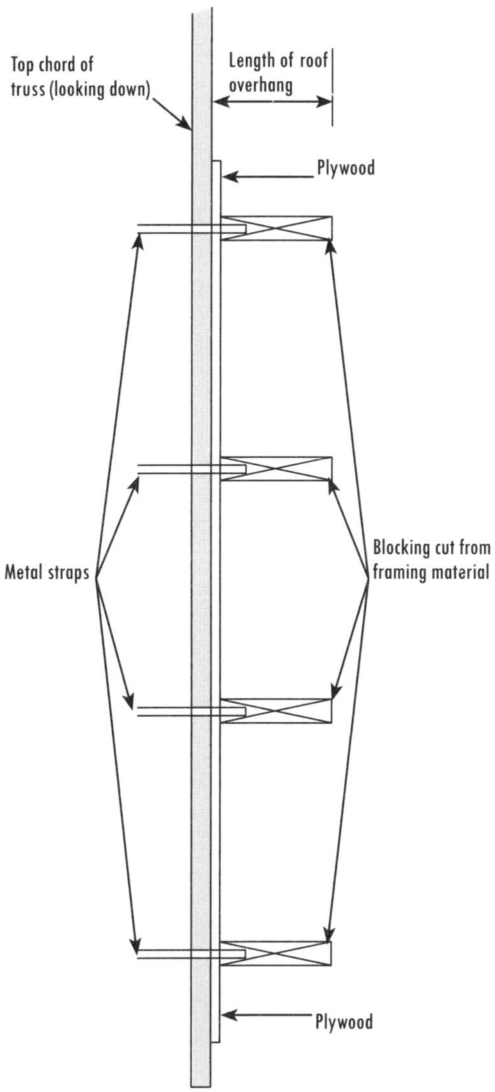

Figure 7-51
Fabricated gable roof overhang

Trusses come in a variety of sizes and shapes and can be used for long or short spans, such as a house roof, or a roof over a porch, walk, or breezeway. Trusses are normally spaced 24 inches on center. When compared to the cost of assembling individual roof members onsite, the use of trusses can cut labor costs substantially. Trusses eliminate the need for the following framing members of the roof system:

Ceiling joists
Rafters
Ridge
Collar beams
Rafter supports
Lookouts
Gable end studs

The bottom chord of the truss is used as a ceiling joist, and the top chord is used as the rafter. The roof overhang at the eave can be fabricated with the truss, as shown in Figure 7-50. Blocking for a gable roof is cut from 2 × 6 or 2 × 8 framing, and is used in the overhang at either 16 or 24 inches on center. Blocking is usually cut into 10-foot-long sections and fastened to the top chord of the truss with metal straps (Figure 7-51).

To estimate the number of roof trusses at 24 inches on center, multiply the length of the wall perpendicular to the truss by 0.50 and add one truss for the starter. Do this for each section of the house where there's a break in the roof line. For example, let's calculate how many trusses we would need for the one- and two-story sections of our house if we chose to use trusses instead of individual framing members.

One-story section

24′0″ × 0.50 = 12 trusses, plus 1 starter = 13

Two-story section

40′0″ × 0.50 = 20 trusses, plus 1 starter = 21

Total: 34 trusses

> **Note:** Even if you use preassembled trusses, you'll need to complete line 7.2.9, Roof Sheathing, and 7.2.10.3, Roof Sheathing Nails, in the *Cost Estimate Worksheet.*

> *Enter your **truss** estimate in section 7.2.11 of your Cost Estimate Worksheet. Since we're not using trusses, we'll enter 0 on line 7.2.11.*

Nails

Use two 10d common nails on each side of the truss at every wall plate. The nail allowance for trusses is considerably less than for ceiling joists and rafters. For 34 trusses, estimate 10 pounds of 10d common nails.

> *Enter your estimate of the number of pounds and size of **nails** on line 7.2.12 of your Cost Estimate Worksheet. If we were using trusses, we'd enter 10 pounds of 10d common nails in our Example Cost Estimate Worksheet. However, since we aren't using trusses, we'll enter 0 on line 7.2.12.*

Porch Shed Roof Framing

When shed roofs are built to cover porches and walkways (Figures 7-52 and 7-53) and they're attached to the house, they're framed as shown in Figure 7-54. These roofs can be built using either ceiling joists and rafters, or trusses.

Using Ceiling Joists and Rafters

If using rafters for a porch roof, you'll estimate the number and lengths needed in the same manner as for the main house, with one exception. For the house roof, half the thickness of the ridge is deducted from the length of a common rafter for the true rafter length. But for the porch shed roof, the full thickness of the rafter nailer (which serves the same purpose as the ridge) is deducted to get the true rafter length, as shown on the worksheet in Figure 7-55. (A copy of a blank worksheet is available on the accompanying CD-ROM.)

Figure 7-56 shows the shed roof dimensions for the terrace at the rear of our house, and Figure 7-57 shows a completed worksheet using those dimensions. The rafters will be cut from 2 × 6 × 14 lumber.

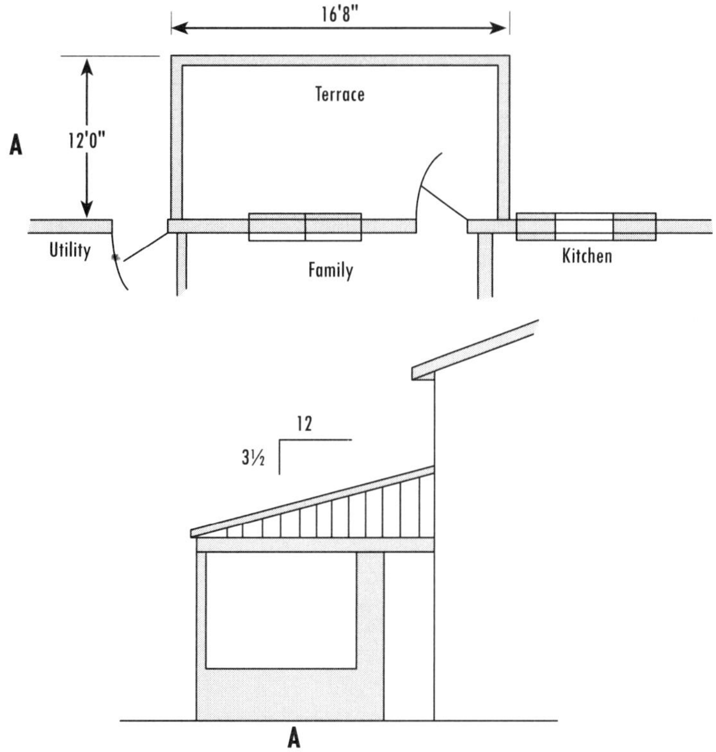

Figure 7-52
Porch shed roof (rear terrace)

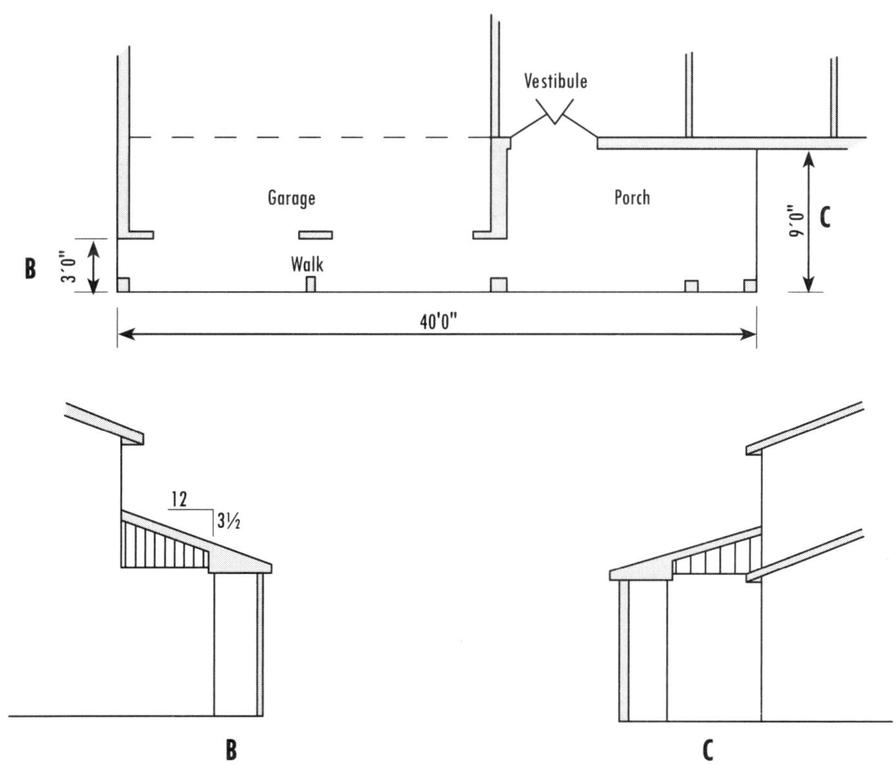

Figure 7-53
Front porch and walkway shed roof

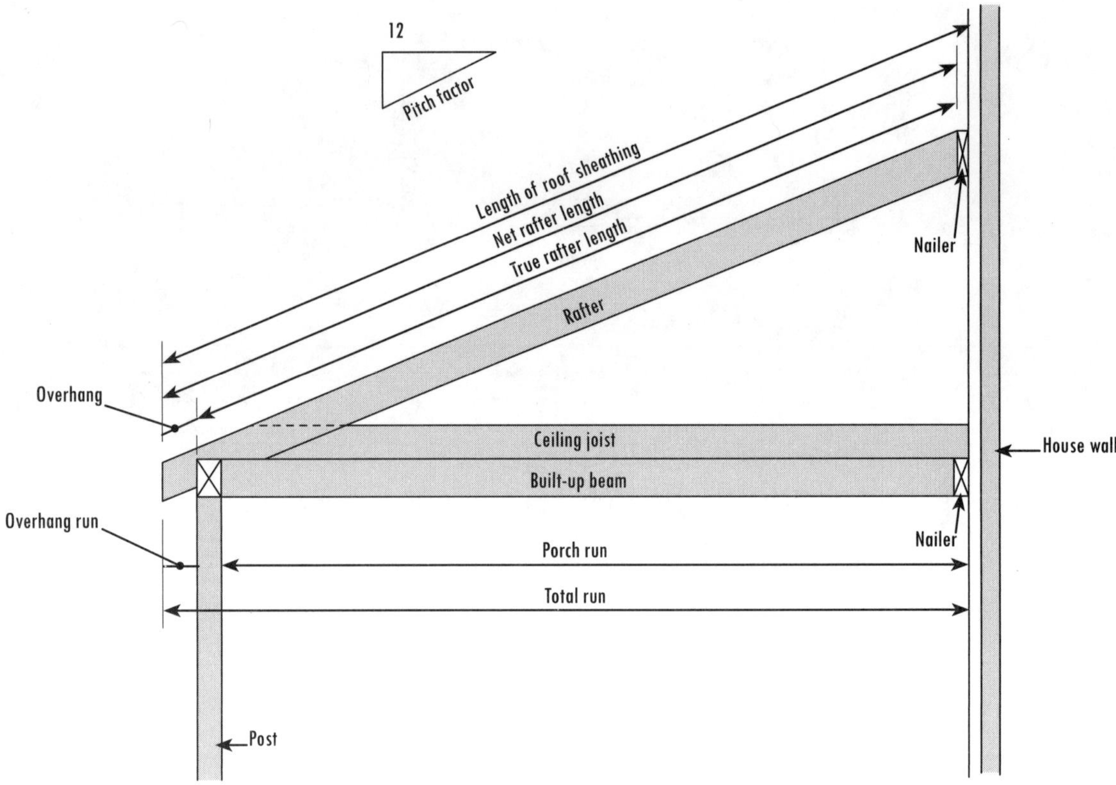

Figure 7-54
Shed roof framing

Worksheet for Common Rafters (Shed Roof)

1 Rafter length to house wall:

_________ ft × _________ = 0.00 inches

Porch run · CR Pitch factor

2 Rafter length to face of nailer:

0.00 minus _______ inches = 0.00 inches

Thickness of nailer · True rafter length

3 Rafter overhang:

_________ ft × 0.00 = 0.00 inches

Overhang run · CR Pitch factor · Rafter overhang

4 Net rafter length:

0.00 plus 0.00 = 0.00 inches

True rafter length · Overhang · Net rafter length

0.00 divided by 12 = 0.00 feet = _______ feet/inches

Net rafter length

5 Length of roof sheathing:

0.00 × 0.00 = 0.00 inches

Total run · CR Pitch factor

0.00 in divided by 12 = 0.00 feet = _______

Figure 7-55
Worksheet for common rafters

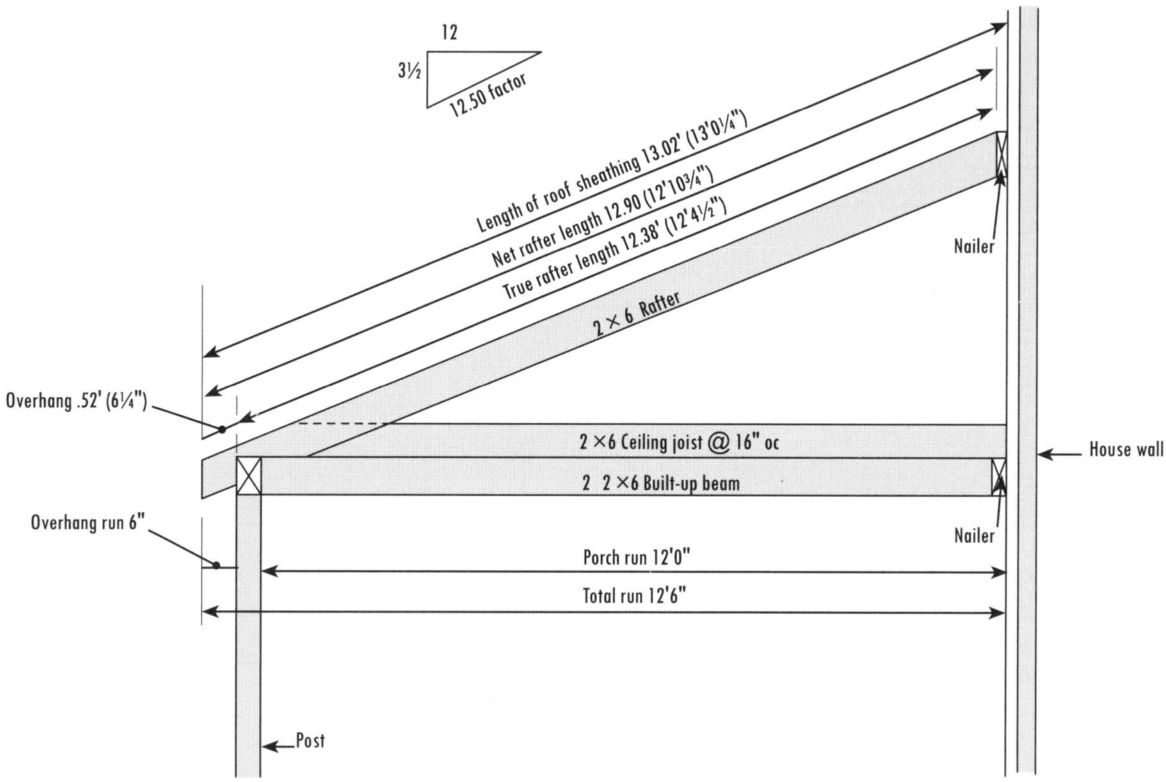

Figure 7-56
Shed roof framing (terrace)

Worksheet for Common Rafters (Shed Roof)

1 Rafter length to house wall:

12.00 ft (Porch run) × 12.50 (CR Pitch factor) = 150.00 inches

2 Rafter length to face of nailer:

150.00 minus 1.50 inches (Thickness of nailer) = 148.50 inches (True rafter length)

3 Rafter overhang:

0.50 ft (Overhang run) × 12.50 (CR Pitch factor) = 6.25 inches (Rafter overhang)

4 Net rafter length:

148.50 (True rafter length) plus 6.25 (Overhang) = 154.75 inches (Net rafter length)

154.75 (Net rafter length) divided by 12 = 12.90 feet = 12'10¾"

5 Length of roof sheathing:

12.50 (Total run) × 12.50 (CR Pitch factor) = 156.25 inches

156.25 in divided by 12 = 13.02 feet = 13'¼"

Figure 7-57
Sample worksheet for common rafters

If ceiling joists and rafters are used for the terrace shed roof in Figure 7-52 (roof overhang of 6 inches for each gable and the eave), these are the materials you'll need:

Built-up beam (or porch rim)		
4	2 × 6 × 12′	48 board feet
2	2 × 6 × 18′	36 board feet
Nailer for ceiling joists		
1	2 × 6 × 18′	18 board feet
Ceiling joists (16″ on center)		
14	2 × 6 × 12′	168 board feet
Rafter nailer		
1	2 × 8 × 18′	24 board feet
Rafters		
15	2 × 6 × 14′	210 board feet
Total framing material:		504 board feet
Roof sheathing (OSB)		
8 pcs	½″ × 4 × 8′	256 square feet
Nails		
16d common nails		7 pounds
6d threaded nails		3 pounds

*Enter your **roof framing and sheathing** estimates on lines 7.3.1 through 7.3.7 of your Cost Estimate Worksheet. We'll enter the numbers on our material list above in our Example Cost Estimate Worksheet.*

Using Trusses

If using trusses at 24 inches on center, you'll need the following materials:

Built-up beam		
4	2 × 6 × 12′	48 board feet
2	2 × 6 × 18′	36 board feet
Truss nailer		
2	2 × 4 × 18′	24 board feet
Trusses		
10		
Roof sheathing (plywood)		
8 pcs	½″ × 4 × 8	256 square feet

Nails	
16d common nails	3 pounds
6d threaded nails	3 pounds

> *If trusses were used, you'd enter your estimates on lines 7.3.8 through 7.3.12 of your Cost Estimate Worksheet. We won't enter any trusses for the rear terrace roof in our Example Cost Estimate Worksheet.*

However, we will use trusses for the porch and walkway shed roof at the front on the house, shown in Figure 7-53. These trusses will be 24 inches on center. The material list will include:

Built-up beam		
1	2 × 6 × 18′	18 board feet
2	2 × 6 × 16′	32 board feet
4	2 × 6 × 12′	48 board feet
1	2 × 6 × 8′	8 board feet
Truss nailer		
1	2 × 4 × 16′	11 board feet
2	2 × 4 × 12′	16 board feet
Trusses		
22		
Roof sheathing (OSB)		
13 pcs	½″ × 4 × 8′	416 square feet
Nails		
16d common nails		6 pounds
6d threaded nails		5 pounds

> *Enter your **truss-related** estimates on lines 7.3.8 through 7.3.12 of your Cost Estimate Worksheet. We'll enter the numbers on our materials list for the front porch roof in our Example Cost Estimate Worksheet.*

Stair Stringers

If a house has more than one floor, stair stringers and temporary tread installation between floors is recommended as soon as possible after the house has a temporary roof. Not only is this more convenient for workmen on the job — it's also safer.

The table in Figure 7-58 has stringer lengths for various floor-to-floor heights that you can use as a guideline for estimating. These lengths are only approximate, since stringer length varies with the number and dimensions of the risers and treads that may be calculated for the same floor-to-floor rise. For example, if the floor-to-floor rise is 8′10⅜″:

> The stringer length will be 13′7½″ if you're using 14 risers at 7⅝″ and 13 treads at 10″; or

> The stringer length will be 17′1$\frac{29}{32}$″, using 17 risers at 6¼″ and 16 treads at 11¼″.

The difference is in how high you want the risers and how deep you want the treads. The minimum tread depth is 10 inches and the maximum riser height is 7¾ inches.

The minimum depth between the bottom of the stair stringer and the cutout for the treads is 3½ inches (see Figure 7-59), unless the stringer is supported by other construction. Consequently, 2 × 12s are recommended for these stringers. If the width of the stairs is greater than 2′6″, a center stringer is required.

Floor-to-Floor Height	Approximate Length of Stringers*
3'0"	6'0"
4'0"	7'0"
4'6"	7'6"
5'0"	9'0"
5'6"	9'6"
6'0"	10'0'
6'6"	11'0"
7'0"	12'0"
7'6"	12'6"
8'0"	13'0"
8'6"	14'0"
9'0"	15'0"
9'6"	15'6"
10'0"	16'0"

*The stair stringer length varies with the same floor-to-floor height depending on the number and dimensions of the risers and treads. The length of the stringer increases as the number of risers increases.

Figure 7-58
Approximate stringer lengths

Figure 7-59 shows the dimensions for the temporary stairs of our sample house (Figures 7-1 and 7-2). What material should we estimate for the stair stringers and temporary treads and risers from the first to the second floor?

The stringer length is 13′7½″ with 14 risers at 7⅝ inches and 13 treads at 10 inches. We'll need 2 × 12s, 14 feet in length. Since the stair width is 3′6″, three stringers are required. There's normally enough waste material from the framing to use as temporary tread material without ordering additional material for these.

Most builders construct stairs to the basement soon after the concrete floor is poured. We'll discuss basement stair stringers later in this book.

Enter your estimate of the numbers and length of ***stringers*** *we discussed on line 7.4 of your Cost Estimate Worksheet. We'll use three 2 × 12 × 14′ stringers in our Example Cost Estimate Worksheet.*

Labor Costs for the Superstructure

Labor costs are always estimates. I can't stress how important cost records from previous jobs are in order to more accurately estimate a new job. A major factor that affects labor costs is weather. Building the superstructure of a house is outside work, so an allowance should be made for lower productivity in cold or very hot weather. To offset the possible severity and duration of poor weather conditions, add 5 to 10 percent to your manhour estimate.

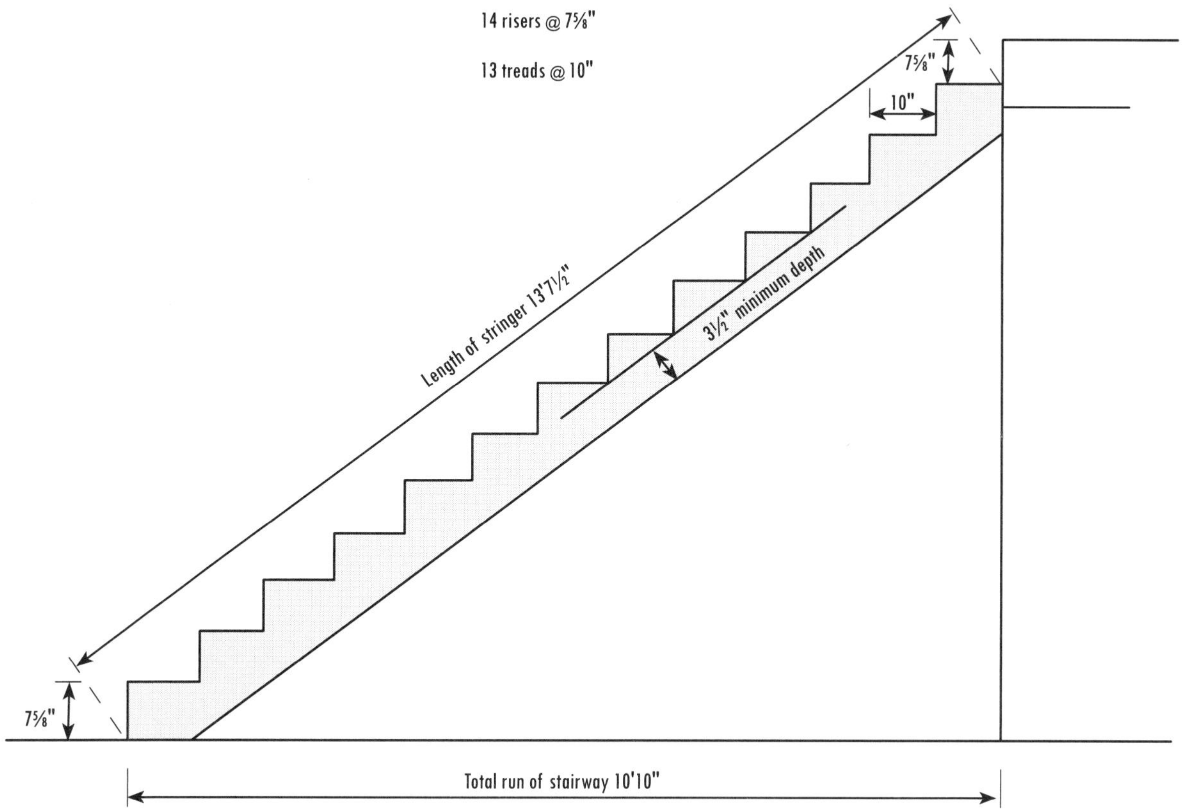

Figure 7-59
Stair layout

Labor Costs

We'll estimate labor costs based on cost records for similarly-built houses, recently constructed.

Custom House Built with Ceiling Joists and Rafters

Our records show that a recently built 2,183 square-foot, two-story house superstructure required 507 manhours to build. Our sample house is similar in design, but with 2,848 square feet of living space, including the garage. We'll use X to designate the estimated number of manhours needed to construct the superstructure of our house, using the following proportion:

$$507 \text{ manhours} : 2{,}183 \text{ sf} = X \text{ manhours} : 2{,}848 \text{ sf}$$

$$\text{Product of extremes} = \text{Product of means}$$

$$507 \times 2{,}848 = 2{,}183 \times X$$

$$X = \frac{507 \times 2{,}848}{2{,}183}$$

$$X = 661.45 \text{ (rounded to 662) manhours}$$

The estimated time to build the superstructure for our house is 662 manhours. We'll use three carpenters and three laborers on this job. The manhours will be divided equally among the six workmen, so each one will work:

$$\frac{\text{Manhours}}{\text{Workmen}} = \frac{662}{6} = 110.33 \text{ (rounded to 111) hours}$$

> *Enter your **labor cost** estimates in section 7.5 of your Cost Estimate Worksheet. We'll use these numbers in our Example Cost Estimate Worksheet.*

If you don't have labor cost records to use as a guideline, the following labor factor can be used to estimate labor for the superstructure of a custom-built house, using ceiling joists and rafters:

Manhour Factor: 0.232 total hours per square foot

Using the manhour factor, let's estimate how many manhours it should take to build the superstructure of our 2,848-square-foot house, using ceiling joists and rafters:

$$2{,}848 \text{ sf} \times 0.232 = 660.74 \text{ (rounded to 661) manhours}$$

Let's compare that to our estimate for how many manhours it should take to build the superstructure of a 2,183-square-foot house, using ceiling joists and rafters:

$$2{,}183 \text{ sf} \times 0.232 = 506.46 \text{ (rounded to 507) manhours}$$

Custom-Built House Using Trusses

Trusses, where usable, require approximately one-third fewer manhours to install than ceiling joists and rafters. Records for a similarly designed custom house of 2,286 square feet show that 415 manhours were required to build the superstructure. Using these records, we can estimate the manhours (X) for our 2,848 square-foot house. Set up the proportion as follows:

$$415 \text{ manhours} : 2{,}286 \text{ sf} = X \text{ manhours} : 2{,}848 \text{ sf}$$

$$\text{Product of extremes} = \text{Product of means}$$

$$415 \times 2{,}848 = 2{,}286 \times X$$

$$X = \frac{415 \times 2{,}848}{2{,}286}$$

$$X = 517.03 \text{ (rounded to 518) manhours}$$

The 518 manhours will be divided equally among 3 carpenters and 3 laborers, so each will work:

$$\frac{\text{Manhours}}{\text{Workmen}} = \frac{518}{6} = 86.33 \text{ (rounded to 87) hours}$$

> *If trusses are used, enter those **labor estimates** on line 7.5 of your Cost Estimate Worksheet.*

If you don't have labor cost records, the following factor can be used for estimating the superstructure of a custom house built using trusses:

Manhour Factor: 0.182 manhours per square foot

Using the manhour factor, let's estimate how many manhours it should take to build the superstructure of our 2,848-square-foot house, using trusses:

2,848 square feet × 0.182 = 518.34 (rounded to 519) manhours

Again, let's compare that to our estimate for how many manhours it should take to build the superstructure of a 2,286-square-foot house, using trusses:

2,286 square feet × 0.182 = 416 manhours

This completes the estimate for the superstructure of a house. In Chapter 8, we'll cover estimating the roofing system.

COST ESTIMATE WORKSHEET FOR SUPERSTRUCTURE

#	Qty	Size	MBF	Cost	Per		Subtotal
7.1	**Exterior and interior walls**						
7.1.1	**Sole and top plates**						
	180 pieces	2 " × 4 " × 12 ft	1.440	@ $697.00	MBF	= $1,003.68	
7.1.2	**Studs**						
	800 pieces	2 " × 4 " × 8 ft	4.267	@ $697.00	MBF	= $2,974.10	
7.1.3	**Headers**						
	2 pieces	2 " × 12 " × 12 ft	0.048	@ $759.00	MBF	= $36.43	
	4 pieces	2 " × 10 " × 10 ft	0.067	@ $725.00	MBF	= $48.58	
	2 pieces	2 " × 8 " × 8 ft	0.021	@ $671.00	MBF	= $14.09	
	7 pieces	2 " × 6 " × 12 ft	0.084	@ $718.00	MBF	= $60.31	
	2 pieces	2 " × 6 " × 10 ft	0.020	@ $718.00	MBF	= $14.36	
	2 pieces	2 " × 6 " × 8 ft	0.016	@ $718.00	MBF	= $11.49	
	9 pieces	2 " × 4 " × 12 ft	0.072	@ $697.00	MBF	= $50.18	
	3 pieces	2 " × 4 " × 10 ft	0.020	@ $697.00	MBF	= $13.94	
	6 pieces	2 " × 4 " × 8 ft	0.032	@ $697.00	MBF	= $22.30	
	pieces	" × " × ft	0.000	@	MBF	= $0.00	
7.1.4	**Temporary wall braces**						
	20 pieces	2 " × 4 " × 12 ft	0.160	@ $697.00	MBF	= $111.52	
7.1.5	**Corner bracing**	Plywood					
	20 pieces	½ " × 4 ft × 8 ft		@ $15.54	piece	= $310.80	
7.1.6	**Wall sheathing**	OSB					
	73 pieces	½ " × 4 ft × 8 ft		@ $8.22	piece	= $600.06	
	pieces	" × ft × ft		@	piece	= $0.00	
7.1.7	**Nails**						
7.1.7.1	138 lbs, 16d common			@ $2.04	lb	= $281.52	
7.1.7.2	7 lbs, 6d common			@ $2.20	lb	= $15.40	
7.1.7.3	24 lbs, 1½" roofing			@ $1.67	lb	= $40.08	
			Line 7.1			$5,608.84	$5,608.84

COST ESTIMATE WORKSHEET FOR SUPERSTRUCTURE, cont.

#	Qty	Size	MBF	Cost	Per			Subtotal
7.2	**Ceiling and roof framing**							
7.2.1	**Ceiling joists**							
7.2.1.1	19 pieces	2 " × 6 " × 14 ft	0.266	@ $592.00	MBF	=	$157.47	
7.2.1.2	19 pieces	2 " × 6 " × 14 ft	0.266	@ $592.00	MBF	=	$157.47	
7.2.1.3	31 pieces	2 " × 8 " × 16 ft	0.661	@ $615.00	MBF	=	$406.52	
7.2.1.4	31 pieces	2 " × 6 " × 12 ft	0.372	@ $592.00	MBF	=	$220.22	
7.2.2	**Ceiling backing**							
	16 pieces	2 " × 6 " × 12 ft	0.192	@ $718.00	MBF	=	$137.86	
7.2.3	**Ceiling joist stiffeners**							
	11 pieces	1 " × 4 " × 12 ft	0.044	@ $1,000.00	MBF	=	$44.00	
7.2.4	**Rafters**							
	106 pieces	2 " × 6 " × 16 ft	1.696	@ $718.00	MBF	=	$1,217.73	
7.2.5	**Ridges**							
	5 pieces	2 " × 8 " × 14 ft	0.093	@ $671.00	MBF	=	$62.40	
7.2.6	**Collar beams**							
	16 pieces	2 " × 4 " × 8 ft	0.085	@ $697.00	MBF	=	$59.25	
7.2.7	**Rafter supports**							
	50 pieces	2 " × 4 " × 12 ft	0.400	@ $697.00	MBF	=	$278.80	
7.2.8	**Lookouts and nailers**							
	20 pieces	2 " × 4 " × 12 ft	0.160	@ $697.00	MBF	=	$111.52	
7.2.9	**Roof sheathing** OSB							
	62 pieces	½ " × 4 ft × 8 ft		@ $8.32	piece	=	$515.84	
7.2.10	**Nails**							
7.2.10.1	9 lbs, 8d common			@ $2.04	lb	=	$18.36	
7.2.10.2	43 lbs, 16d common			@ $2.04	lb	=	$87.72	
7.2.10.3	24 lbs, 6d threaded			@ $3.24	lb	=	$77.76	
7.2.11	**Trusses**							
	units			@	truss	=	$0.00	
	units			@	truss	=	$0.00	
7.2.12	**Nails**							
	lbs			@	lb	=	$0.00	
			Line 7.2				3,552.92	$3,552.92
7.3	**Porch shed roof framing and sheathing (with ceiling joists and rafters)**							
7.3.1	**Built-up beam (or porch rim)**							
	4 pieces	2 " × 6 " × 12 ft	0.048	@ $718.00	MBF	=	$34.46	
	2 pieces	2 " × 6 " × 18 ft	0.036	@ $718.00	MBF	=	$25.85	
	pieces	" × " × ft	0.000	@	MBF	=	$0.00	
7.3.2	**Nailer for ceiling joists**							
	1 piece	2 " × 6 " × 18 ft	0.018	@ $718.00	MBF	=	$12.92	
7.3.3	**Ceiling joists**	16 " oc						
	14 pieces	2 " × 6 " × 12 ft	0.168	@ $718.00	MBF	=	$120.62	
7.3.4	**Rafter nailer (or ridge)**							
	1 piece	2 " × 8 " × 18 ft	0.024	@ $671.00	MBF	=	$16.10	

COST ESTIMATE WORKSHEET FOR SUPERSTRUCTURE, cont.

#	Qty	Size	MBF		Cost	Per			Subtotal
7.3.5	**Rafters**								
	15 pieces	2 " × 6 " × 14 ft	0.210	@	$718.00	MBF	=	$150.78	
	pieces			@		MBF	=	$0.00	
7.3.6	**Roof sheathing** OSB								
	8 pieces	½ " × 4 ft × 8 ft		@	$8.32	piece	=	$66.56	
7.3.7	**Nails**								
	7 lbs, 16d common			@	$2.04	lb	=	$14.28	
	3 lbs, 6d threaded			@	$3.24	lb	=	$9.72	
OR	**Porch shed roof framing and sheathing (with trusses)**								
7.3.8	**Built-up beam (or porch rim)**								
	1 piece	2 " × 6 " × 18 ft	0.018	@	$718.00	MBF	=	$12.92	
	2 pieces	2 " × 6 " × 16 ft	0.032	@	$718.00	MBF	=	$22.98	
	4 pieces	2 " × 6 " × 12 ft	0.048	@	$718.00	MBF	=	$34.46	
	1 piece	2 " × 6 " × 8 ft	0.008	@	$718.00	MBF	=	$5.74	
7.3.9	**Truss nailer**								
	1 piece	2 " × 4 " × 16 ft	0.011	@	$697.00	MBF		$7.67	
	2 pieces	2 " × 4 " × 12 ft	0.016	@	$697.00	MBF		$11.15	
7.3.10	**Trusses**								
	22 units	Custom built		@	$14.75	piece	=	$324.50	
	units			@			=	$0.00	
7.3.11	**Roof sheathing** OSB								
	13 pieces	½ " × 4 ft × 8 ft		@	$8.32	piece	=	$108.16	
	pieces			@			=	$0.00	
7.3.12	**Nails**								
	6 lbs, 16d common			@	$2.04	lb	=	$12.24	
	5 lbs, 6d threaded			@	$3.24	lb	=	$16.20	
			Line 7.3					$1,007.31	$1,007.31
7.4	**Stair stringers**								
	3 pieces	2 " × 12 " × 14 ft		@	$58.00	piece	=	$174.00	
	pieces	" × " × ft		@			=	$0.00	
			Line 7.4					$174.00	$174.00
	Cost of material (add lines from 7.1, 7.2, 7.3 and 7.4)							=	$10,343.07
			Sales tax	@	7.75	%		=	$801.59
			Total cost of material					=	$11,144.66
7.5	**Labor costs for superstructure**								
	111 manhours - carpenter, BC			@	$32.09	hour	=	$3,561.99	
	111 manhours - carpenter, BC			@	$32.09	hour	=	$3,561.99	
	111 manhours - carpenter, BC			@	$32.09	hour	=	$3,561.99	
	111 manhours - laborer, BL			@	$26.64	hour	=	$2,957.04	
	111 manhours - laborer, BL			@	$26.64	hour	=	$2,957.04	
	111 manhours - laborer, BL			@	$26.64	hour	=	$2,957.04	
			Total cost of labor					$19,557.09	$19,557.09
	Total superstructure cost (entered on line 7 of Form 100)								**$30,701.75**

Chapter 8
Roofing

ROOFING INSTALLED OVER THE roof deck prevents moisture from entering the structure. In residential construction, the majority of roofs are covered with asphalt and fiberglass shingles. The rest are covered with tiles made of clay, concrete, or slate, metal sheets, wood shingles or built-up roofing. Wood shingles are rarely used today because of their limited life span and high fire risk. There are a number of Web sites with information on the advantages and disadvantages of different roofing types.

We'll discuss asphalt and fiberglass shingle roofing in this chapter, as well as the materials associated with its application, including:

- Felt underlayment
- Flashing and plastic roof cement
- Metal drip edge
- Nails

These materials are important to any roofing installation. Regardless of the quality of the roof covering, if the underlayment and the flashing aren't properly installed and the roofing isn't attached according to the manufacturer's recommendation, the roof won't be waterproof.

Roof Covering

Installing a good roof is a sound investment, so initial cost shouldn't be the deciding factor in choosing materials. Materials vary in price, and while one may save you money at the time of installation, a more expensive roofing material could be more cost-effective in the long run. When making a selection, both builder and owner should keep this in mind. Other factors to consider are the type of building, roof pitch, and local weather conditions.

Roofing Felt

Roofing felt is an underlayment made of either fiberglass or polyester fleece impregnated with a bituminous material (tar or asphaltic bitumen). Sand is often applied to one side of the roll to keep the material from sticking together when the felt is rolled up. Roofing felt, weighing 15 pounds per square (100 square feet), typically comes in rolls 3 feet wide by 144 feet long. One roll covers 432 square feet, including edge overlap. Double-thickness roofing felt, weighing 30 pounds per square, is also available. It come in rolls 3 feet wide by 72 feet long, and covers 216 square feet. The advantage of using double-thickness roofing felt is that it can be placed in a single layer where added thickness is required, rather than having to install two layers of 15-pound felt. The roof pitch determines whether you need to use double- or single-layer felt, or if any underlayment is necessary at all (see Figure 8-1).

	Minimum Roof Slope	
Underlayment	Double Coverage Shingle*	Triple Coverage Shingle*
Not required	7 in 12	4 in 12
Single	4 in 12	3 in 12
Double	2 in 12	2 in 12

* Double coverage for a 12" × 36" shingle is considered to be an exposure of 5". Triple coverage is considered to be an exposure of 4".

Note: The headlap for single coverage of underlayment should be 2" and 19" for double coverage.

Figure 8-1
Underlayment for asphalt and fiberglass shingles

To estimate single-layer roofing felt (15 pound), calculate the total roof area, divide by 400, and round up to the next highest whole number to find the number of rolls required.

Using the roof area calculations from Chapter 7 for the sample house, and the porch dimensions in Figures 7-52 and 7-53, let's calculate how many rolls of 15-pound roofing felt will be required for the house.

Area of second-story roof section	1,225.39 sf
Area of first-story roof section	729.40 sf
Area of rear terrace roof (17.67′ × 13.00′ × 1.042 CR pitch ratio)	239.36 sf
Area of front porch and walkway roof cover (41.00′ × 9.50′ × 1.042 CR pitch ratio)	405.89 sf
Total roof area:	2,600.04 sf

$$\frac{2{,}600.04\text{sf}}{400\text{ sf per roll}} = 6.50\text{(rounded to 7) rolls}$$

*Enter your estimate of 15-pound **roofing felt** rolls on line 8.1.1 of your Cost Estimate Worksheet. We'll use 7 rolls of 15-pound roofing felt in our Example Cost Estimate Worksheet.*

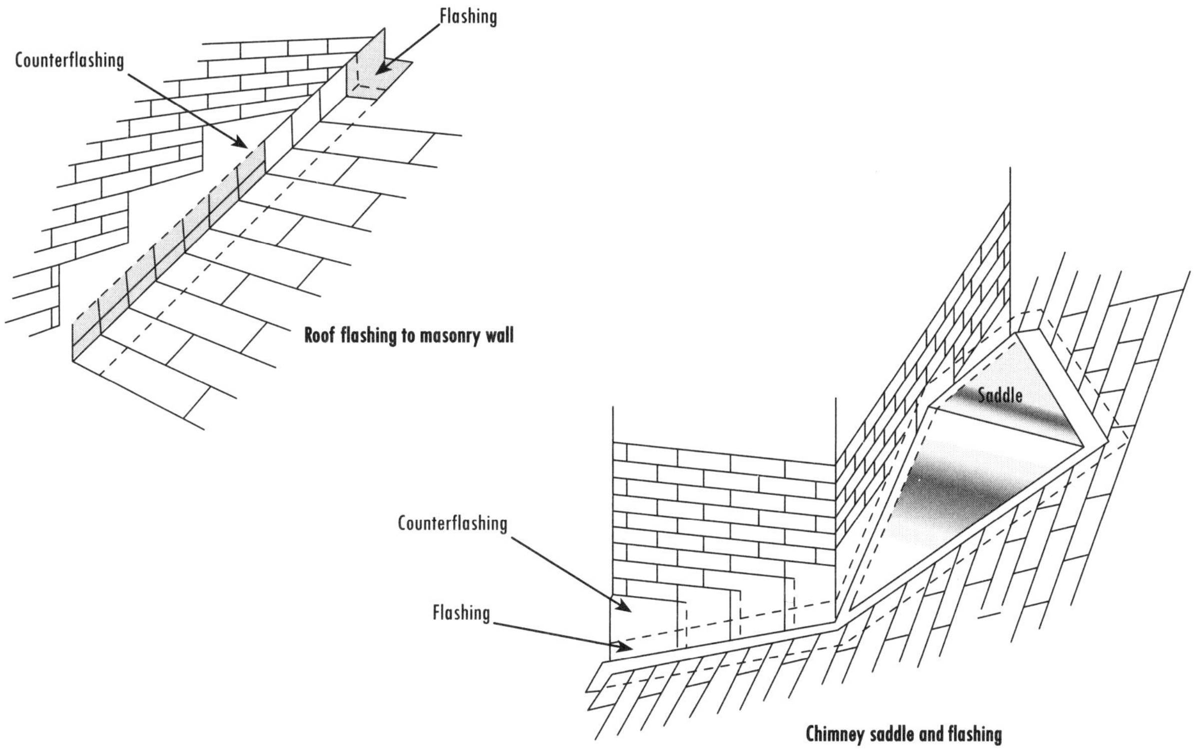

Figure 8-2
Roof and chimney flashing

Nails for Roofing Felt

Allow 2 pounds of 1-inch roofing nails per roll of 15-pound roofing felt underlayment. So, for our sample house roof, we'll need to order the following:

7 rolls × 2 lbs of nails per roll = 14 lbs of 1-inch roofing nails

> *Enter your estimate of the number of pounds of 1-inch **roofing nails for felt** on line 8.1.2 of your Cost Estimate Worksheet. We'll use 14 pounds of 1-inch roofing nails in our Example Cost Estimate Worksheet.*

Flashing

Flashing provides waterproofing at all critical joints in roof construction. It's installed in valleys and at all wall and chimney intersections. Counterflashing is an additional layer of flashing that overlaps, providing extra protection. It's used primarily where roof and masonry walls meet, and around chimneys and chimney saddles (Figure 8-2). Aluminum, asphalt, copper, galvanized sheet metal, and stainless steel are acceptable flashing materials.

If prior job records are available, use them to help you estimate. If not, use the following guideline:

- Order 60 feet of 14-inch-wide by 0.0175-inch-thick commercial grade aluminum flashing per 1,000 square feet of plain gable or shed roof area.
- Increase your order by 50 percent if there are hips and valleys in the roof structure.

Our sample house has plain gable and shed roofs. Let's estimate how much flashing material we'll need. We calculated the total roof area of the two gable roofs and the two shed roofs at 2,600.04 square feet. There are no hips or valleys, so we'll use the guideline just mentioned.

Using X to represent the amount of flashing for our 2,600.04-square-foot roof area, set up the proportion as follows:

$$\text{Feet of flashing} : \text{Square feet of roof} = \text{Feet of flashing} : \text{Square feet of roof}$$

$$60 \text{ ft} : 1{,}000 \text{ sf} = X \text{ ft} : 2{,}600 \text{ sf}$$

$$\text{Product of extremes} = \text{Product of means}$$

$$60 \times 2{,}600 = 1{,}000 \times X$$

$$X = \frac{60 \times 2{,}660}{1{,}000}$$

$$X = 156 \text{ (rounded to 160) feet}$$

We'll need 160 feet of 14- by 0.0175-inch aluminum flashing. It typically comes in 10-foot rolls, so divide the total feet of flashing by 10 to find how many rolls we'll order:

$$\frac{160 \text{ ft}}{10 \text{ ft per roll}} = 16 \text{ rolls}$$

> *Enter your estimate of the number of 10-foot rolls of **flashing** on line 8.1.3 of your Cost Estimate Worksheet. We'll use 16 rolls of 14-inch-wide × 10-foot-long × 0.0175-inch-thick aluminum flashing in our Example Cost Estimate Worksheet.*

Ridge Vents, Joint Straps, and End Connector Plugs

Ridge vents, installed in conjunction with under-eave soffit vents, provide effective ventilation for the attic (Figure 8-3). They can reduce summer attic temperatures by as much as 50 degrees, decreasing the cost of air conditioning in the home's living areas. When used, ridge vents should be installed before the roof shingles and included in the roofing materials estimate.

To estimate ridge vents, divide the linear feet of the ridge by 10 (ridge vents typically come in 10-foot sections), and round up.

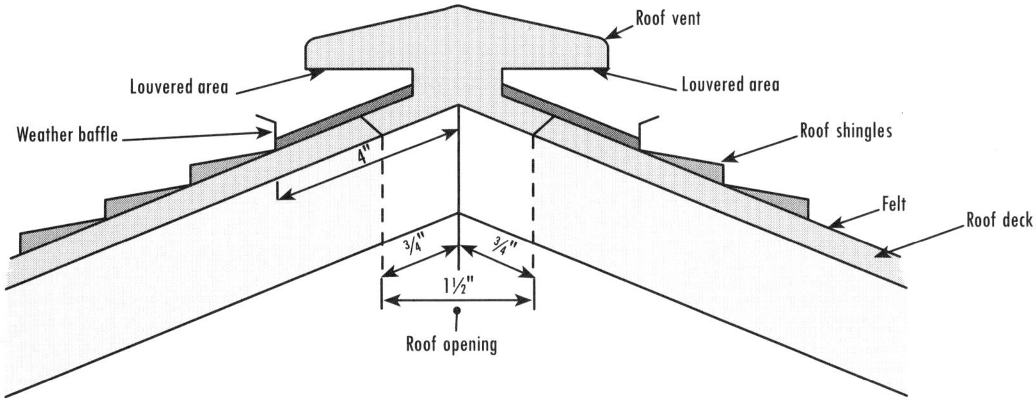

Figure 8-3
Ridge vent detail

The total linear feet of the two gable roofs, including overhangs, for our sample house is:

Two-story section	42 lf
Single-story section	25 lf
Total:	67 lf

$$\frac{67\text{ ft}}{10\text{ ft per piece}} = 6.7\text{ (rounded to 7) pieces of ridge vent}$$

Enter your estimate of the number of 10-foot pieces of ***ridge vent*** *on line 8.1.4 of your Cost Estimate Worksheet. We'll use 7 pieces of 10-foot ridge vent in our Example Cost Estimate Worksheet.*

Wherever sections of the ridge vent meet, you need a joint strap. You also need one end connector plug at each ridge vent opening. We now need to calculate how many joint straps and end connector plugs we'll need for the two gable roofs on our house.

The two-story section is 42 feet, and each ridge vent piece is 10 feet long. The number of ridge vent pieces used will be:

$$\frac{42}{10} = 4.2\text{ (rounded to 5) pieces}$$

The 5 pieces of ridge vent will have 4 joints, so we'll need 4 joint straps for the 2-story section. The ridge vent for this section is open at both ends, so we'll need two end connector plugs.

The single-story section is 25 feet in length; each ridge vent piece is 10 feet long.

$$\frac{25}{10} = 2.5\text{ (rounded to 3) pieces}$$

The 3 pieces will have 2 joints, so we'll need 2 joint straps on the first-floor section. Since the ridge vent for this roof section abuts the second-story wall, we only need one end connector plug for the first-floor roof vent section.

We'll need a total of 6 joint straps and 3 end connector plugs.

> *Enter your estimate of the number of ridge vent **joint straps** on line 8.1.5 of your Cost Estimate Worksheet. We'll use 6 ridge vent joint straps in our Example Cost Estimate Worksheet.*

> *Enter your estimate of the number of ridge vent **end connector plugs** on line 8.1.6 of your Cost Estimate Worksheet. We'll use 3 ridge vent end connector plugs in our Example Cost Estimate Worksheet.*

Soffit Vents

You'll need to install under-eave soffit vents to allow air to enter and escape through the ridge vent for adequate attic ventilation. Typically, you need 1 square foot of ventilation area per 300 square feet of attic floor space, provided there's a vapor retarder in the attic insulation. If your roof pitch is 7/12 to 10/12, add 20 percent to your calculation to satisfy ventilation requirements; add 30 percent for roof pitches of 11/12 and steeper. Soffit vents usually come in 8-foot lengths, providing 9 square inches of net free ventilation area per linear foot.

How many 8-foot lengths of soffit ventilation strip will we need for our sample house? Calculate the attic area in the two sections of the house, and divide by 300 for the square feet of soffit ventilation required.

Two-story attic area (40.00′ × 26.00′)	1,040 sf
Single-story attic area (23.83′ × 26.00′)	620 sf
Total attic area:	1,660 sf

The required soffit ventilation is 1 square foot of soffit per 300 square feet of attic floor space:

$$\frac{1{,}660\,\text{sf}}{300} = 5.53\,\text{sf}$$

There are 9 square inches of ventilation per linear foot of soffit vent. Multiply by 144 to convert square feet to square inches.

$$5.53 \times 144 = 796.32 \text{ square inches (rounded up to 797)}$$

$$\text{Soffit ventilation required} = \frac{797\text{si}}{9\text{si per lf}} = 88.55 \text{ (rounded to 89) ft}$$

We need 89 feet of soffit ventilation. Soffit vents come in 8-foot lengths.

$$\text{Number of 8-foot sections of soffit ventilation required} = \frac{89\text{ lf}}{8\text{ ft per section}} = 11.13 \text{ (rounded to 12) sections}$$

We need 12 sections of 8-foot soffit ventilation.

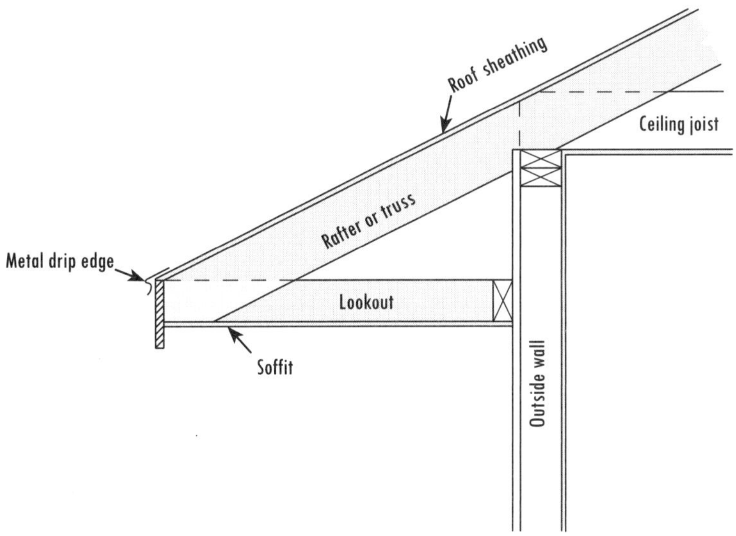

Figure 8-4
Metal drip edge

> *Enter your estimate of the number of* ***soffit ventilation*** *strips on line 8.1.7 of your Cost Estimate Worksheet. We'll enter 12 soffit ventilation strips in our Example Cost Estimate Worksheet.*

Metal Drip Edge

A metal drip edge (Figure 8-4) is installed over the roof sheathing and roofing felt along the fascia, under the roof shingles. It directs water away from the fascia. When estimating drip edge, divide the total linear feet of eave by 10 (10-foot lengths), and round up for the number of pieces required.

Let's calculate how many pieces of 6-inch by 10-foot drip edge we'll require for our sample house. Add the eave sections for each part of the house to get the total eave length in linear feet. Include the 1-foot overhangs for each roof section and the 6-inch overhangs for the porch and rear terrace.

Two-story section (front), 40.00′ + 2.00′ overhang	42.00
Two-story section (rear), 40.00′ + 2.00′ overhang	42.00
One-story section (front), 24.00′ + 1.00′ overhang	25.00
One-story section (rear), 24.00′ + 1.00′ overhang	25.00
Porch and walkway roof, 40.00′ + 1.00′ overhang	41.00
Rear terrace roof, 16.67′ + 1.00′ overhang	17.67
Total eave length:	192.67 lf

$$\frac{192.67\text{ lf}}{10'\text{ lengths}} = 19.27 \text{ (rounded to 20) pieces}$$

We'll need 20 pieces of drip edge in 10-foot lengths.

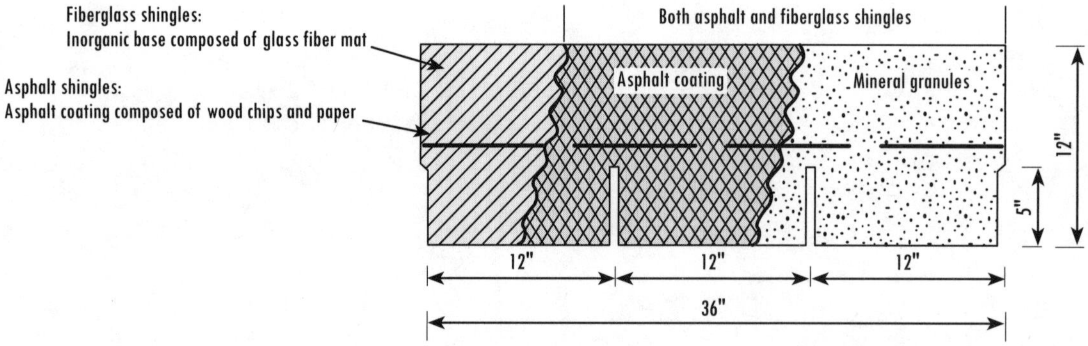

Figure 8-5
Asphalt and fiberglass shingle

*Enter your estimate of the number of 10-foot lengths of **drip edge** on line 8.1.8 of your Cost Estimate Worksheet. We'll use 20 pieces of drip edge in 10-foot lengths in our Example Cost Estimate Worksheet.*

Asphalt and Fiberglass Shingles

Asphalt shingles are constructed of either organic materials or glass fiber (fiberglass), as shown in Figure 8-5. The organic type is made of asphalt-saturated felt material, with ceramic granules embedded in the surface. Organic asphalt shingles provide exceptional waterproofing.

The most widely used asphalt shingle comes in 12-inch-wide by 36-inch-long strips. The most common weight is 235-240 pounds per square, though higher-quality shingles can exceed 450 pounds per square. Shingles are packaged three bundles per square, and one square covers 100 square feet of roof area.

With a 5-inch weather exposure, it takes 80 12- by 36-inch shingles to cover 100 square feet, calculated as follows:

$$5\text{" exposure} \times 36\text{" shingle length} = 180 \text{ si per shingle}$$

$$100 \text{ sf} \times 144 = 14{,}400 \text{ si}$$

$$\frac{14{,}400}{180 \text{ si}} = 80 \text{ shingles}$$

To check your figures:

$$5\text{" exposure} \times 36\text{" shingle length} \times 80 \text{ shingles} = 14{,}400 \text{ square inches}$$

$$\frac{14{,}400}{144} = 100 \text{ sf (1 square)}$$

Note: 1 square foot = 144 square inches.

Fiberglass Shingles

In recent years, fiberglass shingles have gained popularity in what had been a strictly asphalt-shingle roofing market. The two types of shingle look very much alike, but the fiberglass version is nearly 30 percent lighter, and easier to install.

Fiberglass shingles consist of a glass fiber reinforcing mat coated with asphalt that contains mineral fillers. By itself, the fiberglass mat isn't waterproof. Its purpose is simply to reinforce. It's the asphalt coating that makes fiberglass shingles waterproof. Fiberglass shingles weigh 180 to 230 pounds per square. Laminated, or architectural, fiberglass shingles use two distinct layers bonded together with asphalt sealant. These shingles are heavier, more expensive, and more durable than traditional three-tab shingle designs. They also have a more varied, contoured surface that's visually pleasing.

Estimating Asphalt and Fiberglass Shingles

To estimate asphalt or fiberglass shingles, calculate the roof area the same way that you calculate for roof sheathing (see Chapter 7). You can either multiply rafter length by eave length; or using the ratios in Figure 7-31, multiply the CR pitch factor by the number of square feet of floor area to get the roof area. Add an allowance for the starter strip, ridge, and hip caps. Then, divide the total roof area by 100 to get the number of roofing squares you need.

Some estimators allow one extra foot for the starter course and an extra foot for each ridge and hip. When using 12- by 36-inch shingles, you can also use the following factors for estimating the extra shingles needed.

Starter Courses:

$$0.0042 \times \text{Total Eave Length} = \text{Number of Squares}$$

Ridge and Hip Caps (5-inch exposure):

$$0.010 \times \text{Linear Feet of Ridge and Hips} = \text{Number of Squares}$$

Putting the Estimate Together

Now we're ready to estimate all the materials needed for the roof area of our sample house. Let's start by estimating the number of fiberglass asphalt shingle roofing squares we need for the house, porch/walkway and terrace roofs shown in Figures 7-1, 7-2, 7-9, 7-52 and 7-53. We'll use 12- by 36-inch shingles with a 5-inch exposure.

We estimated the following square feet of roof area earlier in the chapter:

Two-story section roof area	1,225.39 sf
Single-story section roof area	729.40 sf
Rear terrace	239.36 sf
Front porch and walkway cover	405.89 sf
Total roof area:	2,600.04 sf

Divide the total square foot roof area by 100 to determine the number of squares we need.

$$\frac{2{,}600.04\text{ sf}}{100} = 26.00\text{ squares}$$

Now figure how many squares of shingles we need for the starter course on our house. Our house has a total eave length of 192.67 linear feet.

$$0.0042 \times \text{Total Eave Length} = \text{Number of Squares}$$

$$0.0042 \times 192.67\text{ lf} = 0.81\text{ square}$$

Next, calculate how many squares we need for the ridge caps on our house. There are no hip caps required. One shingle makes three caps (see Figure 8-6). The total ridge length is 67 linear feet (40 feet + 24 feet plus three 1-foot overhangs).

$$0.010 \times \text{Linear Feet of Ridge and Hips} = \text{Number of Squares}$$

$$0.010 \times 67\text{ lf} = 0.67\text{ square}$$

The number of fiberglass asphalt roofing squares we need is:

Main roof areas	26.00
Starter courses	0.81
Ridges	0.67
Subtotal	27.48
Add 3 percent for waste	0.82
Total:	28.30 (rounded to 29) squares of shingles

*Enter your estimate of the number of squares and type of **roofing shingles** on line 8.1.9 of your Cost Estimate Worksheet. We'll use 29 squares of fiberglass asphalt shingles in our Example Cost Estimate Worksheet.*

Nails to Attach Asphalt Roofing

We'll estimate 1¾ pounds of 1¼-inch roofing nails per square. For the ridge and hip caps that require 1½-inch roofing nails, multiply the total linear feet of ridge and hip by a nail factor of 0.027 to get the number of pounds of nails needed.

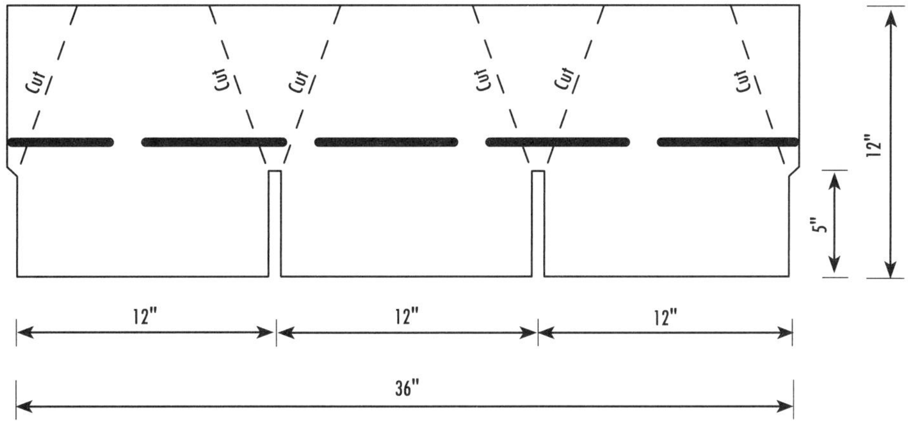

Figure 8-6
Ridge and hip caps

Let's figure how many pounds of 1¼-inch roofing nails we need for 29 squares of asphalt roofing. Multiply the number of roof squares by 1¾ (1.75):

29 squares × 1.75 lbs = 50.75 (rounded to 51) lbs of 1¼" roofing nails

Now let's calculate how many pounds of 1½-inch roofing nails we need for the ridges. Multiply the linear feet of ridge by 0.027 (nail factor):

67 lf × 0.027 (nail factor) = 1.81 (rounded to 2) lbs of 1½" roofing nails

> *Enter your estimate of the number of pounds and size of* ***roofing nails*** *in section 8.1.10 of your Cost Estimate Worksheet. We'll use 51 pounds of 1¼-inch roofing nails and 2 pounds of 1½-inch roofing nails in our Example Cost Estimate Worksheet.*

Labor Costs for Roofing

Labor costs to install roofing vary with the roof pitch, the type of roofing material used, and the skill of the workmen. The costs can range from less than one manhour per square to install asphalt shingles on a plain gable roof (roof pitch 6 in 12 or less), to 6 or more manhours per square for installing heavy tiles on steep roofs, or where foot rests are required. On mansard roofs (with two pitches on all four sides, from the eave to the ridge), the lower pitch may exceed 20 in 12. The manhours per square for installation on these roofs will be considerably higher.

Generally, roofing contractors can install roofing more economically than in-house tradesmen. That also allows the in-house tradesmen to focus on other work that might otherwise be delayed. It's a win-win situation for both the general and the roofing contractor. Roofing contractors work on a per-job basis, and will either furnish the material and labor, or labor only, for a fixed amount per square.

A complete specification sheet should be signed by the roofing contractor before you award the contract. Here's a guideline for items that should be included in a spec sheet:

- Felt paper: Weight
- Metal drip edging
- Roof or ridge vents: Type and size
- Flashing: Type and size, where it's to be used, and how it's to be installed. Will valleys be open with the flashing showing, will they be closed, or woven where the roof shingles overlap?
- Roof shingles: Type and weight
- Nails: Size and number per shingle
- Insurance: The roofing contractor must have workers' compensation and liability insurance for his employees

Installation of the roofing felt, flashing, vents, drip edge, and shingles for our house can also be calculated using the individual labor costs given in the current edition of the *National Construction Estimator*.

> *Enter the **subcontractor estimate** or your **estimate of the individual labor costs** in section 8.2 of your Cost Estimate Worksheet. We'll use the current individual National Construction Estimator labor costs in our Example Cost Estimate Worksheet.*

COST ESTIMATE WORKSHEET FOR ROOFING

#	Qty	Size		Cost	Per			Subtotal
8.1	**Roof Covering**							
8.1.1	**Felt**							
	7 rolls	15 lb	@	$20.00	roll	=	$140.00	
8.1.2	**Nails for felt**							
	14 lbs	1" roofing	@	$1.88	lb	=	$26.32	
8.1.3	**Flashing**							
	16 rolls	Aluminum 14" × 10' × 0.0175"	@	$11.50	roll	=	$184.00	
8.1.4	**Ridge vents**							
	7 pieces	10' lengths	@	$13.90	piece	=	$97.30	
8.1.5	**Joint straps**							
	6 pieces		@	$1.22	piece	=	$7.32	
8.1.6	**End connector plugs**							
	3 pieces		@	$1.94	piece	=	$5.82	
8.1.7	**Soffit ventilation**							
	12 pieces	2" × 8'	@	$5.12	piece	=	$61.44	
8.1.8	**Metal drip edge**							
	20 pieces	6" × 10'	@	$7.54	piece	=	$150.80	
8.1.9	**Roofing shingles**							
	29 squares		@	$46.00	square	=	$1,334.00	
	Fiberglass asphalt shingles, 25 year							
8.1.10	**Nails**							
	51 lbs	1¼" roofing	@	$1.96	lb	=	$99.96	
	2 lbs	1½" roofing	@	$1.67	lb	=	$3.34	
	lbs		@		lb	=	0.00	
		Total for line 8.1					$2,110.30	$2,110.30
		Sales tax	@	7.75	%		=	$163.55
		Total cost of material					=	$2,273.85
8.2								
	Subcontractor estimate	29 squares	@		square	=	0	
	or							
	Felt	29 squares	@	$11.24	square	=	$325.96	
	Flashing	16 rolls	@	$14.26	roll	=	$228.16	
	Ridge Vents	7 pieces	@	$20.86	piece	=	$146.02	
	Soffit ventilation	12 pieces	@	$11.76	piece	=	$141.12	
	Metal drip edge	20 pieces	@	$14.26	piece	=	$285.20	
	Shingles	29 squares	@	$57.11	square	=	$1,656.19	
				Total for line 8.2			$2,782.65	$2,782.65
	Total cost of roof covering (entered on line 8 of Form 100)							**$5,056.50**

Chapter 9
Electrical, Plumbing, Heating, and Air Conditioning

ELECTRICAL, PLUMBING, HEATING, AND air conditioning installations are best handled by specialized tradesmen. Most localities require this work to be performed by licensed tradesmen or contractors, with permits issued by the local building department. The work must meet all local building codes, and the construction site visited by one or more inspectors to ensure compliance. Normally, this work is done by subcontractors.

Rough-in for work by these trades usually follows installation of the roof, and precedes finishing the walls and ceilings.

Electrical

The electrical system provides the power a house needs to function efficiently. Before letting the contract out for bid, take the blueprints to the local power company to get their recommendation regarding the electrical layout of the house. It will be time well spent. The electrical layout shown on the plans may not be adequate for the type of appliances and equipment intended for the house. Information gathered from the power company can be very helpful, and may save the owner the expense of adding to the system at a later date.

An electrical contractor normally bids on labor and material for the rough-in work such as wiring, boxes for outlets, switches and fixtures, the entrance panel, circuit breakers and cover plates for the outlets and switches. His bid normally includes installing hook-ups for appliances (water heater, range, oven, and the dishwasher shown on the plans and specifications) and hanging the light fixtures. The light fixtures and the appliances themselves are typically provided by the contractor. Any additional equipment required, heating, cooling and ventilating systems hook-up, installation of smoke detectors, intercom systems, central vacuum systems, sump pumps, door bells, telephone service line and boxes, cable service line and boxes should be included in the bid request.

Make sure you have a clear understanding with the electrical contractor regarding where his responsibility begins and ends. It normally begins at the meter. If the service line to the house is underground, the electric company will probably charge for this service. If the closest available source of power is a distance from the house, you can expect additional expense based on that distance.

Here's a checklist for items to look for in the electrical contractor's bid:

- Workers' compensation and liability insurance, provided by the electrical contractor for his employees
- Cost of service line to house
- Cost of hook-up for temporary electric service
- Labor and material for rough-in. This includes wiring, outlets, boxes and plates, boxes for the fixtures, switches, connectors, the entrance panel and circuit breakers
- Installation of light fixtures
- Hook-up of kitchen appliances
- Outlets for laundry appliances
- Provision of hook-ups for heating, cooling and ventilating equipment
- Provision and hook-up of special equipment – central vacuum, smoke detectors, etc. (the electrical contractor may subcontract these to specialist companies*)
- Telephone boxes and provisions for routing wiring through the house (service and hook-ups are usually provided by the telephone company*)
- Cable television boxes and service to house (may be subcontracted to a cable company*)
- Cable for home computer network (may be subcontracted to a specialist*)

*You may opt to subcontract with specialist suppliers yourself, rather than go through your electrical subcontractor. That could save you the mark-up on the materials he'd include in his bid. However, your savings could be offset by material and supervision costs incurred when you hire independent subcontractors. Experience and records can help you determine the most expedient and cost-effective approach for your situation.

Remember to include these items in your cost estimate:

Electrical contractor's bid	Smoke detectors*
Light fixtures	Cable TV installation*
Kitchen appliances	Computer cable installation*
Central vacuum	Other special equipment

*if not included in subcontractor's bid

*Enter the **electrical contractor's bid and other costs** in Section 9.1 of your Cost Estimate Worksheet. We'll use our electrical contractor's bid for everything except the lighting fixtures and kitchen appliances in our Example Cost Estimate Worksheet.*

Plumbing

The plumbing system must be designed to provide adequate water and proper drainage and venting for the dwelling. Most jurisdictions require a license and permit for plumbing work. All plumbing must comply with the local building codes, and pass worksite inspection checks. If the house is built in an area without public water service or a public sewer system, the local health department will usually have final authority over the private water supply and sewage discharge system required.

The material take-off for plumbing rough-in should be done by a plumbing contractor. Fixtures such as the kitchen sink, garbage disposal, bathtubs, shower stalls, water closets, lavatories and faucets are normally selected by the owner and installed by the plumbing contractor.

Before putting the plumbing contract out for bid, be sure the specifications state the type and size of both the water supply and the drain pipes. If the plumber is contracted to furnish any fixtures (including the water heater), the specifications should make clear the type and size fixture required. Fixtures furnished by the owner should be included in the specifications, and the responsibility for installing them made clear.

The limits of responsibility for the water supply lines and sewer lines outside the house should also be included in the specifications. Some plumbing contractors will bid on the work up to a specified distance outside the foundation (usually 5 or 10 feet). The cost of extending the water supply and sewer lines to a more distant point of connection and discharge will be extra. The responsibility for digging the water and sewer line trenches and covering them should be defined in the specifications. A plumbing contractor will seldom extend his area of responsibility beyond the property line without a separate bid.

Use this checklist to identify the costs to be included in the plumbing contractor's bid:

- ➢ Plumbing contractor to provide workers' compensation and liability insurance for his employees
- ➢ Cost of the water supply line from point of connection to the house, including the trench for the water pipe
- ➢ Temporary water service hook-up
- ➢ Cost of the sewer line from the house to the point of discharge, including the trench for the sewer pipe
- ➢ Cost of labor and material for rough-in. This includes type and size of water supply pipes, type and size of drain pipes, type and size of vent pipes, number and location of outside water faucets, and all material required to make the necessary connections
- ➢ Cost of plumbing fixtures (if not owner/builder provided)
- ➢ Installation of plumbing fixtures
- ➢ Cost of water heater (if not owner/builder provided)
- ➢ Installation of water heater

- Installation of special equipment such as a water softener, pressure relief valves, sump pump, shower doors, etc. Ensure purchase cost is included in plumber's bid (if not owner/ builder provided)
- Cost of labor and material to rough-in gas supply from the meter to the heating system furnace, water heater, clothes dryer, range/oven and fireplace outlets, as required
- Hook-up of appliances (owner/builder or contractor provided) if not performed by an electrical contractor. Make sure the contracts stipulate which contractor is responsible for gas hook-up and testing
- Cost of gas supply line to the house and meter (may be subcontracted to the gas company)

Your past experience can help you determine the most expedient and cost-effective approach for your situation. Remember to include the following items in your cost estimate:

Plumbing contractor's bid
Water heater*
Pressure relief valves*
Plumbing fixtures
Water softener*
Sump pump*
* If not included in bid

> *Enter the **plumbing contractor's bid and other costs** in Section 9.2 of your Cost Estimate Worksheet. We'll use our plumbing contractor's bid for everything except the plumbing fixtures, water heater, and water softener in our Example Cost Estimate Worksheet.*

HVAC

A HVAC contractor is the best person to advise you on the heating and air conditioning needs for your house design, and to bid on equipment and installation costs. His calculation, which takes into account the size of the house, the amount of insulation in the walls and attic, the insulation value of windows, and other criteria, determines the optimum furnace capacity. Not all homes need air conditioning, especially along the west coast of the U.S. So, for our example here, we'll separate the heating and air conditioning quotes.

Heating

There are many types of heating systems designed for residential buildings, including oil, gas, electric and biofueled furnaces, space heaters, electric radiant heat with heating cables in the ceilings or floors, heat pumps that double as air conditioners and, in increasing numbers, solar heaters.

The heating system installed in a house should be efficient, safe and convenient to use. Regardless of the system you choose, it should be designed and installed by a reputable HVAC contractor, and meet all applicable building codes. If the heating unit is too small or not installed properly, it won't provide sufficient heating; if the heating unit is too large, it'll be unnecessarily expensive to purchase and operate.

The size of the heating unit depends on the size of the house and its calculated heat loss. Good insulation in the walls and attic and dual pane windows will reduce heat loss, resulting in a savings in heating cost and reducing the size of the heating unit you need. For example, if electric radiant heat is to be installed, the size of the heating unit will be computed at approximately 6 to 10 watts per square foot of living area, depending on heat loss. A room with 200 square feet and a high heat loss factor will probably require 2,000 watts of heat. In contrast, the same room with a low heat loss factor will probably be computed between 1,200 and 1,500 watts of heat. This will result in savings both in the cost of the heating unit and its operating costs over time.

The most common residential heating system uses a gas-fired furnace with a heat exchanger, blowing hot air through ducts in the walls and attic. The same ducts can be used for the air conditioning system. All that needs to be added to install air conditioning with this system are a compressor and a heat exchanger. That's the type of system we'll install here.

Use this checklist to ensure that all of the following are covered in the HVAC contractor's heating bid:

- Furnace heater and hook-up
- Thermostat and system testing
- Ductwork and installation
- Humidifier (if required)
- Contractor to supply workers' compensation and liability insurance for his employees

Remember to include the HVAC contractor's heating system bid and the humidifier (if required) in your cost estimate. We'll subcontract the complete heating system for our sample house.

*Enter the HVAC contractor's **heating system bid** on line 9.3.1 of your Cost Estimate Worksheet. We'll use our HVAC contractor's heating system bid and other costs in our Example Cost Estimate Worksheet.*

Air Conditioning

Air conditioning should be safe, quiet, economical and effective. The size of the air conditioning system, as with the heating system, depends on the size of the house and the calculated heat loss. It should meet all applicable building codes.

Central air conditioning is measured in tons of refrigeration (one ton equals 12,000 BTUs per hour) and is designed to cool the entire building, or a part of it (one floor, for example).

In recent years, the heat pump has gained increased acceptance for use in residential construction. It supplies heat in the winter and air conditioning in the summer, by reversing its cycle. It has a built-in auxiliary resistance heating unit that automatically switches on when the outside temperature is too cold for the heat pump to operate efficiently. Heat pumps are more efficient to operate than resistant-type heaters.

Heat pumps and other central air conditioning should be designed and installed by a reputable HVAC contractor. Your contractor can advise you on the AC needs for your house, do the load calculation, and bid on equipment and installation costs. For a preliminary estimate, you can use this rule of thumb to determine the size of the heat pump required for you house:

1 ton of refrigeration (12,000 BTUs) per 600 square feet of living space

Our sample house is approximately 2,238 square feet (excluding the garage and basement) and will require a 4-ton central air conditioning system.

Use this checklist to cover the following items in the HVAC contractor's air conditioning bid:

- AC compressor and heat exchanger and hook-up
- Thermostat and system testing (if not part of the heating system)
- Ductwork and installation (if not part of the heating system)
- Condensate removal tray, pump and pipe work
- Workers' compensation and liability insurance to be provided for his employees

*Enter the HVAC contractor's **air conditioning bid** in line 9.3.3 of your Cost Estimate Worksheet. We'll use our HVAC contractor's AC bid in our Example Cost Estimate Worksheet.*

COST ESTIMATE WORKSHEET FOR ELECTRICAL, PLUMBING, HEATING AND AIR CONDITIONING

#	Item	Cost	Subtotal
9.1	**Electrical**		
9.1.1	Contractor's bid	$16,500.00	
9.1.2	Light fixtures	$2,000.00	
9.1.3	Kitchen appliances	$3,500.00	
9.1.4	Central vacuum*		
9.1.5	Sump pump*		
9.1.6	Smoke detectors*		
9.1.7	Phone installation*		
9.1.8	Cable TV installation*		
	Total for line 9.1	$22,000.00	$22,000.00
9.2	**Plumbing**		
9.2.1	Contractor's bid	$14,500.00	
9.2.2	Plumbing fixtures	$5,500.00	
9.2.3	Water heater*	$450.00	
9.2.4	Water softener*	$1,300.00	
9.2.5	Pressure relief valves*		
9.2.6	Sump pump*		
	Total for line 9.2	$21,750.00	$21,750.00
9.3	**Heating and air conditioning**		
9.3.1	Furnace and ductwork	$7,000.00	
9.3.2	Humidifier		
9.3.3	AC compressor and heat exchanger	$2,500.00	
	Total for line 9.3	$9,500.00	$9,500.00
	Labor for carpenters**		$0.00
	Total (entered on line 9 of Form 100)		**$53,250.00**

* Note: If not included in contractor's bid.

** Note: Coordinate work with the subcontractors for cutouts, openings, supports, etc.

Chapter 10

Estimating Brickwork

BRICKWORK IS GENERALLY STARTED after the window and door frames are installed. All materials needed for the brickwork should either be on the job site or scheduled for delivery as the masons require them. These materials include the bricks, masonry cement mix, clean water, brick lintels, wall ties, antifreeze (if needed), and scaffolding. The mason's work area should be cleaned of debris and the ground leveled.

The builder or the customer generally selects the brick type to use. It's not uncommon for an owner to be undecided about which brick he wants to use at the time you're doing the estimate. If that's the case, make an allowance based on the cost of the brick per thousand (M) for each different type the owner may be interested in. As with most materials, the cost of brick varies according to type and quality.

Common Types of Brick

Standard brick is commonly used in residential construction. This brick is 2¼ inches high (the *face*) by 3¾ inches wide (the *bed)* by 8 inches long. Other popular types used in residential construction are *Norman* bricks (2¼ × 3¾ × 11½ inches) and *Roman* bricks, which have a long lean dimension (1½ × 3¾ × 11½ inches).

Another brick originally intended for home building, but now more commonly used in single-story commercial buildings, is *SCR* brick. This brick is 2¼ × 5½ × 11½ inches, and can be used to build a single wall with a 6-inch nominal width. These bricks don't need a frame or masonry backup, but are furred for drywall or plaster application. The advantage to SCR is that 4.40 units cover a square foot, making the work go quickly. The downside to SCR brick is that it provides less insulation than other types of brick.

Brick and Mortar Mix Factors for a Single Wythe Wall

Type of brick	Brick factor per square foot*		Mortar mix factor per brick (60 lb bags)**		Mortar mix factor per brick (80 lb bags)**	
	3/8" joint (a)	1/2" joint (b)	3/8" joint (c)	1/2" joint (d)	3/8" joint (e)	1/2" joint (f)
Standard 2¼ × 3¾ × 8	6.55	6.16	0.026	0.035	0.019	0.026
Roman 1½ × 3¾ × 11½	6.47	6.00	0.030	0.043	0.023	0.033
Norman 2¼ × 3¾ × 11½	4.60	4.36	0.033	0.046	0.024	0.034
SCR 2¼ × 5½ × 11½	-	4.36	-	0.065	-	0.048

* **Note:** This brick factor *does not* include an allowance for waste. An additional allowance of 3 to 5 percent should be made for brick waste.

Example: 1,000 square feet × 6.55 (factor) 6,550.00 bricks
Add 5 percent for waste 327.50 bricks
6,877.50 (estimate 6,900 bricks)

** **Note:** Mortar mix factors include 20 percent for waste.

Figure 10-1
Factors for single wythe wall

Brick, even of the same type, may have slight size differences due to variations in the clay used and the firing temperature. Overfired bricks will be smaller than underfired bricks. A difference of even ¼ inch in size can change the number of bricks required. Keep this in mind when estimating material. Colors also vary, so make sure the homeowner is happy with the brick when it's delivered, before your masons begin work.

Estimating Brick Materials

The chart in Figure 10-1 shows brick factors per square foot of a single wythe wall, and the mortar mix factor per brick for Standard, Roman, Norman, and SCR bricks. Use this chart to help estimate the number of bricks and quantity of mortar mix you'll need.

Mortar is made from masonry cement (purchased by the bag) and sand (purchased by the cubic yard or ton). However, preblended mortar mixes are being used more and more frequently. These have specific amounts of masonry cement and dry sand for an optimum mortar mix, ensuring consistent quality. Preblended mortar mixes (such as Quikrete® or Spec Mix®) are more expensive, but they eliminate the need to purchase masonry cement and sand separately. The additional cost is offset by the product's dependability, reduced mixing cost, and ease of application.

Brick Factors

As mentioned earlier, using the factors in Figure 10-1, you can estimate your brick and mortar mix requirements. But where do these factors come from? Let's calculate the factor for a Standard brick with a ½-inch joint (Column b). The table gives us a factor of 6.16 bricks per square foot of single wythe wall. That factor is derived in the following manner.

Standard brick height is 2¼ inches; adding a ½-inch joint gives you a height of 2¾ inches (2.75 inches). Standard brick length is 8 inches; adding a ½-inch joint gives you a length of 8½ inches (8.5 inches). To find the total area of the brick plus joint, multiply height by length, then divide by 144 to get the area in square feet.

$$2.75'' \times 8.5'' = 23.375 \text{ square inches (si)}$$

$$\frac{23.375 \text{ si}}{144 \text{ (si to sf)}} = 0.1623 \text{ sf}$$

One brick covers 0.1623 square foot of a single wythe wall, so how many bricks are required per square foot? Set up a proportion as follows:

$$1 \text{ brick} : 0.1623 \text{ sf} = X \text{ bricks} : 1 \text{ sf}$$

$$\text{Product of extremes} = \text{Product of means}$$

$$1 \text{ brick} \times 1 \text{ sf} = 0.1623 \text{ sf} \times X \text{ brick}$$

$$X = \frac{1 \text{ brick} \times 1 \text{ sf}}{0.1623 \text{ sf}}$$

$$X = 6.16 \text{ bricks per sf}$$

That's how we got our Standard brick factor. The other factors were computed in a similar manner for each brick type and mortar joint thickness shown.

Mortar Mix Factors

The Standard brick mortar mix factor for a ½-inch joint (Column d) is 0.035 bag per brick for a single wythe wall, using 60-pound bags. Let's see now how we calculate the mortar mix factor for Standard brick. The other mortar mix factors were computed in a similar manner for each brick type and the thickness of the mortar joints, as shown.

Standard brick height is 2¼ (2.25) inches, the width is 3¾ (3.75) inches, and the brick length is 8 inches. Extending the mortar bed by ½ inch for the joint brings the length of the mortar bed for each brick to 8½ (8.50) inches. A ½-inch mortar joint is then trowled onto one end of the brick and it's laid in place.

The volume of the mortar per brick, in cubic inches, is:

Bottom of brick	8.50″ L × 3.75″ W × 0.5″ joint	15.94 ci
End of brick	2.25″ H × 3.75″ W × 0.5″ joint	4.22 ci
Total mortar per brick:		20.16 ci

One 60-pound bag of mortar mix makes 685 cubic inches of mortar. We can set up the following proportion to calculate how much of a 60-pound bag of mortar (X) is required for one Standard brick requiring 20.16 cubic inches of mortar for ½-inch joints:

$$685 \text{ ci} : 1 \text{ bag of mortar mix} = 20.16 \text{ ci} : X \text{ bag of mortar mix}$$

$$\text{Product of extremes} = \text{Product of means}$$

$$685 \text{ ci} \times X \text{ bag} = 1 \text{ bag} \times 20.16 \text{ ci}$$

$$X = \frac{1 \text{ bag} \times 20.16 \text{ ci}}{685 \text{ ci}}$$

$$X = 0.029 \text{ bag}$$

$$0.029 \times 1.20 \ (20\% \text{ waste}) = 0.035 \text{ bag}$$

Height of Wall Multiplied by the Following Factors Equals the Number of Brick Courses		
Type of brick	⅜" joint	½" joint
Standard 2¼ × 3¾ × 8	4.571	4.364
Roman 1½ × 3¾ × 11½	6.400	6.000
Norman 2¼ × 3¾ × 11½	4.571	4.364
SCR 2¼ × 5½ × 11½	4.571	4.364

Figure 10-2
Brick courses per wall height factors

Brick Courses per Wall Height

Figure 10-2 shows the factors to use for estimating the number of courses for any wall height, using Standard, Roman, Norman, or SCR bricks. Using these factors, we can make the calculations shown in Figure 10-3.

The number of courses per foot of height using Standard brick with ½-inch mortar joints is illustrated in Figure 10-4. If you look at the example for 1-foot and 3-foot wall heights in Figure 10-3, you'll see that there are 4.36 courses per foot and 13.09 courses in 3 feet of wall height, as shown in the illustration.

Figure 10-5 shows the inch and fractions of an inch conversion to feet and decimal factors used in Figure 10-3. For example, 8⅞ inch = 0.740 foot (double boxed in Figure 10-5).

Brick veneer may cover all or part of the exterior walls on residential construction. Check the elevation section of the blueprints and the specs for this information. Brickwork is expensive, so estimate as accurately as possible. Estimating too high could mean you lose the contract; estimating too low means you get the contract, but at a lower profit margin than you were expecting.

Examples: Wall height	× Factor	= Number of brick courses
Standard bricks with ½" joints		
2¾" (0.229')	4.364	1.00 course
1'0" (1.00')	4.364	4.36 courses
3'0" (3.00')	4.364	13.09 courses
7'0" (7.00')	4.364	30.55 courses
14'0" (14.00')	4.364	61.10 courses
Roman bricks with ⅜" joints		
1⅞" (0.156')	6.400	1.00 course
3'9" (3.75')	6.400	24.00 courses
7'6" (7.50')	6.400	48.00 courses
8'0" (8.00')	6.400	51.20 courses
11'3" (11.25')	6.400	72.00 courses
Norman bricks with ½" joints		
2¾" (0.229')	4.364	1.00 course
11" (0.917')	4.364	4.00 courses
7'4" (7.333')	4.364	32.00 courses
11'0" (11.00')	4.364	48.00 courses

Figure 10-3
Brick courses per wall height

Be careful not to overlook the cost for tooling the mortar joints when estimating brickwork. Few homeowners even think about how they want

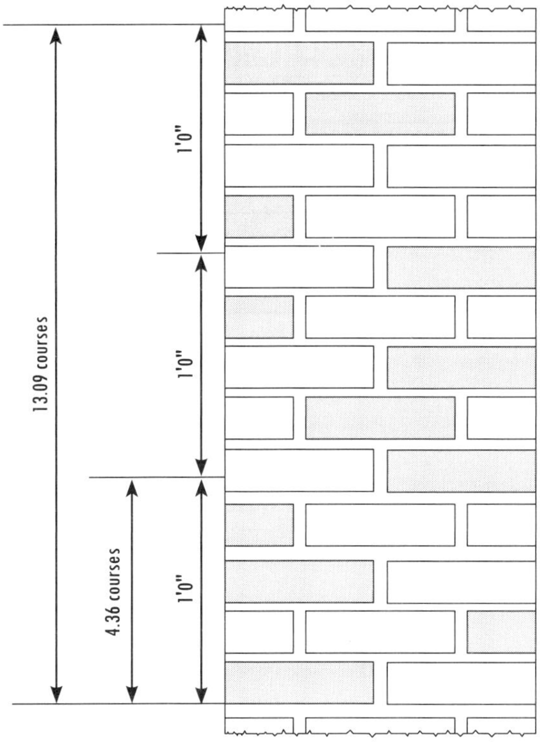

Figure 10-4
Standard brick with ½" joints

the mortar joints finished. However, once the work's been started, it may be too late for them to decide. Figure 10-6 shows the most common tooled mortar joints. Even though mortar joint finishing has no significant effect on the cost estimate, it could affect customer satisfaction. Determine how the homeowner wants the joints tooled before brickwork begins, and pass the information on to the brick masons.

Fireplaces and Chimneys

In residential construction, masonry fireplaces (that start at a footing and terminate with the chimney cap above the roof) have largely been replaced by prefabricated fireplaces. A prefab fireplace (shown in Figure 10-7) is energy-efficient, easy to install, and can be finished with the same mantle, hearth, and chimney top above the roof as a masonry fireplace — but at a fraction of the cost.

Regardless of which type of fireplace you use, the chimney must extend at least 2 feet above the roof ridge or any other part of the roof that's within 10 feet of the chimney (Figure 10-8).

1/2	4th	8th	16th	0"	1"	2"	3"	4"	5"	6"	7"	8"	9"	10"	11"
				0.000	0.083	0.167	0.250	0.333	0.417	0.500	0.583	0.667	0.750	0.833	0.917
			1	0.005	0.089	0.172	0.255	0.339	0.422	0.505	0.589	0.672	0.755	0.839	0.922
		1		0.010	0.094	0.177	0.260	0.344	0.427	0.510	0.594	0.677	0.760	0.844	0.927
			3	0.016	0.099	0.182	0.266	0.349	0.432	0.516	0.599	0.682	0.766	0.849	0.932
	1			0.021	0.104	0.188	0.271	0.354	0.438	0.521	0.604	0.688	0.771	0.854	0.938
			5	0.026	0.109	0.193	0.276	0.359	0.443	0.526	0.609	0.693	0.776	0.859	0.943
		3		0.031	0.115	0.198	0.281	0.365	0.448	0.531	0.615	0.698	0.781	0.865	0.948
			7	0.036	0.120	0.203	0.286	0.370	0.453	0.536	0.620	0.703	0.786	0.870	0.953
1				0.042	0.125	0.208	0.292	0.375	0.458	0.542	0.625	0.708	0.792	0.875	0.958
			9	0.047	0.130	0.214	0.297	0.380	0.464	0.547	0.630	0.714	0.797	0.880	0.964
		5		0.052	0.135	0.219	0.302	0.385	0.469	0.552	0.635	0.719	0.802	0.885	0.969
			11	0.057	0.141	0.224	0.307	0.391	0.474	0.557	0.641	0.724	0.807	0.891	0.974
	3			0.063	0.146	0.229	0.313	0.396	0.479	0.563	0.646	0.729	0.813	0.896	0.979
			13	0.068	0.151	0.234	0.318	0.401	0.484	0.568	0.651	0.734	0.818	0.901	0.984
		7		0.073	0.156	0.240	0.323	0.406	0.490	0.573	0.656	0.740	0.823	0.906	0.990
			15	0.078	0.161	0.245	0.328	0.411	0.495	0.578	0.661	0.745	0.828	0.911	0.995

Figure 10-5
Decimal equivalents of fractional parts of a foot

Figure 10-9 shows different sizes and shapes of chimneys and their flues, and Figure 10-10 gives the number of bricks required for each type (see corresponding letters), per foot of chimney height.

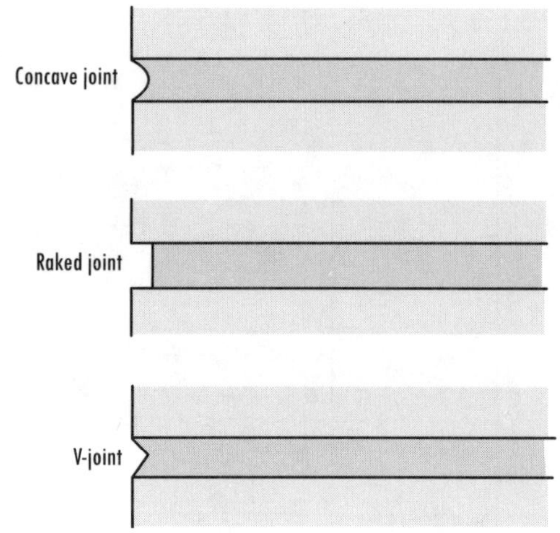

Figure 10-6
Mortar joints

Estimating Brick for Our Sample House

Now let's estimate the brickwork material we'll need for our sample house. We'll use Standard brick with ½-inch joints. The brick factor we need, from Figure 10-1, Column b, is 6.16 bricks per square foot, not including waste.

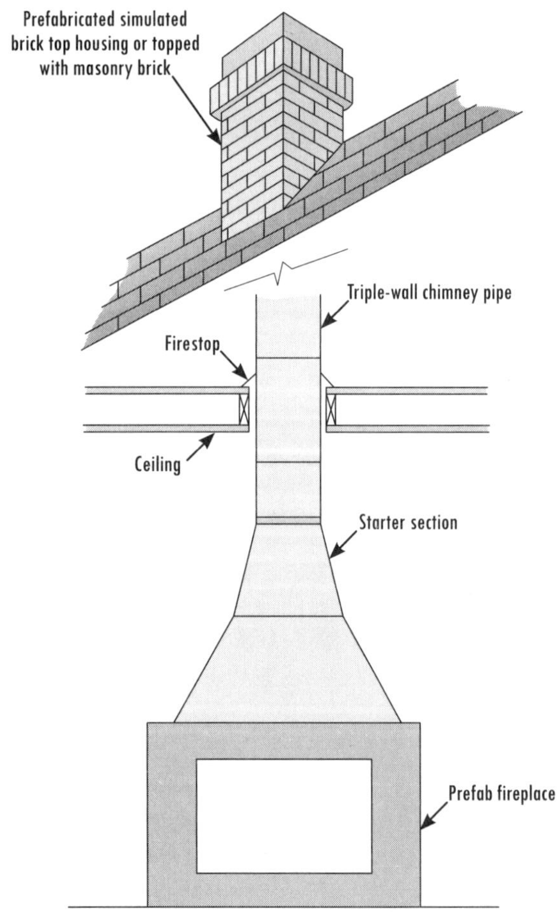

Figure 10-7
Chimney for prefab fireplace

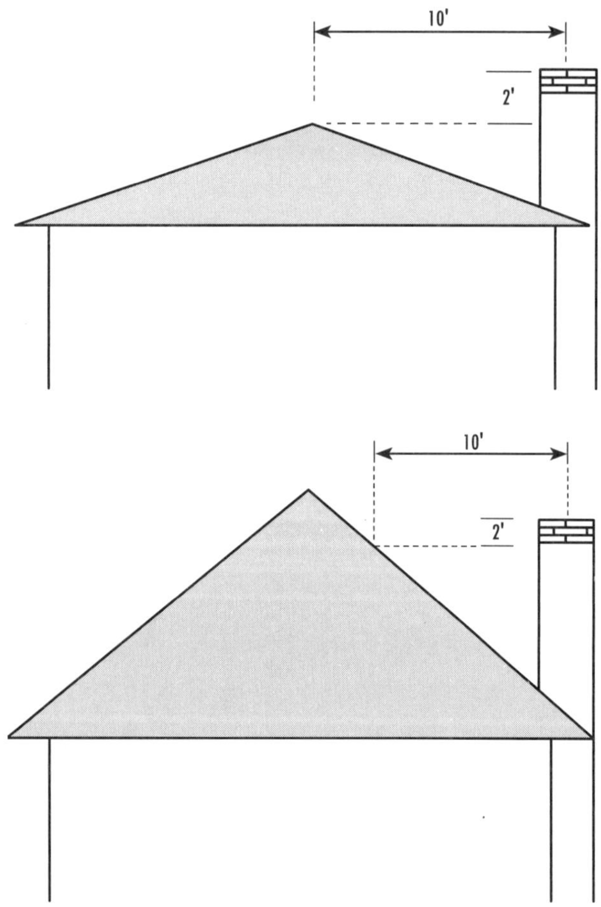

Figure 10-8
Chimney heights

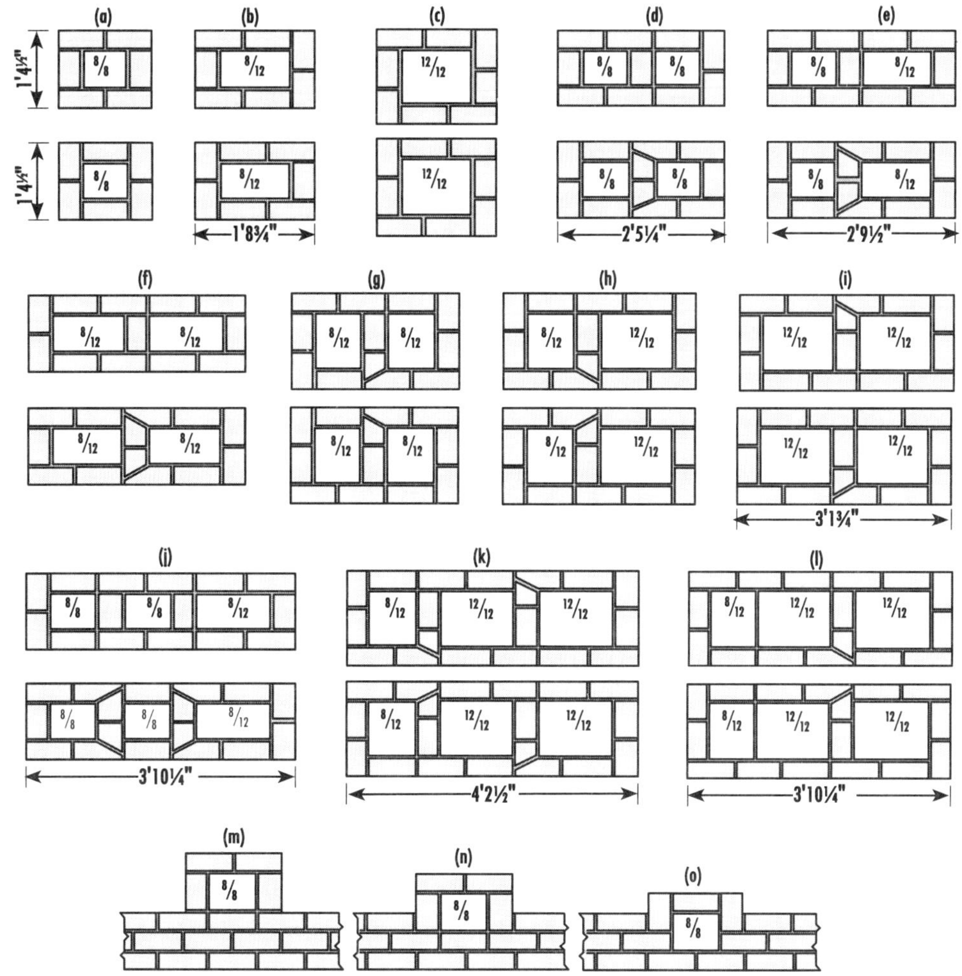

Figure 10-9
Chimney sizes and brick requirements

	Number and Size of Flues		Number of Standard Bricks
(a)	1	8" × 8" flue	27
(b)	1	8" × 12" flue	31
(c)	1	12" × 12" flue	35
(d)	2	8" × 8" flues	46
(e)	1	8" × 8" flue and 1 8" × 12" flue	51
(f)	2	8" × 12" flues	55
(g)	2	8" × 12" flues	53
(h)	1	8" × 12" flue and 1 12" × 12" flue	58
(i)	2	12" × 12" flues	62
(j)	2	8" × 8" flues and 1 8" × 12" flue	70
(k)	1	8" × 12" flue and 2 12" × 12" flues	83
(l)	1	8" × 12" flue and 2 12" × 12" flues	70
(m)	1	8" × 8" flue extending 12" from face of wall	18
(n)	1	8" × 8" flue extending 8" from face of wall	9
(o)	1	8" × 8" flue extending 4" from face of wall	0

Figure 10-10
Number of bricks required in chimneys per foot of chimney height

Exterior Wall Opening	Room	Size (inches) Width		Height	Type	Area (si)	Area (sf)
Windows		**Width**		**Height**			
A	Living Room	108	×	62	Horizontal sliding, Low E	6,696	46.50
B	Bathroom	36	×	24	Horizontal sliding, Low E	864	6.00
C	Family Room	72	×	48	Horizontal sliding, Low E	3,456	24.00
D	Kitchen	84	×	36	Horizontal sliding, Low E	3,024	21.00
E	Dining Room	60	×	48	Horizontal sliding, Low E	2,880	20.00
					First Floor Total Window Area	16,920	117.50
F	Bedroom # 2	36	×	50	Double-hung, Low E	1,800	12.50
G	Bedroom # 2	36	×	50	Double-hung, Low E	1,800	12.50
H	Master Bedroom	36	×	50	Double-hung, Low E	1,800	12.50
I	Master Bedroom	36	×	50	Double-hung, Low E	1,800	12.50
J	Master Bathroom	36	×	24	Horizontal sliding, Low E	864	6.00
K	Family Bathroom	36	×	24	Horizontal sliding, Low E	864	6.00
L	Bedroom # 3	36	×	50	Double-hung, Low E	1,800	12.50
					Second Floor Total Window Area	10,728	74.50
Doors							
5/4	Front	64	×	80		5,120	35.56
2/8	Storage	32	×	80		2,560	17.78
2/8	Utility	32	×	80		2,560	17.78
2/8	Family	32	×	80		2,560	17.78
					Total Door Area	12,800	88.89

Figure 10-11
Window and door openings

House Exterior

We're using brick veneer around the perimeter of the house and garage (192 feet), starting about 2 feet below grade, and going up to the second floor. That's 11 feet of wall height covered with brick veneer. The exterior wall openings for the doors and the windows on the ground floor are shown in Figure 10-11. We'll subtract them from our brick area, as they obviously don't need brick. Let's calculate the total exterior area we'll cover with brick veneer:

192.00′ × 11.00′ =	2,112.00 sf
Minus two (9′ × 7′) garage door openings	(126.00) sf
Minus window openings*	(117.50) sf
Minus door openings*	(88.89) sf
Total for the main house:	1,779.61 sf

* From Figure 10-11

Now we need to multiply by our brick factor to find the number of bricks required for the house exterior:

1,779.61 sf × 6.16 (brick factor) = 10,962.40 bricks

Retaining Wall and Steps

There will be a double-wythe retaining wall and 10 brick steps built from the basement door to the backyard. The retaining wall is 14½ feet long and 7 feet high, and the steps are 3 feet wide. First calculate the amount of brick required for the double-wythe wall:

14.50' × 7.00' × 2 = 203.00 sf

203.00 sf × 6.16 (brick factor) =	1,250.48 bricks

Next, we need 58 bricks per step:

10 steps × 58 bricks per step =	580.00 bricks
Total for wall and steps:	1,830.48 bricks

Chimney and Fireplace

The chimney for the prefab fireplace will be topped with brick. The brickwork will extend from below the roof to 2 feet above the ridge, for a total height of 5 feet. Using the numbers from Figure 10-10 (line c) for a 12-inch-square flue, our brick factor is 35 bricks per foot of chimney height. Calculate the number of bricks needed:

5.00' × 35 (brick factor) = 175 bricks

We'll also use brick on the wall surrounding the fireplace. The wall is 6 feet wide and 8 feet high, and the fireplace opening (requiring no bricks) is 40 × 33 inches. Calculate the amount of brick required for the wall:

6.00′ × 8.00′	48.00 sf
Minus 3.33′ (40″) × 2.75′ (33″)	(9.16) sf
	38.84 sf

38.84 sf × 6.16 (brick factor) = 239.25 bricks

The hearth in front of the fireplace will also be brick. It'll be raised, two courses plus a single course on edge, 6 feet wide and 16 inches deep. Estimate 4 bricks per square foot for Standard brick laid flat, and 6.16 bricks per square foot laid on edge. We need to first calculate the area of the hearth:

6.00' × 1.33' (16") = 7.98 (rounded to 8) sf

Now calculate the number of bricks for the courses:

8 sf × 4 bricks/sf × 2 courses	64.00
8 sf × 6.16 bricks/sf × 1 course	49.28
Total for hearth:	113.28

Let's total the brick we'll need for our sample house:

House exterior	10,962.40
Retaining wall and steps	1,830.48
Chimney	175.00
Fireplace wall	239.25
Hearth	113.28
Subtotal:	13,320.41
Add 5 percent for waste	666.02
Total:	13,986.43 (rounded to 14,000) bricks

> *Enter your estimate of the number, type and size of* ***bricks*** *on line 10.1 of your Cost Estimate Worksheet. We'll enter 14,000 Standard 2¼" × 3¾" × 8" bricks in our Example Cost Estimate Worksheet.*

Prefabricated Fireplace

We're installing a zero-clearance factory-built fireplace, including metal fireplace body, refractory interior, and fuel grate. We'll need to add the cost for flue, doors, blower, and spark arrestor to the fireplace estimate.

> *Enter your estimates for the* ***fireplace*** *items in section 10.2 of your Cost Estimate Worksheet. We'll get our numbers from the current National Construction Estimator for our Example Cost Estimate Worksheet.*

Mortar Mix

We'll use 60-pound bags of mortar mix, calculating our requirements using the factor for Standard bricks with ½-inch joints (0.035 bag per brick), found in Figure 10-1. When calculating mortar mix, there's no need to include mortar for waste bricks.

$$13{,}320 \text{ bricks} \times 0.035 \text{ mortar mix factor} = 466.2 \text{ (rounded to 470) bags}$$

> *Enter your estimate of the number of 60-pound bags of* ***mortar mix*** *on line 10.3 of your Cost Estimate Worksheet. We'll use 470 60-pound bags of mortar in our Example Cost Estimate Worksheet.*

Brick Lintels

Concrete lintels are needed to support the brickwork above each wall opening. There must be at least 4 inches of lintel support extending past the edge of each wall opening. Lintels typically come in increments of 6 inches. Figure 10-12 shows the lintel sizes we'll need for the first floor wall openings (see Figure 10-11).

> *Enter your estimate of the number and size of* ***lintels*** *required in section 10.4 of your Cost Estimate Worksheet. We'll use the number and size of lintels shown in Figure 10-12 in our Example Cost Estimate Worksheet.*

Wall Ties

Wall ties anchor the brick veneer to the wall sheathing underneath. Use no less than 22-gauge ties, with each supporting no more than 2.67 square feet of veneer wall. The total area of veneer brickwork we calculated for the exterior of the main house was 1,779.61 square feet. We'll need the following number of wall ties:

$$\frac{1{,}779.61 \text{ sf}}{2.67 \text{ sf per tie}} = 666.52 \text{ ties}$$

Exterior Wall Opening	Room	Opening Width	Minimum Lintel Width*		Lintel Width, Rounded Up to the Next Highest 6"	Total Numbers of Lintels Required	
Windows		Inches	Inches	Feet/Inches	Feet/Inches	Size	Qty
A	Living Room	108	116	9' 8"	10' 0"	10' 0"	1
B	Bathroom	36	44	3' 8"	4' 0"	8' 0"	1
C	Family Room	72	80	6' 8"	7' 0"	7' 0"	1
D	Kitchen	84	92	7' 8"	8' 0"	6' 0"	2
E	Dining Room	60	68	5' 8"	6' 0"	4' 0"	1
						3' 6"	3
Doors							
5/4	Front	64	72	6' 0"	6' 0"		
2/8	Storage	32	40	3' 4"	3' 6"		
2/8	Utility	32	40	3' 4"	3' 6"		
2/8	Family	32	40	3' 4"	3' 6"		

* Lintels must have a minimum of 4" support at each end

Figure 10-12
Lintel number and sizes

Wall ties are typically sold in boxes of 1,000, so we'll estimate one box.

Enter your estimate of the number of boxes of ***wall ties*** *on line 10.5 of your Cost Estimate Worksheet. We'll estimate 1 box of 1,000 wall ties in our Example Cost Estimate Worksheet.*

Cleaning Material

The final step of a brickwork job is to use muriatic acid to clean the finished masonry. The total area to be cleaned, in square feet (sf), is:

Main house	1,779.61 sf
Retaining wall and brick steps	203.00 sf
Chimney (12″ × 5′ high × 4 sides)	20.00 sf
Fireplace surround	38.84 sf
Hearth	8.00 sf
Total brick area:	2,049.45 sf

Estimate 1 gallon of muriatic acid (mixed 1 pint to 1 gallon of water) for each 300 square feet of brick wall to be cleaned.

$$\frac{2{,}049.45\text{ sf}}{300\text{ sf per gallon}} = 6.83\text{ (rounded to 7) gallons}$$

Enter your estimate of the number of gallons of muriatic acid under ***cleaning material*** *on line 10.6 of your Cost Estimate Worksheet. We'll estimate 7 gallons of muriatic acid in our Example Cost Estimate Worksheet.*

Labor Costs for Brickwork

If you subcontract brickwork, be sure to discuss with the mason how the mortar joints are to be tooled, and who'll be responsible for cleaning the brick when the job's finished.

Brickwork labor varies depending on the efficiency of the bricklayer(s), weather conditions, and the size and shape of the building. The number of corners, window and door openings, the height of the walls, type of brick, type of bond and job conditions are also factors to consider when you're estimating how much brick a mason can lay in an eight-hour day.

Bricklaying

It generally takes a bricklayer and one helper about 34 manhours to lay 1,000 bricks, regardless of whether the bricks are Standard, Roman, Norman or SCR. You can get a closer approximation by using your own records, if they're accurate. As always, your best bet is to use records from previous jobs. However, if you don't have those records, or they're not up-to-date, having a manhour figure will help you make a reasonable estimate.

Let's figure how many manhours (X) it will take to lay the 13,320 bricks in our house. For estimating purposes, it's not necessary to allow for waste. We estimated enough brick to allow for some brick to be broken, damaged or rejected for any reason. Set the problem up as a proportion:

$$1{,}000 \text{ bricks} : 34 \text{ manhours} = 13{,}320 \text{ bricks} : X \text{ manhours}$$

$$\text{Product of extremes} = \text{Product of means}$$

$$1{,}000 \text{ bricks} \times X \text{ manhours} = 34 \text{ manhours} \times 13{,}320 \text{ bricks}$$

$$X = \frac{34 \text{ manhours} \times 13{,}320 \text{ bricks}}{1{,}000 \text{ bricks}}$$

$$X = 452.88 \text{ (rounded to 453) manhours}$$

*Enter your estimate in manhours, under **labor costs for brickwork: bricklaying,** using a bricklayer plus one helper, on line 10.7 of your Cost Estimate Worksheet. We'll use 453 manhours in our Example Cost Estimate Worksheet.*

Brick Cleaning

The brickwork will be cleaned by a laborer. A typical laborer working on a surface that's less than 12 feet high can clean 100 square feet per hour. Add 20 percent more for work above 12 feet.

Let's figure how many manhours it will take to clean the 2,049.45 square feet of brick area of our sample house. Again, we'll set the problem up as a proportion:

$$100 \text{ sf} : 1 \text{ manhour} = 2{,}049.45 \text{ sf} : X \text{ manhours}$$

$$\text{Product of extremes} = \text{product of means}$$

$$100 \text{ sf} \times X \text{ manhours} = 1 \text{ manhour} \times 2{,}049.45 \text{ sf}$$

$$X = \frac{1 \text{ manhour} \times 2{,}049.45 \text{ sf}}{100 \text{ sf}}$$

$$X = 20.49 \text{ (rounded to 21) manhours}$$

> *Enter your estimate of the number of laborer manhours for* ***brick cleaning*** *on line 10.8 of your Cost Estimate Worksheet. We'll estimate 21 manhours using a laborer (BL) in our Example Cost Estimate Worksheet.*

COST ESTIMATE WORKSHEET FOR BRICKWORK

#	Qty	Size		$ Cost	Per			Subtotal
10.1	**Bricks**							
	14,000	Standard 2¼" × 3¾" × 8"	@	$425.00	1,000	=	$5,950.00	$5,950.00
10.2	**Fireplace**							
	1 insert	43" wide		$667.00	each	=	$667.00	
	20 ft flue	8" diam		$23.00	ft	=	$460.00	
	2 flue spacers (one per ceiling)			$13.30	each	=	$26.60	
	1 doors (set of 2)	43" wide		$266.00	set	=	$266.00	
	1 blower			$119.00	each	=	$119.00	
	1 spark arrestor			$109.00	each	=	$109.00	
							$1,647.60	$1,647.60
10.3	**Mortar mix**							
	470 bags	60 lb bag	@	$3.91	each	=	$1,837.70	$1,837.70
10.4	**Brick lintels**							
	1 piece	10.00 ft	@	$13.20	ft	=	$132.00	
	1 piece	8.00 ft	@	$13.20	ft	=	$105.60	
	2 pieces	7.00 ft	@	$13.20	ft	=	$184.80	
	1 piece	6.00 ft	@	$7.06	ft	=	$42.36	
	1 piece	4.00 ft	@	$7.06	ft	=	$28.24	
	3 pieces	3.50 ft	@	$7.06	ft	=	$74.13	
	pieces	ft	@		ft	=	$0.00	
	pieces	ft	@		ft	=	$0.00	
							$567.13	$567.13
10.5	**Wall ties**							
	1 box of 1,000 ties	22 gauge	@	$60.00	box	=	$60.00	$60.00
10.6	**Cleaning material**							
	7 gallons muriatic acid		@	$8.57	gallon	=	$59.99	$59.99
				Cost of material				$10,122.42
		Sales tax	@	7.75	%			$784.49
								$10,906.91
	Labor costs for brickwork							
10.7	**Bricklaying**							
	453 manhours, B9 crew (bricklayer + helper)		@	$28.74	hour	=	$13,019.22	
10.8	**Cleaning**							
	21 manhours, BL crew (laborer)		@	$26.64	hour	=	$559.44	
				Total cost of labor			$13,578.66	$13,578.66
	Cost of brickwork (entered on line 10 of Form 100)							**$24,485.57**

Chapter 11

Saving Energy

THE COST OF HEATING and cooling homes is steadily increasing, a trend that experts expect to continue. For homes with insufficient insulation, that could mean sky-high energy bills.

It's the nature of heat to flow to colder areas, whether those places are up, down, or on the other side of a wall. Heat loss is measured in *British Thermal Units* per hour (Btu/hr or BTUH). One Btu is the amount of heat it takes to raise one pound of water one degree Fahrenheit. In other words, if a wall loses heat at 1,500 Btu/hr, it's losing heat at the same rate it takes to heat 1,500 pounds of water 1 degree Fahrenheit every hour. The heating system must replace that heat to maintain the desired temperature within a living space. Ideally, heat loss must be kept to a minimum for maximum fuel conservation.

Insulation

The single most important factor in energy conservation is proper insulation. Insulation inhibits the flow of heat, both out of the home in wintertime and into the home during the summer. Insulation significantly reduces the energy required for heating and cooling. The ability of an insulating material to resist heat transfer is its *resistance value*, or *R-value*. Industry standard R-values are numbered according to their effectiveness; the higher the R-value number, the more effective the insulation. By insulating with energy-efficient materials, you'll reduce future utility bills. Good insulating materials will pay for themselves many times over during the life of the home.

Let's compare the cost of heating/cooling a 1970s-vintage house, before and after energy upgrades are made. When the original house was built, insulation was only installed in the ceiling — none in the walls or floors.

	Batts or Blankets		Loose Fill (poured in)			
	Glass fiber	Rockwool	Glass fiber	Rockwool	Cellulose fiber	
R-11	3½" - 4"	3"	5"	4"	3"	R-11
R-19	6" - 6½"	5¼"	8" - 9"	6" - 7"	5"	R-19
R-22	6½"	6"	10"	7" - 8"	6"	R-22
R-30	9½" - 10½" *	9" *	13" - 14"	10" - 11"	8"	R-30
R-38	12" - 13" *	10½" *	17" - 18"	13" - 14"	10" - 11"	R-38

* Two batts or blankets required

Figure 11-1
R-values and insulation thickness

The home had single-paned windows, and no storm windows or storm doors. During the fuel crisis some years ago, an additional four inches of insulation was blown into the attic. The walls were insulated, and the floors (over an unheated basement) were insulated with full thick batts. Storm windows and storm doors were also installed on the windows and doors.

The same family lived in the house for eleven years, and had saved the energy bills during that time. Following the upgrades, fuel consumption was reduced by an average of 46 percent a year — almost half the original energy usage. And, this dividend on the owners' investment will continue to grow as long as the house stands.

Types of Insulation

Although there are many types of insulation, we'll discuss only the most commonly-used materials. Figure 11-1 shows R-values and thicknesses of common types of insulation.

Glass Fiber and Rockwool

Glass fiber and rock wool are available in batts, blankets or as loose fill (poured in). Batts and blankets are cut in sections 15 inches wide for 16-inch on-center stud or joist spacing, and 23 inches for 24-inch on-center spacing. Batts and blankets are fire- and moisture-resistant, and come with or without vapor retarders.

Cellulose Fiber, Vermiculite and Perlite

The materials typically used as loose fill insulating in attics are cellulose fiber, vermiculite or perlite. Loose fill (poured in) insulation doesn't have any vapor retarder. If you opt for loose fill, you must purchase and install the vapor retarder separately. Remember to add that cost to your estimate. Loose fill insulations are best suited for nonstandard or irregular joist spacing, or where there are obstructions between the joists. If the material is fire resistant, it'll be clearly labeled. Check the insulation packaging for federal specifications.

Rigid Insulation Boards

Rigid insulation boards are made from organic fiber, polystyrene foam, or urethane. Boards made from organic fiber are generally used for wall sheathing. Polystyrene and urethane rigid insulation boards have a higher R-value than organic fiber boards. Follow the manufacturer's instructions when installing rigid insulation boards. Some manufacturers use aluminum foil facers for an effective vapor retarder. The standard size used is 4 by 8 feet, with nominal thicknesses of 3/8 inch through 1 7/8 inches, but boards can be special ordered in thicknesses up to 2¼ inches.

Other Insulation Factors to Consider

When choosing insulation, heat-transfer material is only a part of the installation package. Vapor retarders, previously mentioned, are an integral component. Another component used to maximize the effectiveness of insulation is weatherstripping.

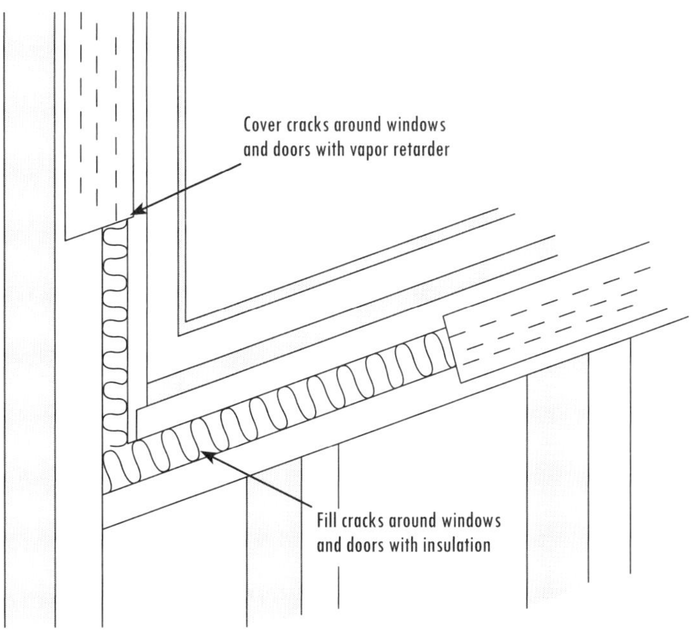

Figure 11-2
Infiltration and vapor retarders

Vapor Retarders

Vapor retarders restrict the moisture produced in warm air from entering walls, ceilings, and floors. Their installation is vital for the simple reason that wet insulation does not insulate. Worse still, if moisture becomes trapped, it can cause a structure to rot. By limiting vapor, you prolong the life of your insulation.

Weatherstripping

It's common for air to leak into a structure through cracks around windows and doors, especially when it's windy. Cold air seeping into a room puts an additional load on the heating system. By weatherstripping windows and doors and sealing around cracks, as shown in Figure 11-2, you can minimize air infiltration.

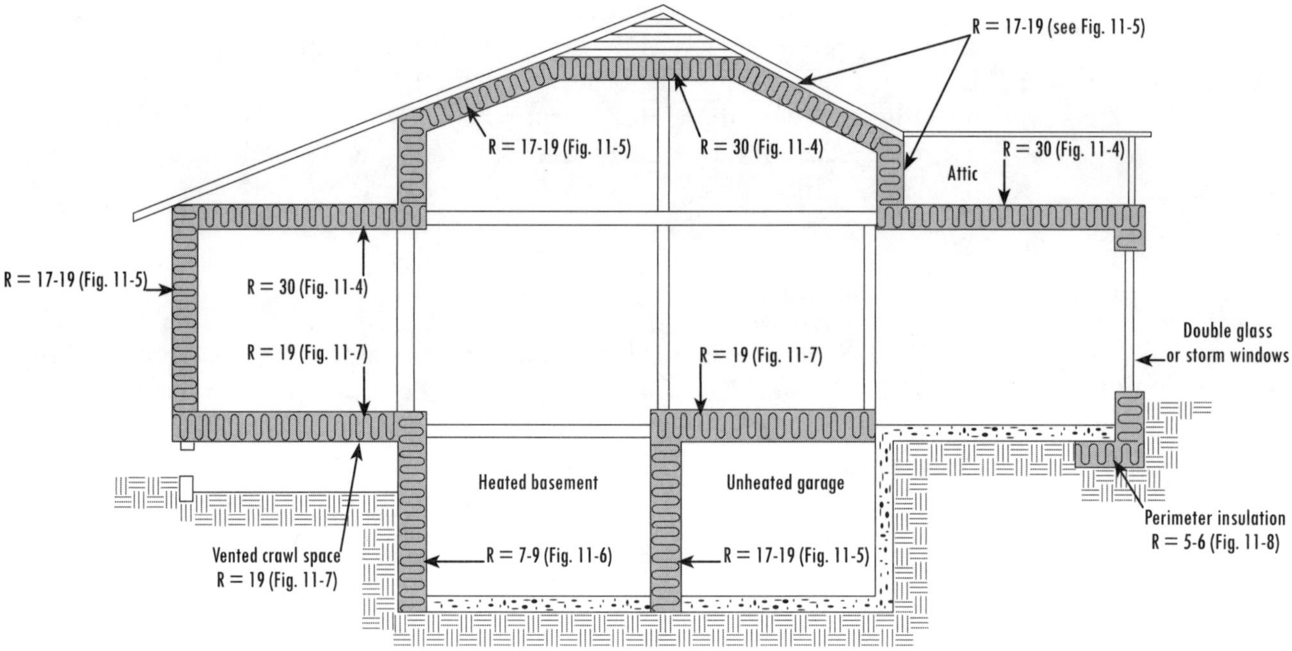

Figure 11-3
House insulation

Where to Insulate

A few years ago, children in a local elementary school were asked how they'd ease the energy shortage. One child said, "When we're not in bed, I'd take our blankets and put them around the room to keep it warm." That's the right idea, right down to the blanket reference. Figure 11-3 shows a house with a blanket of insulation correctly installed around it for efficient energy conservation. R-values vary from location to location; use R-19 for ceilings and R-11 for walls and floors in warmer regions; R-38 for ceilings, R-19 for walls and R-22 for floors in colder areas. Regional fuel costs also need to be taken into account when selecting R-value.

Ceilings

Ceiling insulation comes in batts, blankets, loose fill (poured in) or a combination of any two of those materials — whatever it takes to achieve the desired R-value (see Figures 11-1 and 11-4). If you can install insulation after the ceiling's finished, it will save labor to place the insulation in the attic space before installing the drywall and install the insulation on the ceiling boards later. If you can't install insulation after the ceiling is finished, it'll have to be stapled in place between the joists before the drywall is installed. Be careful not to block air flow from the soffit vents to gable or roof vents when installing the ceiling insulation.

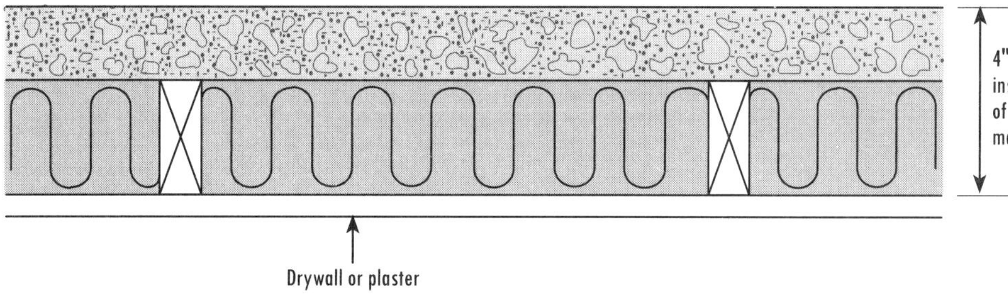

Figure 11-4
Ceiling insulation

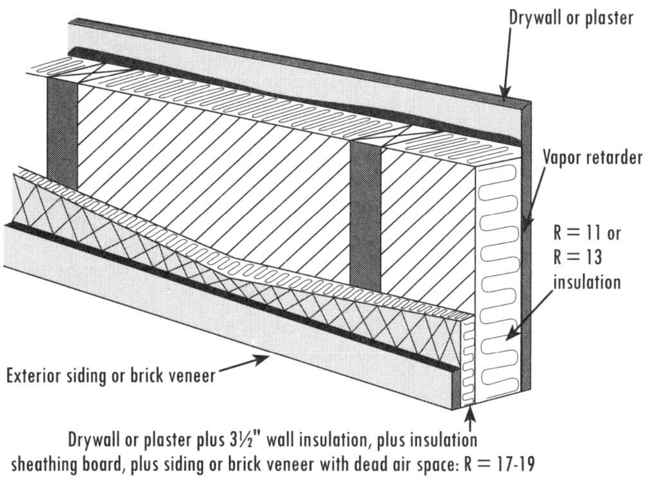

Figure 11-5
Wall insulation

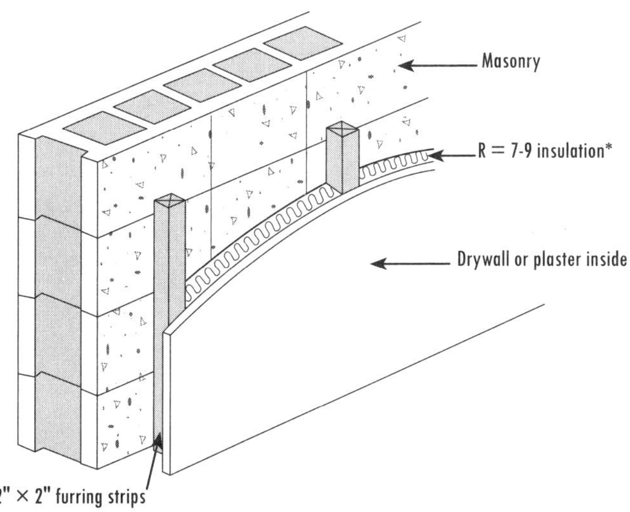

*Only insulating products whose thermal efficiency is not affected by moisture should be used

Figure 11-6
Masonry wall insulation

Ceiling Heat Loss Comparisons

The following are insulation heat-loss figures for some insulation thicknesses and R-values:

- 8-inch (R-25) insulation has 21 percent less heat loss than 6 inch (R-19).
- 10-inch (R-31) insulation has 35 percent less heat loss than 6 inch (R-19).
- 12-inch (R-38) insulation has 46 percent less heat loss than 6 inch (R-19).

Walls

The exterior walls (excluding windows and doors) of a house are the areas that roughly account for 15 percent of total heat loss. When R-11 or R-13 batt or blanket insulation (which are easier to install in walls than loose fill) are added to other building materials, as shown in Figure 11-5, you can achieve an R-17 to R-19 value. A brick veneer wall with dead air space, without extra insulating material, can insulate to R-3 or so. Adding R-11 insulation between the studs can reduce the heat loss up to 70 percent.

Masonry walls with no insulation lose heat rapidly, but adding furring strips and R-7 to R-9 insulation between the strips (Figure 11-6) can reduce that loss by 80 percent.

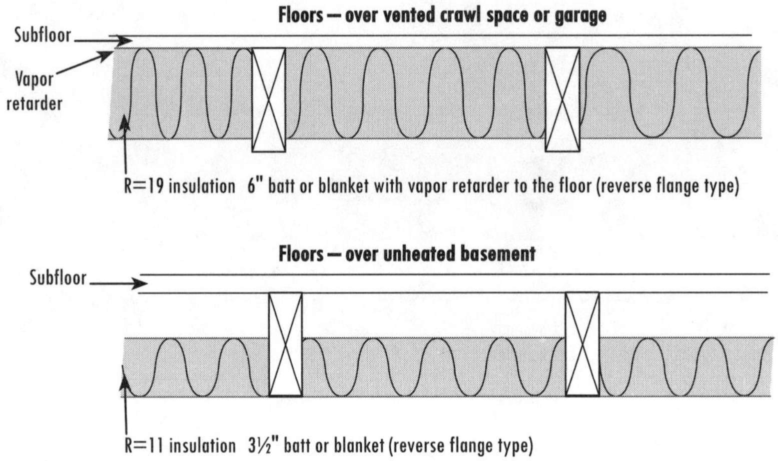

Figure 11-7
Floor insulation

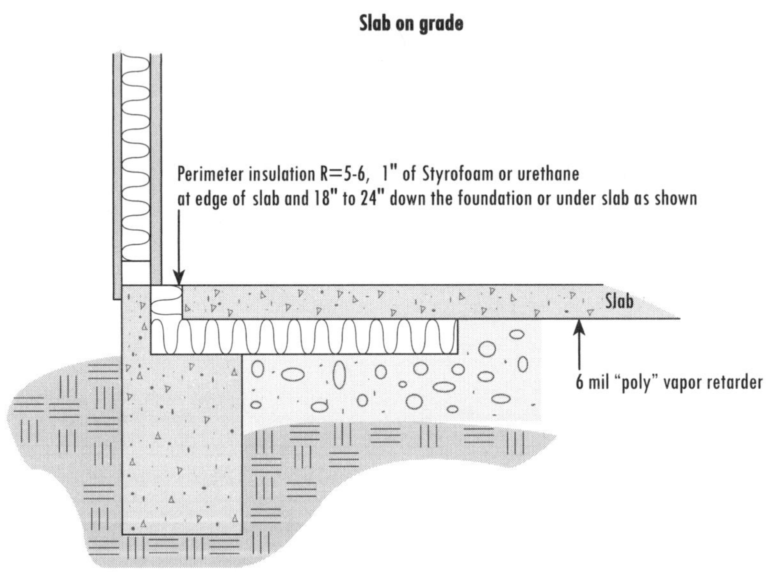

Figure 11-8
Perimeter insulation under slabs

Floors

Floors over heated rooms or basements (Figure 11-3) need no insulation. However, you must insulate floors over unheated basements, garages or vented crawl spaces. A value of R-19 is generally recommended, depending on climate (see Figures 11-3 and 11-7). For slabs on grade, as seen in Figures 11-3 and 11-8, insulate the perimeter with 1 inch of Styrofoam or urethane (R-5 or R-6). It should be installed at the edge of the slab, and 18 to 24 inches down the foundation or under the slab.

Window Condensation (inside temperature 70°F)		
Window type	Outside temperature (low wind velocity)	Condensation occurs at:
Single glass	0°F 30°F	12% humidity 33% humidity
Insulating glass	0°F 30°F	35% humidity 55% humidity
Single glass with storm window	0°F 30°F	40% humidity 61% humidity
Insulating glass with storm window	0°F 30°F	53% humidity 70% humidity

Figure 11-9
Window condensation

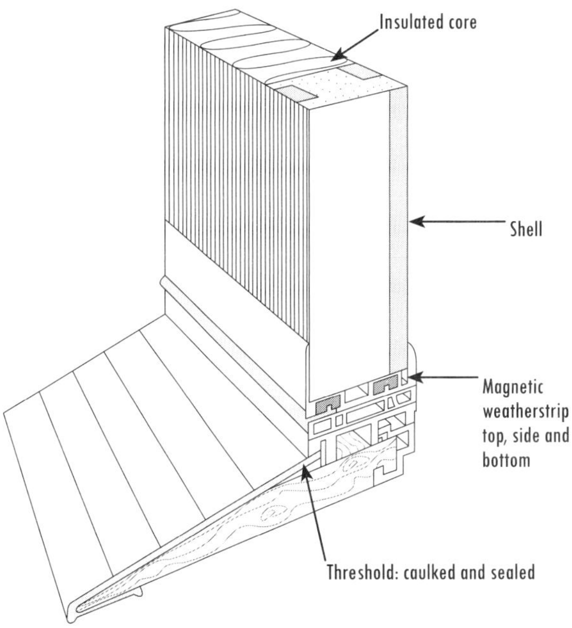

Figure 11-10
Insulated door

Windows

Windows and sliding glass doors account for roughly 40 percent of total heat loss in a house, depending on the type of windows and doors installed. Cracks around windows should be filled with insulation and covered with a vapor retarder, as shown in Figure 11-2.

Window Heat Loss

Here's a comparison of heat loss between single pane and insulated windows:

- Single pane windows, with storm windows added, lose approximately 50 percent less heat than single pane windows alone.
- Insulating glass loses approximately 46 percent less heat than single pane windows with *no* storm windows.

Figure 11-9 is a window condensation guide. In cold climates, you need to counteract moisture that can form on the inside of windows. A dehumidifier inside the house can help control condensation.

Doors

Entry and exit doors (excluding sliding glass doors) account for less than 5 percent of the total heat loss in a house. Fill any cracks around the perimeter with insulation and cover with a vapor retarder.

Door Heat Loss

The following is a comparison of the heat loss between plain doors and weatherstripped doors:

- Solid core (1¾-inch) doors with storm doors retain approximately 35 percent more heat than solid core doors without storm doors.
- A steel/urethane insulated door with an R-13.8 value (see Figure 11-10) retains approximately 85 percent more heat than a solid core (1¾-inch) door. This same steel/urethane door (with no storm door) retains approximately 76 percent more heat than a solid core door *with* a storm door.

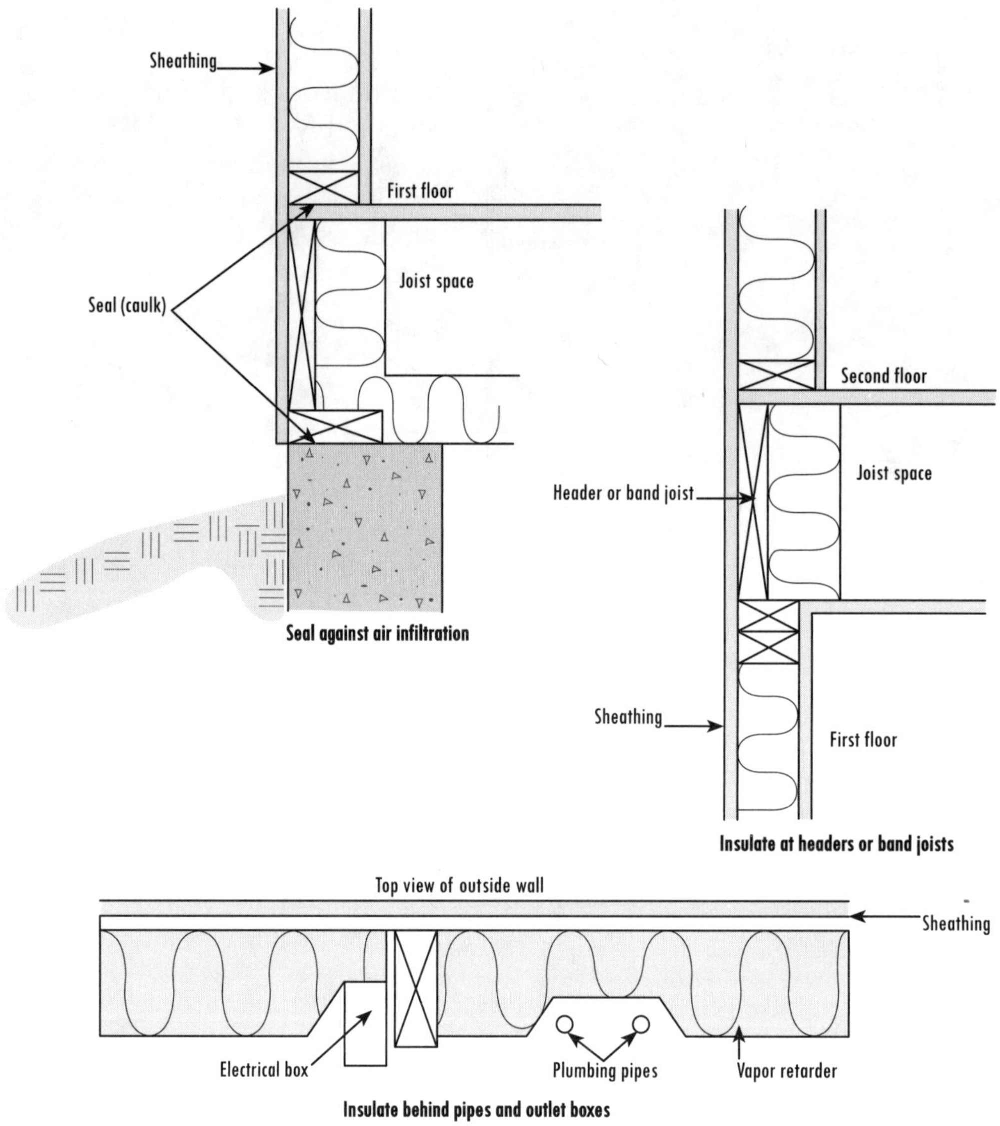

Figure 11-11
Remember little spaces

Other Sources of Heat Loss

There are several places where air can infiltrate a building if they're not properly sealed off with caulking or insulation. Some of the small spaces that allow heat loss are shown in Figure 11-11.

The true efficiency of insulation hinges on its installation. R-values are effective only if their densities are maintained. Compressing batts or blankets increases their density, but reduces their R-value. When there's too much air in the blown insulation mixture, you'll get a "fluffy" installation — less dense than the product is intended for optimal R-value. Over time, settling occurs, which also reduces thickness and R-value.

Other common insulating problems can be caused by:

- Not filling small spaces around window and door frames with insulation, and neglecting to install a vapor retarder;
- Installing batts or blankets incorrectly, leaving voids between framing members;
- Stapling batt or blanket flanges improperly, resulting in a poor vapor retarder;
- Installing the batt or blanket under floors over crawl spaces with the vapor retarder facing the crawl space instead of the floor;
- Improperly placing and fitting batts or blankets at cross bracing (bridging) between joists, or where purlin and/or knee bracing is placed between studs.

Estimating Insulation Materials

You'll need the total floor area from the house plans to measure for ceiling and floor insulation. For the walls, multiply the perimeter of the outside walls (plus any interior walls where insulation is required) by the wall height. Deduct any window and door openings that are 50 square feet or larger, to get the net area. Divide the area to be insulated by the amount of insulation in each bag or roll, and round up for the number of bags or rolls you need.

To estimate the insulation we need for our sample house (shown in Figures 7-1 and 7-2), figure the following *ceiling areas* separately:

- Kitchen, dining, and living rooms: R-30 insulation, 624 sf
- Second floor area: R-30 insulation, 1,040 sf
- Unheated garage and storage area: R-19 insulation, 610 sf

Add these miscellaneous areas to your insulation estimate:

- Net exterior walls, first and second floors: R-13 insulation, 1,968 sf
- Floor area over the *unheated* basement: R-13 insulation, 1,040 sf

Additional costs for storm windows, storm doors, or installation materials should be entered separately.

*Enter your estimates for **insulation** in section 11.1 of your Cost Estimate Worksheet. We'll use these areas and R-values in our Example Cost Estimate Worksheet.*

> *Enter your estimates for* ***storm windows, storm doors****, or any other materials on lines 11.2 through 11.4 of your Cost Estimate Worksheet. We have nothing additional to add to our Example Cost Estimate Worksheet.*

Estimating Labor

A carpenter generally installs insulation as the framing and sheathing is being completed. R-13 insulation will typically cost less to install than R-19 through R-30, since it's less bulky and easier to handle.

> *Enter the estimate of* ***labor costs,*** *square footage, R-values, and costs per square foot just calculated in section 11.5 of your Cost Estimate Worksheet. We'll use these areas, R-values, and costs from the National Construction Estimator in our Example Cost Estimate Worksheet.*

COST ESTIMATE WORKSHEET FOR ENERGY SAVING MATERIALS

#	Item	Location	R-Value	Area, sf		Cost	Per			Subtotal
11.1	**Insulation**									
		Ceiling - 1st floor area	30	624	@	$0.84	sf	=	$524.16	
		Ceiling - 2nd floor area	30	1,040	@	$0.84	sf	=	$873.60	
		Ceiling - garage and storage	19	610	@	$0.56	sf	=	$341.60	
		Exterior walls	13	1,968	@	$0.32	sf	=	$629.76	
		Basement ceiling	13	1,040	@	$0.32	sf	=	$332.80	
					@		sf	=	$0.00	
11.2	**Storm windows** (List on separate sheet and enter cost here)							=	$0.00	
11.3	**Storm doors** (List on separate sheet and enter cost here)							=	$0.00	
11.4	**Other material** (List on separate sheet and enter cost here)							=	$0.00	
				Subtotal				=	$2,701.92	$2,701.92
				Sales tax	@	7.75	%			$209.40
				Total cost of material						$2,911.32
11.5	**Labor costs**									
			R-13	3,008		$0.16	sf	=	$481.28	
			R-19 - R-30	2,274		$0.19	sf	=	$432.06	
				Total cost of labor				=	$913.34	$913.34
	Cost of energy saving materials (entered on line 11 of Form 100)									**$3,824.66**

Chapter 12

Interior Wall and Ceiling Finish

SEVERAL CONSTRUCTION PHASES NEED to be completed before interior walls and ceilings get their final finish. Electrical, plumbing, heating, and air conditioning must be roughed in, and insulation installed. Drop ceilings over kitchen cabinets and bathroom vanities and similar carpentry work should be completed before the wall finish is applied.

Thin-Coat Plaster

Lath and plaster application has largely been replaced by thin-coat plaster for plastered walls in residential construction. Plaster contains a large amount of water and takes a long time to dry.

Thin-coat plaster is applied over a plaster base of gypsum core with a special face paper. These plaster base sheets come in ½- or ⅝-inch thicknesses, 4 feet wide and usually 12 feet long for residential construction. The plaster is spread in a 1/16- to 3/32-inch-thick coat over the plaster base sheets, drying in 24 hours or less, and providing a hard and durable finish. Carpenters can begin the interior trim in the same drying time as they would for gypsum drywall applications.

The plaster base material is estimated by the square foot, the same as drywall (explained later in this chapter). Thin-coat is applied by plasterers specializing in the trade, who bid the contract based on either labor and material, or labor only, on a per square yard basis. If you, the builder, furnish the material, follow the manufacturer's recommendation when selecting the type and quantity of material to estimate.

Width in Feet \ Length in Feet	2'	3'	4'	5'	6'	7'	8'	9'	10'	11'	12'	13'	14'	15'	16'	17'	18'	19'	20'
2'	68	86	104	122	140	158	176	194	212	230	248	266	284	302	320	338	356	374	392
3'	86	105	124	143	162	181	200	219	238	257	276	295	314	333	352	371	390	409	428
4'	104	124	144	164	184	204	224	244	264	284	304	324	344	364	384	404	424	444	464
5'	122	143	164	185	206	227	248	269	290	311	332	353	374	395	416	437	458	479	500
6'	140	162	184	206	228	250	272	294	316	338	360	382	404	426	448	470	492	514	536
7'	158	181	204	227	250	273	296	319	342	365	388	411	434	457	480	503	526	549	572
8'	176	200	224	248	272	296	320	344	368	392	416	440	464	488	512	536	560	584	608
9'	194	219	244	269	294	319	344	369	394	419	444	469	494	519	544	569	594	619	644
10'	212	238	264	290	316	342	368	394	420	446	472	498	524	550	576	602	628	654	680
11'	230	257	284	311	338	365	392	419	446	473	500	527	554	581	608	635	662	689	716
12'	248	276	304	332	360	388	416	444	472	500	528	556	584	612	640	668	696	724	752
13'	266	295	324	353	382	411	440	469	498	527	556	585	614	643	672	701	730	759	788
14'	284	314	344	374	404	434	464	494	524	554	584	614	644	674	704	734	764	794	824
15'	302	333	364	395	426	457	488	519	550	581	612	643	674	705	736	767	798	829	860
16'	320	352	384	416	448	480	512	544	576	608	640	672	704	736	768	800	832	864	896
17'	338	371	404	437	470	503	536	569	602	635	668	701	734	767	800	833	866	899	932
18'	356	390	424	458	492	526	560	594	628	662	696	730	764	798	832	866	900	934	968
19'	374	409	444	479	514	549	584	619	654	689	724	759	794	829	864	899	934	969	1004
20'	392	428	464	500	536	572	608	644	680	716	752	788	824	860	896	932	968	1004	1040

Figure 12-1
Square feet of walls and ceiling with 8-foot ceiling (no openings deducted)

Drywall

Gypsum wallboard (*drywall*) is the most commonly used material for walls and ceilings. It has a high fire-resistance rating, and comes in a standard 4-foot width, with lengths from 8 to 16 feet. Thicknesses are usually ½, ⅝ and ¾ inch, but ¼- and ⅜-inch thicknesses are also available for special applications, such as for curved finishes. Drywall work is often subcontracted for labor and material, or for labor only.

Estimating Wallboard

To estimate drywall material, you need to know the square footage of the walls and ceilings in each room, closet, hallway, and stairway that you'll be covering. The chart in Figure 12-1 shows the total square footage for walls and ceilings of different sized rooms with 8-foot ceiling heights. When calculating square footage, round up any fractions to the whole foot, for both width and length.

Estimating square feet using Figure 12-1 is easy. Just find the point where the width row and length column intersect. This gives you a rough estimate of how much wallboard you'll need for the walls and ceiling of any particular room. For a room that's 10 feet 8 inches wide by 12 feet 3 inches long (and 8 feet high, for this chart), you simply round up to the next highest whole numbers, 11 feet and 13 feet, and see where they intersect. The point of intersection (shaded in Figure 12-1) is 527 square feet.

Worksheet for Area of Walls and Ceiling

Room	________						
	Width			**Length**			
	ft in	decimal		ft in	decimal		
Size	____	____	×	____	____		
	ft in	decimal					
Ceiling height	____	____					
Perimeter of walls	0.00	width	+	0.00	length	= 0.00 × 2 =	0.00 lf
				Area			
Perimeter walls	0.00	lf	×	0.00	ceiling height	=	0.00 sf
Ceiling	0.00	width	×	0.00	length	=	0.00 sf
	Gross square feet of walls and ceiling					=	**0.00** sf
Openings > 50 sf	**Width**			**Length**			
	ft in	decimal		ft in	decimal		
	____	____	×	____	____		0.00
	____	____		____	____		0.00
	____	____		____	____		0.00
	Net square feet of walls and ceiling					=	**0.00** sf

Figure 12-2
Worksheet for square feet of walls and ceiling

Size	Approximate weight, lbs (each)	Bending radii, feet Lengthwise	Bending radii, feet Widthwise
¼" × 4' × 8'	35	5	15
⅜" × 4' × 8'	50	7½	25
⅜" × 4' × 12'	75	7½	25
½" × 4' × 8'	67	10 *	-
½" × 4' × 12'	100	10 *	-
⅝" × 4' × 8'	90	-	-
⅝" × 4' × 12'	135	-	-

* Bending two pieces of ¼" board permits lengthwise radius shown for ½" board

Figure 12-3
Gypsum board weight and bending radii

Use the worksheet in Figure 12-2 to help you estimate the square footage of walls and ceilings for rooms larger than 20 feet long or wide, or for rooms with ceiling heights other than 8 feet. There's a copy of Figure 12-2 on the CD-ROM inside the back cover of this book. Enter the dimensions (length, width, and height) in feet and inches. Use Figure 7-29, in Chapter 7, to convert each entry into decimals.

When estimating, consider all surfaces as solid areas, except when there are openings of more than 50 square feet. This will help cover any material waste. Total all the wall and ceiling areas you plan to cover. If you're using 4 × 12s, divide the total area by 48, and round up to the next whole number to find the number of boards to order. If you're using 4 × 8s, divide the total area by 32, and round up. Keep in mind that 4 × 12 boards are heavier, so they're more difficult to handle. If you use the longer boards, any money you anticipate saving by having fewer joints could be offset by higher labor cost. Figure 12-3 gives the approximate weight and bending radii of various sizes of gypsum board.

Estimating Drywall Materials

In addition to the gypsum boards, you'll need the following materials to complete a gypsum wallboard installation. Use the guidelines below to help you estimate these materials:

Nails, 1¼-inch bright ring shank*	6 pounds per 1,000 square feet
Adhesive, cartridge	10.5 ounces per 170 square feet
Tape	500 foot roll per 1,300 square feet
Joint compound	5 gallons per 1,300 square feet
Joint compound (for texturizing)	5 gallons per 400 square feet

*Use with adhesive

You'll also need metal corner bead for all outside corners. Add the total linear feet of outside corners and divide by 8, for 8-foot lengths of corner bead; or by 10 or 12 for those lengths, which are used with higher ceilings.

Worksheet for Area of Walls and Ceiling									
Room	Dining Room								
	Width			**Length**					
	ft in	decimal		ft in	decimal				
Size	12'2"	12.17	×	13'0"	13.00				
	ft in	decimal							
Ceiling height	8'0"	8.00							
Perimeter of walls	12.17	width	+	13.00	length =	25.17	× 2 =	50.34	lf
				Area					
Perimeter walls	50.34	lf	×	8.00	ceiling height		=	402.72	sf
Ceiling	12.17	width	×	13.00	length		=	158.21	sf
		Gross square feet of walls and ceiling					=	**560.93**	sf
Openings > 50 sf	**Width**			**Length**					
	ft in	decimal		ft in	decimal				
			×					0.00	
								0.00	
								0.00	
		Net square feet of walls and ceiling					=	**560.93**	sf

Figure 12-4
Completed worksheet for walls and ceiling

Calculating Drywall Costs

Now let's practice calculating drywall costs based on square feet (rather than using Figure 12-1), and include individual estimates for material costs. We'll begin with a simple rectangular room — the dining room of our sample house.

To determine the finish materials needed for dining room walls and ceiling (refer to Figure 7-1), we'll need to calculate wallboard, corner bead, nails, adhesive, tape, and joint compound (the ceiling will be plain finished).

Wallboard

The dining room dimensions are 12′2″ × 13′0″. None of the openings is greater than 50 square feet, so we won't deduct them from our estimate, which gives us an allowance for waste. We'll use the form shown in Figure 12-2 (also found on the CD-ROM) to calculate the wallboard we'll need. Figure 12-4 shows the completed estimate for this room. We'll need 560.93 square feet of wallboard to finish the walls and ceiling of the dining room.

Metal Corner Bead

We'll use corner bead on the opening to the kitchen and the doorway to the living room, counting only the corner bead on the dining room side of the opening. The corner bead on the kitchen and living room sides will be counted when we estimate for those two rooms.

Opening to kitchen, 2 corners, 8′0″ (96″) each	192″
Doorway to living room, 2 corners, 6′8″ (80″) each	160″
Total:	352″

$$\frac{352}{96} = 3.66 \text{ (rounded to 4)}$$

Using 8-foot (96-inch) lengths, we'll need 4 pieces of corner bead, totaling 32 linear feet.

Nails

6 pounds of 1¼-inch bright ring shank nails are needed per 1,000 square feet of wallboard. Set up a proportion to find how many nails we'll need for our 560.93-square-foot dining room.

$$6 \text{ lbs nails} : 1{,}000 \text{ sf} = X \text{ lbs nails} : 560.93 \text{ sf}$$
$$\text{Product of extremes} = \text{Product of means}$$
$$6 \text{ lbs nails} \times 560.93 \text{ sf} = 1{,}000 \text{ sf} \times X \text{ lbs of nails}$$
$$X = \frac{6 \text{ lbs nails} \times 560.93 \text{ sf}}{1{,}000 \text{ sf}}$$
$$X = 3.37 \text{ lbs of } 1\tfrac{1}{4}\text{" bright ring shank nails}$$

Adhesive

One 10.5-ounce cartridge of adhesive is needed per 170 square feet of wallboard, using a ⅜-inch bead on 16-inch on-center framing. Set up a proportion to find how many cartridges of adhesive (X) we'll need for the dining room.

$$1 \text{ cartridge} : 170 \text{ sf} = X \text{ cartridges} : 560.93 \text{ sf}$$
$$\text{Product of extremes} = \text{Product of means}$$
$$1 \text{ cartridge} \times 560.93 \text{ sf} = 170 \text{ sf} \times X \text{ cartridges}$$
$$X = \frac{1 \text{ cartridge} \times 560.93 \text{ sf}}{170 \text{ sf}}$$
$$X = 3.30 \text{ (rounded to 4) cartridges}$$

We'll need 4 (10.5-ounce) cartridges of adhesive.

Tape

One 500-foot roll of tape is needed for every 1,300 square feet of wallboard. Set up the following proportion to find how many rolls of tape (X) we'll need.

$$1 \text{ roll} : 1{,}300 \text{ sf} = X \text{ rolls} : 560.93 \text{ sf}$$
$$\text{Product of extremes} = \text{Product of means}$$
$$1 \text{ roll} \times 560.93 \text{ sf} = 1{,}300 \text{ sf} \times X \text{ rolls}$$
$$X = \frac{1 \text{ roll} \times 560.93 \text{ sf}}{1{,}300 \text{ sf}}$$
$$X = 0.43 \text{ roll of tape (500')}$$

We'll need about 215 feet (0.43 of a 500-foot roll) of drywall tape.

Joint Compound

One 5-gallon can of joint compound is needed per 1,000 square feet of wallboard. Set up the following proportion to find how much joint compound (X) we'll use for our dining room.

$$1 \text{ can} : 1{,}000 \text{ sf} = X \text{ cans} : 560.93 \text{ sf}$$
$$\text{Product of extremes} = \text{Product of means}$$
$$1 \text{ can} \times 560.93 \text{ sf} = 1{,}000 \text{ sf} \times X \text{ cans}$$
$$X = \frac{1 \text{ can} \times 560.93 \text{ sf}}{1{,}000 \text{ sf}}$$
$$X = 0.56 \text{ can (5 gallon)}$$

We'll need about 2.8 gallons (0.56 of a 5-gallon can) of joint compound.

Labor

Labor to install these materials can vary by as much as 30 percent, so we'll calculate labor for each item separately. We'll discuss labor costs in detail for the complete house estimate in the next section.

L-Shaped Room

Let's do another practice wallboard estimate based on square footage. This time, we'll calculate the ceiling/wall areas for the L-shaped living/dining room arrangement shown in Figure 12-5.

How do we calculate wallboard for this room? We treat the two areas as if they were two separate rooms. Calculate the area of the "adjoining wall" and multiply by 2, since this area was calculated in the wall area for each of the two rooms. Then, subtract that from the originally calculated area for the walls of the L-shaped room. This step is shown in Figure 12-6, which has all the correct calculations for this room. You can use the blank worksheet on the accompanying CD-ROM to do your own figures.

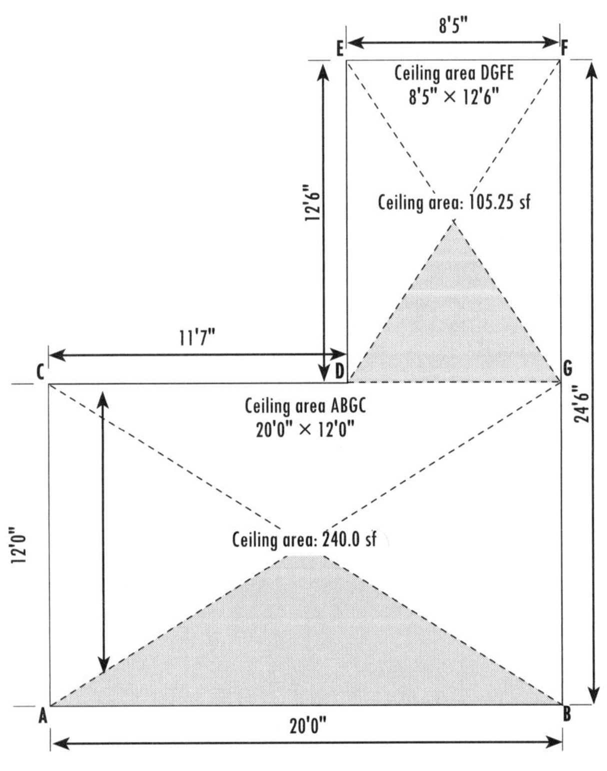

Figure 12-5
Offset rooms

Whole House Wallboard Estimate

Now let's do a per-square-foot estimate for wallboard and accessories for an entire house.

Wallboard

We'll say that we've done our calculations for each room and found that we'll need 10,016 square feet of wallboard. We're going to use 4 × 8 boards, so we need to divide our square feet by 32 and round up to the nearest whole board. In our case, however, we have an even multiple of 32, so that step won't be necessary.

$$\frac{10{,}016 \text{ sf}}{32 \text{ sf per board}} = 313 \text{ boards}$$

Worksheet for Walls and Ceilings in Offset Rooms										
Rooms	Living Room/Dining Room									
	Width			**Length**						
	ft in	decimal		ft in	decimal					
Size	12'0"	12.00	×	20'0"	20.00					
	Width			**Length**						
	ft in	decimal		ft in	decimal					
Size	12'6"	12.50	×	8'5"	8.42					
	ft in	decimal								
Ceiling height	8'0"	8.00								
Perimeter of walls	12.00	width	+	20.00	length	=	32.00	× 2 =	64.00	lf
	12.50	width	+	8.42	length	=	20.92	× 2 =	41.84	lf
		Total perimeter of the two "separate" rooms							105.84	lf
					ft in		decimal			lf
Minus length of "adjoining wall not present"					8'5"		8.42	× 2 =	16.84	lf
									89.00	lf
				Area						
Perimeter walls	89.00	lf	×	8.00	ceiling height			=	712.00	sf
Ceiling	12.00	width	×	20.00	length			=	240.00	sf
Ceiling	12.50	width	×	8.42	length			=	105.25	sf
		Gross square feet of walls and ceiling						=	**1,057.25**	sf
Openings > 50 sf	**Width**			**Length**						
	ft in	decimal		ft in	decimal					
			×						0.00	
									0.00	
									0.00	
		Net square feet of walls and ceiling						=	**1,057.25**	sf

Figure 12-6
Worksheet for walls and ceilings in offset rooms

Of the total 10,016 square feet, 2,850 square feet are ceiling area, leaving 7,166 square feet of wall area. Because ceiling applications are more labor intensive, we'll need a separate break down to estimate the labor per square foot for the ceiling and wall areas. If you choose not to calculate them separately for your estimate, you can figure that the ceiling is approximately 28.5 percent of the total area to be covered.

> *Enter your estimate for the square footage of **wallboard** on line 12.1 of your Cost Estimate Worksheet. We'll use 10,016 square feet in our Example Cost Estimate Worksheet*

Metal Corner Bead

There are 196 linear feet of outside corners for the interior walls that need corner bead. We'll use 8-foot lengths.

$$\frac{196}{8} = 24.5 \text{ (rounded to 25) pieces}$$

We'll need 25 pieces of 8-foot metal corner bead, totaling 200 linear feet.

> *Enter your estimate of the linear feet of **corner bead** on line 12.2 of your Cost Estimate Worksheet. We'll use 200 linear feet of corner bead in our Example Cost Estimate Worksheet.*

Nails

We've determined that 6 pounds of 1¼-inch bright ring shank nails are needed per 1,000 square feet of wallboard. Let's find the number of pounds we'll need for the 10,016 square feet of wallboard in our house. Set up the following proportion:

$$6 \text{ lbs nails} : 1{,}000 \text{ sf} = X \text{ lbs nails} : 10{,}016 \text{ sf}$$

$$\text{Product of extremes} = \text{Product of means}$$

$$6 \text{ lbs nails} \times 10{,}016 \text{ sf} = 1{,}000 \text{ sf} \times X \text{ lbs nails}$$

$$X = \frac{6 \text{ lbs nails} \times 10{,}016 \text{ sf}}{1{,}000 \text{ sf}}$$

$$X = 60.01 \text{ (rounded to 60) lbs}$$

We'll need 60 pounds of 1¼-inch bright ring shank nails.

> *Enter your estimate for 1¼-inch bright ring shank **nails** on line 12.3 of your Cost Estimate Worksheet. We'll use 60 pounds of 1¼-inch bright ring shank nails in our Example Cost Estimate Worksheet.*

Adhesive

One 10.5-ounce cartridge of adhesive is needed for every 170 square feet of wallboard, using a ⅛-inch bead on 16-inch on-center framing. Set up a proportion to estimate the amount of adhesive we'll need for the interior walls and ceilings of our house.

$$1 \text{ cartridge} : 170 \text{ sf} = X \text{ cartridges} : 10{,}016 \text{ sf}$$

$$\text{Product of extremes} = \text{Product of means}$$

$$1 \text{ cartridge} \times 10{,}016 \text{ sf} = 170 \text{ sf} \times X \text{ cartridges}$$

$$X = \frac{1 \text{ cartridge} \times 10{,}016 \text{ sf}}{170 \text{ sf}}$$

$$X = 58.92 \text{ (rounded to 59) cartridges}$$

We'll need 59 10.5-ounce cartridges of adhesive.

> *Enter your estimate for the number of cartridges of **adhesive** on line 12.4 of your Cost Estimate Worksheet. We'll enter 59 cartridges of adhesive in our Example Cost Estimate Worksheet.*

Tape

We determined that one 500-foot roll of tape is required per 1,300 square feet of wallboard. Set up a proportion to find the number of rolls of tape (X) we'll need for 10,016 square feet of wallboard.

$$1 \text{ roll} : 1{,}300 \text{ sf} = X \text{ rolls} : 10{,}016 \text{ sf}$$

$$\text{Product of extremes} = \text{Product of means}$$

$$1 \text{ roll} \times 10{,}016 \text{ sf} = 1{,}300 \text{ sf} \times X \text{ rolls}$$

$$X = \frac{1 \text{ roll} \times 10{,}016 \text{ sf}}{1{,}300 \text{ sf}}$$

$$X = 7.70 \text{ (rounded to 8) rolls}$$

We'll need 8 (500-foot) rolls of drywall tape.

> *Enter your estimate for number of rolls of **tape** on line 12.5 of your Cost Estimate Worksheet. We'll use 8 rolls of tape in our Example Cost Estimate Worksheet*

Joint Compound

One 5-gallon can of joint compound is required per 1,000 square feet of wallboard. How many cans of compound will we need for the interior walls and ceilings of our house? Set up the following proportion:

$$1 \text{ can} : 1{,}000 \text{ sf} = X \text{ cans} : 10{,}016 \text{ sf}$$

$$\text{Product of extremes} = \text{Product of means}$$

$$1 \text{ can} \times 10{,}016 \text{ sf} = 1{,}000 \text{ sf} \times X \text{ cans}$$

$$X = \frac{1 \text{ can} \times 10{,}016 \text{ sf}}{1{,}000 \text{ sf}}$$

$$X = 10.01 \text{ (rounded to 10) 5-gallon cans}$$

We'll need 10 (5-gallon) cans of joint compound.

> *Enter your estimate for the number of 5-gallon cans of **joint compound** on line 12.6 of your Cost Estimate Worksheet. We'll use 10 5-gallon cans of joint compound in our Example Cost Estimate Worksheet.*

Estimating Labor

Now we need to estimate the labor for hanging the wallboard, installing the corner bead, applying the adhesive, and finishing the joints with tape and compound. Hanging wallboard and finishing joints costs about 30 percent more for ceilings than for walls. We'll use separate entry lines in the Cost Estimate Worksheet for each accessory and type of installation, based on square feet or linear feet of material installed.

> *Enter your **labor** costs in section 12.8 of your Cost Estimate Worksheet — the worksheet will automatically transfer the material quantities. We'll use the calculated quantities of wallboard, metal corner bead, adhesive and tape, and National Construction Estimator labor costs in our Example Cost Estimate Worksheet.*

COST ESTIMATE WORKSHEET FOR INTERIOR WALL AND CEILING FINISH

#	Item	Qty	Size		Cost	Per			Subtotal
12.1	**Gypsum wallboard**								
		10,016 sf							
		sf							
		sf							
		10,016 sf		@	$0.33	sf	=	$3,305.28	
12.2	**Metal corner bead**								
		200 lf		@	$0.54	lf	=	$108.00	
12.3	**Nails**								
		60 lbs	1¼" ring shank	@	$9.08	5 lb	=	$108.96	
12.4	**Adhesive**								
		59 cartridges	10.5 ounce	@	$2.69	each	=	$158.71	
12.5	**Tape**								
		8 rolls	500 ft	@	$3.40	roll	=	$27.20	
12.6	**Joint compound**								
		10 cans	5 gallon	@	$12.70	5 gal	=	$127.00	
12.7	**Other material**						=	$0.00	
	(List on separate sheet and enter cost here)								
			Subtotal				=	$3,835.15	$3,835.15
			Sales tax	@	7.75	%			= $297.22
			Cost of material						= $4,132.37
12.8	**Labor**								
	Gypsum wallboard - wall		7,166 sf	@	$0.56	sf	=	$4,012.96	
	Gypsum wallboard - ceiling		2,850 sf	@	$0.74	sf	=	$2,109.00	
	Metal corner bead		200 lf	@	$0.84	lf	=	$168.00	
	Adhesive		10,016 sf	@	$0.09	sf	=	$901.44	
	Tape - wall joints		7,166 sf	@	$0.31	sf	=	$2,221.46	
	Tape - ceiling joints		2,850 sf	@	$0.40	sf	=	$1,140.00	
			Cost of labor				=	$10,552.86	$10,552.86
	Cost of wall and ceiling finish (entered on line 12 of Form 100)								**$14,685.23**

Chapter 13
Exterior Trim

THE EXTERIOR TRIM OF a house is installed at the same time the electrical, plumbing, heating, and air conditioning are roughed-in and the interior walls are being finished. Exterior trim materials should be weather-resistant, and able to hold paint and retain their appearance for many years. Ideally, the number of joints should be kept to a minimum.

Windows and exterior doors should be in place before the interior walls are finished, but they're estimated with the exterior trim.

Exterior Trim Materials

The trim materials we'll be discussing in this chapter are:

- Windows, including weatherstripping, locks, and window pulls
- Exterior doors, including frames, weatherstripping, trim, hardware and locks
- Siding, including accessories and nails
- Fascia, frieze and rake boards
- Soffit and porch ceilings
- Porch column posts
- Flashing

Windows

Refer to the blueprints for the number, type, and size of windows for the building. Windows can be framed with wood, vinyl, or metal (or a combination). They can be double-hung, casement, awning, sliding, fixed, or a combination of fixed and sliding. The glass can be single pane with storm windows and screens, or dual-pane (even triple-pane) insulated glass with screens. Check the specifications for this information.

Exterior Doors

The blueprints and specifications give the number, type, and size of the doors required. The frame for the main entrance door is often a special size and design, so be sure to note that on your estimate. Doors may be hollow or solid-core or insulated and metal-clad. Sliding doors are framed with wood, metal or vinyl, and may be single pane or insulated glass. Screens are normally included with sliding doors.

Siding

Siding comes in many styles, shapes, patterns, and materials. It may be used as the only covering on the exterior of the house, with a brink veneer, or you may have a combination of two or more types of siding. Several combinations are possible. The most commonly used siding materials are wood, hardboard, aluminum, and vinyl.

Choose good quality siding that holds paint or one that has a permanent finish, like vinyl or anodized aluminum. Be careful when choosing your material; money saved by purchasing inexpensive siding is wasted if the material's appearance deteriorates quickly.

A few of the most common siding styles are:

- Beveled and/or lapped
- Board and batten
- Shiplap
- Tongue and groove
- Panel
- Wood shingles

Some siding is designed specifically for horizontal installation, while others must be installed vertically. Hardboard siding comes in strips and panels. It's made of wood chips and fiber, machine-pressed together to form a durable material.

Lapped siding is 3/8-inch or 7/16-inch thick, and comes in widths up to 12 inches. It's available in lengths ranging from 8 to 16 feet, in increments of 2 feet. Panel siding is 1/4-, 3/8-, or 7/16-inch thick, and can be purchased in 4-foot-wide pieces of various lengths; the longest is usually 16 feet.

Wood Siding Estimating Factors	
Nominal width (inches)	**Multiply net wall area by:**
Wood bevel siding (lapped or rabbetted)	
8"	1.34
10"	1.26
12"	1.21
Tongue and groove	
8"	1.37
10"	1.33
12"	1.31
Vertical board siding	
8"	1.10
10"	1.08
12"	1.07

Figure 13-1
Wood siding factors

Aluminum siding comes in V-groove (vertical), board and batten, and beveled styles. The paint on aluminum siding is baked on, making it very durable.

Vinyl siding is a product of modern science. It never needs painting, and is easily cleaned with soap and water. It can be molded into various shapes and given a number of textures.

Redwood and cedar are the most common woods used for siding due to their durability and desirable appearance. The styles most used for wood siding are beveled, board and batten, shiplap, and tongue and groove. Sidewall shingles are generally manufactured from select cedar logs.

To estimate any type of siding, calculate the wall area, including gables. As with interior measurements, don't deduct openings of less than 50 square feet. The extra material is used to offset waste. Refer to the wall sections and floor plans of the blueprints for dimensions. After the area is calculated and large openings deducted, use the information following to determine the material quantities you'll need.

Wood Siding

Use the factors in Figure 13-1 to calculate the amount of material you'll need for the various types of wood siding.

Aluminum and Vinyl Siding

Aluminum and vinyl siding are also sold by the square. Divide your net wall area by 100 and add the waste factor recommended by the manufacturer. Round up for the number of squares required.

Panel Siding

Divide your net wall area by the area covered by one sheet, and round up to find the number of sheets you'll need. For example:

$$\frac{1{,}200\text{ sf}}{32\text{ sf per sheet}} = 37.50\text{ (rounded to 38) } 4 \times 8\text{ sheets}$$

Cedar Sidewall Shingles

Cedar shingles come in 16-, 18-, and 24-inch lengths, and are sold by the square. Due to the nature of wood graining, no two cedar shingles are exactly alike. Cedar shingles are extremely durable and require no painting or staining.

One square of shingles with the following exposures covers 100 square feet:

- 5-inch exposure for 16-inch-long shingles
- 5½-inch exposure for 18-inch-long shingles
- 7½-inch exposure for 24-inch-long shingles

Divide your net wall area by 100 for the number of squares you'll need (using the recommended exposures), then multiply by the number of bundles per square, and round up.

Let's calculate how many bundles of shingles we'd need to order to cover 1,265 square feet of wall area with 24-inch cedar shingles having an exposure of 7½ inches. We know that one square of these shingles will cover 100 square feet, and our shingles are packed 4 bundles to a square.

Divide the wall area by 100 and multiply by the number of bundles per square (4, in this case).

$$\frac{1{,}265 \text{ sf}}{100 \text{ sf per square}} = 12.65 \text{ squares}$$

$$12.65 \text{ squares} \times 4 \text{ bundles per square} = 50.6 \text{ (rounded up to 51) bundles}$$

If you don't use the recommended exposures, consult the manufacturer's specification sheet to determine how many square feet are covered by one square.

The number of accessory pieces (outside and inside corners, etc.) you need to order will vary with the type and size of siding. Consult the manufacturer's specifications for accessory piece requirements.

Fascia, Frieze and Rake Boards

A *fascia* is the vertical board nailed on the ends of rafters or trusses, see Figure 13-2. The fascia is part of the cornice (the part of the roof that projects from the wall). The *frieze* is also a vertical board, but it's under the cornice or porch ceiling, adjacent to the wall or beam (shown in Figure 13-3).

Figure 13-4 shows the *rake* and the *rake board*. The rake is the sloping edge of a gable roof, and the *rake board* is part of the gable trim.

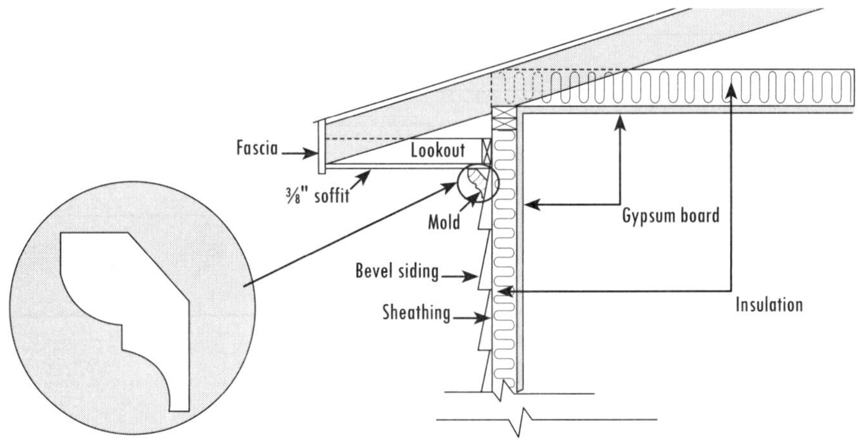

Figure 13-2
Siding and trim

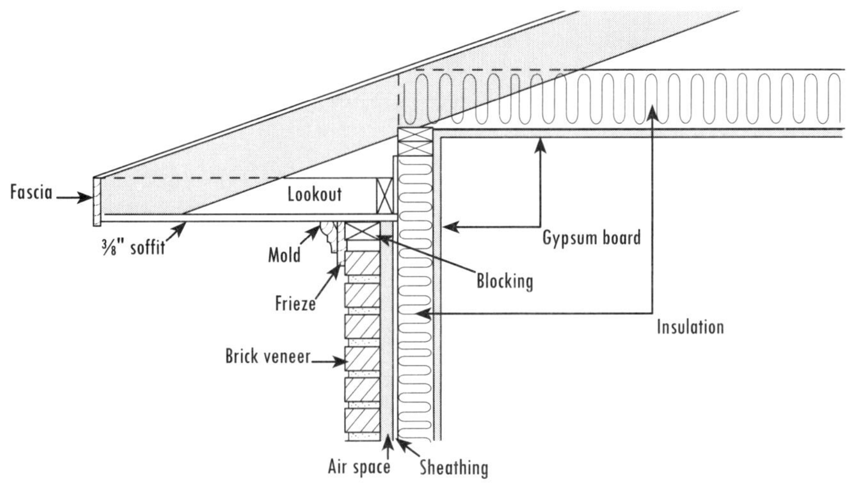

Figure 13-3
Brick veneer and trim

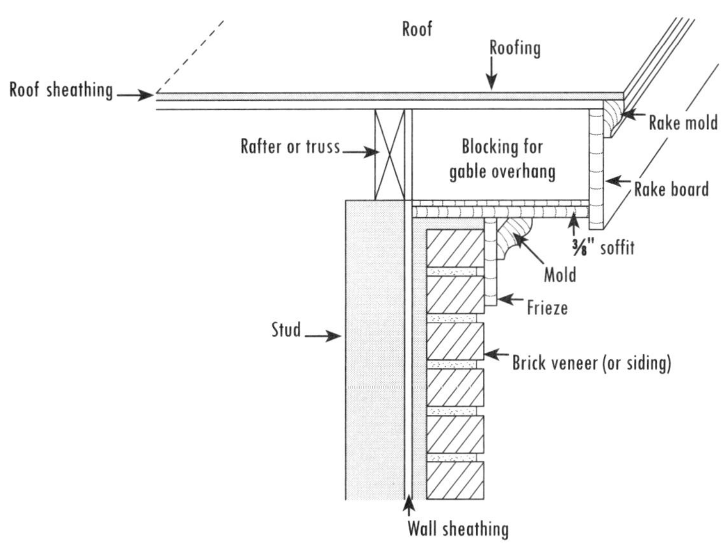

Figure 13-4
Gable overhang and trim

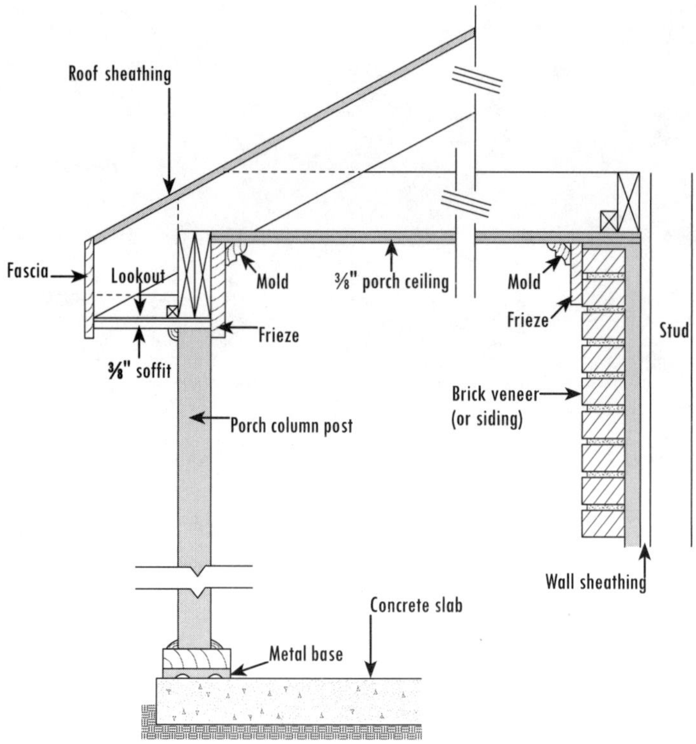

Figure 13-5
Porch trim

The elevations and cornice detail on the blueprints give the size and grade of the material required to estimate fascia, frieze, and rake boards. Calculate the linear feet (include porches) of each member. Add 10 percent for waste to the total footage of each member and round up to the next multiple of 10 feet.

Let's look at an example. If the blueprints show a 1- × 6-inch fascia requirement of 136 linear feet, how much fascia board should we estimate? Remember to add 10 percent for waste, and round up to the next multiple of 10.

$$136 + 13.60 \text{ (10 percent)} = 149.60$$

Rounding up, we'd estimate 150 linear feet of 1- × 6-inch fascia boards.

Soffit and Porch Ceilings

The most commonly used material for constructing soffits and porch ceilings is A-C exterior grade plywood. There are a couple of reasons why it's favored. First, it resists warping. And second, it's inexpensive to install and maintain — only one side is visible, so only that side needs finishing. The most common thicknesses are ¼, ⅜, ½, ⅝, and ¾ inch. Figure 13-5 shows an example of a soffit and porch ceiling.

When estimating plywood for soffits and porch ceilings, calculate the area to be covered and divide by the number of square feet of each piece of plywood used. So, if our blueprints show that the soffits and porch ceilings on our house (296 square feet) are to be covered with 3⁄8-inch-thick, 4- × 8-foot A-C plywood, how many pieces of plywood should we estimate?

Divide the area to be covered by 32 (for 4 x 8 sheets) and round up to the next whole number.

$$\frac{296 \text{ sf}}{32 \text{ sf}} = 9.25 \text{ (rounded up to 10) pieces}$$

We'd estimate 10 pieces of 3⁄8-inch, 4 × 8 A-C plywood.

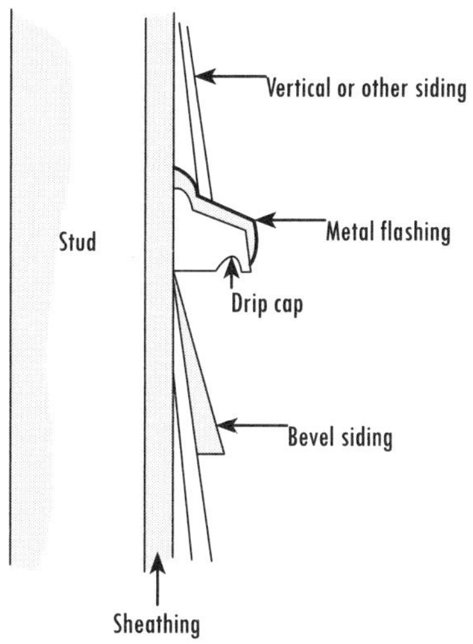

Porch Column Posts

Porch column posts can be made of wood, wrought iron, or other metal. The design, type, and spacing of posts are shown on the elevation section of the plans. Post spacing is determined by the architect to support a specific load. Since posts are weight-bearing, the span between them must never be increased from the plan specifications. Wood posts should rest on metal (or concrete) bases to avoid wood rot (see Figure 13-5).

Refer to the elevation section of the blueprints and specifications to get the number and size of posts you'll need for your estimate.

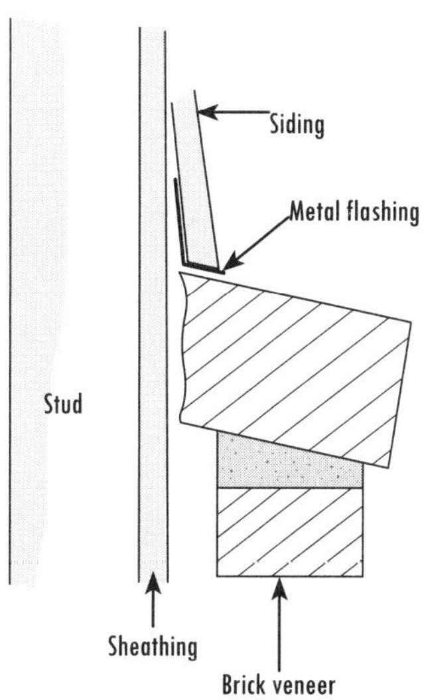

Figure 13-6
Intersection of different materials

Flashing

Flashing should be installed over all wall openings and intersections of different materials, as shown in Figure 13-6. Flashing is made of corrosion-resistant sheet metal (typically aluminum), which comes in 14-inch-wide rolls of 0.0175-inch-thick material, usually 10 feet long.

When estimating metal flashing in rolls, allow 7 inches for the width of all wall openings and intersections. Figure the total linear feet of flashing needed by using the dimensions on the plan's elevation section. If 14-inch aluminum flashing is selected, divide the total by 2 (for the 7-inch wall openings) to get the amount of 14-inch-wide flashing you'll need.

If, for example, we need 135 linear feet of flashing for wall openings and intersections, how many rolls of 14-inch-wide aluminum flashing would we estimate?

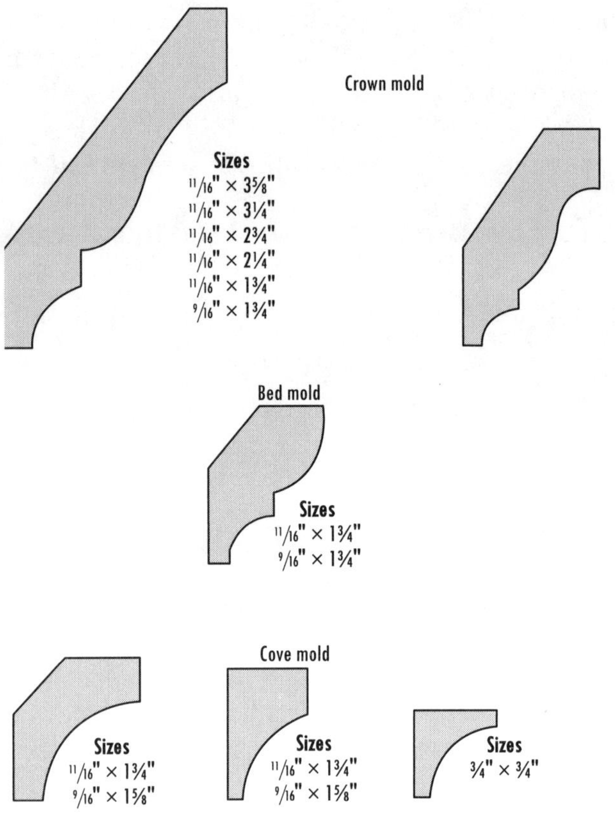

Figure 13-7
Crown, bed and cove molding

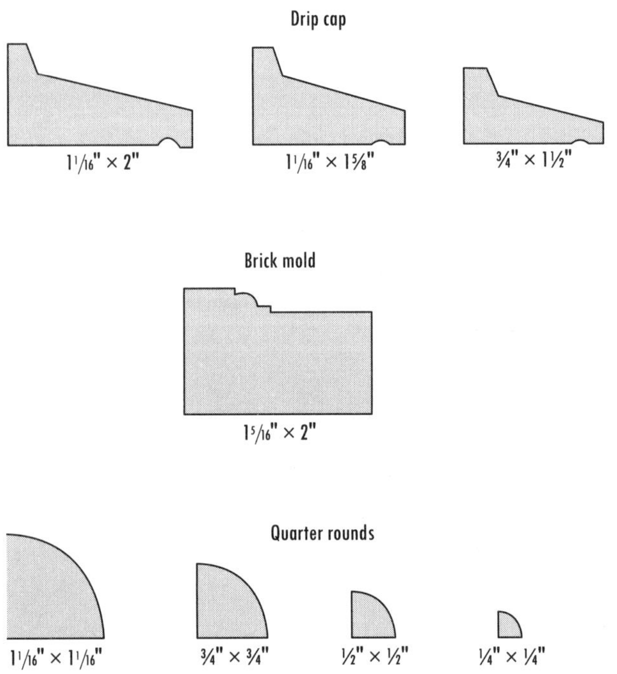

Figure 13-8
Drip cap, brick and quarter round molding

Divide the number of linear feet (135) by 2 to get the linear feet of 14-inch aluminum; then divide by 10, for the number of 10-foot rolls we need. Round up to the next whole number of rolls.

$$\frac{135 \text{ lf}}{2 \times 10'} = 6.75 \text{ (rounded up to 7) rolls}$$

We'd estimate 7 rolls of 14-inch aluminum flashing.

Moldings

Figures 13-7 and 13-8 show different types of moldings (and their sizes) commonly used for exterior trim. Moldings are estimated in linear feet. Allow 10 percent for waste, and round up to the next multiple of 10 feet.

Nails

Nails used for exterior trim should be corrosion-resistant, aluminum or galvanized steel. Figure 13-9 shows nail estimates for different applications.

Exterior Trim Labor

The labor to install siding can vary significantly, depending on the type of siding used. When estimating, take into account the height of the walls to be sided. Setting up and dismantling scaffolding takes time, and needs to be included in the labor costs. Here again, the most accurate labor estimates come from your previous job records. In the absence of those records, you can use the following guidelines to help you estimate.

- Window and door frames: 2 manhours per unit
- Fascia boards: 14 manhours per 100 linear feet

Nail Quantity Estimates	
Nail size	**Quantity**
Horizontal siding	
6d	6 lbs per 1,000 square feet
8d	7 lbs per 1,000 square feet
Panel siding	
6d	17 lbs per 1,000 square feet
8d	21 lbs per 1,000 square feet
Cedar shake siding	
6d	1 lb per square
8d	1 ½ lbs per square
Soffit and porch ceiling	
6d	1 ¼ lbs per 100 square feet
8d	1 ½ lbs per 100 square feet
Cornice	
6d	1 lb per 100 linear feet
8d	2 lbs per 100 linear feet

Figure 13-9
Estimating nail quantities

- Rake boards: 16 manhours per 100 linear feet
- Porch ceilings, including the frieze boards and molding: 10 manhours per 100 square feet

Estimating Our Exterior Trim

Let's do a materials estimate for the exterior trim on our sample house (look back to Chapter 7, Figures 7-1, 7-2, and 7-9).

Windows

The windows, including insulated glass, window screens, and all trim are listed in Figure 10-11 (refer back to Chapter 10). No storm windows are needed. We'll use the *National Construction Estimator* costs for our estimate at the end of this chapter.

> *Enter your estimate for* ***windows*** *in section 13.1 of your Cost Estimate Worksheet. We'll use the windows listed in Chapter 10, Figure 10-11 in our Example Cost Estimate Worksheet.*

Exterior Doors

All exterior doors (except the basement door) are also listed in Figure 10-11. They're prehung and include weatherstripping, hardware, locks, and trim. We'll use mahogany 6-panel doors, with no storm doors.

> *Enter your estimate for* ***exterior doors*** *on line 13.2 of the Cost Estimate Worksheet. We'll use the list of exterior doors in Chapter 10, Figure 10-11 in our Example Cost Estimate Worksheet.*

Siding

All walls and gables on the second floor, plus the front and back porch gables, will have 0.040-inch vinyl siding, with a 4-inch traditional lap profile. Remember, vinyl siding is sold by the square (100 square feet). The total area to be covered is 1,176 square feet. Calculate the number of squares we need for our estimate as follows:

$$\frac{1{,}176 \text{ sf}}{100 \text{ sf (1 square)}} = 11.76 \text{ (rounded to 12) squares of 0.040" vinyl siding}$$

> *Enter your estimate for **siding** on line 13.3 of your Cost Estimate Worksheet. We'll use 12 squares of 0.040-inch vinyl siding in our Example Cost Estimate Worksheet.*

Fascia, Frieze and Rake Boards

We'll need 193 linear feet of 1- × 6-inch lumber for the fascia boards. Allowing 10 percent for waste, the total is 212.30 linear feet [193 + 19.30 (10 percent for waste) = 212.30]. Rounding up to the next multiple of 10 gives us 220 linear feet of 1- × 6-inch fascia boards.

There are 176 linear feet of 1- × 6-inch frieze boards. Adding 10 percent for waste, the total is 193.60 linear feet [176 + 17.60 (10 percent for waste) = 193.60]. Rounding up to the next multiple of 10 we get 200 linear feet, which is what we'll estimate.

There are 137 linear feet of 1- × 6-inch rake boards. Allowing 10 percent for waste, the total is 150.70 linear feet [137 + 13.70 (10 percent for waste) = 150.70]. Rounding up to the next multiple of 10 gives us 160 linear feet.

Our totals are:

Fascia boards	220 linear feet
Frieze boards	200 linear feet
Rake boards	160 linear feet

> *Enter your estimate for **fascia, frieze, and rake boards** on lines 13.4.1 through 13.4.3 of your Cost Estimate Worksheet. We'll use these numbers in our Example Cost Estimate Worksheet.*

Soffit and Porch Ceilings

We'll use 3/8-inch, 4 × 8 A-C exterior plywood for the soffit and porch ceilings. There are 804 square feet to be covered. Divide the square footage by 32 square feet (for 4 × 8s) to find the number of pieces we'll need for our estimate:

$$\frac{804 \text{ sf}}{32 \text{ sf}} = 25.13 \text{ (rounded to 26) pieces}$$

We'll estimate 26 pieces of 3/8-inch 4 × 8 A-C exterior plywood.

> *Enter your estimate for **soffit and porch ceiling** plywood on line 13.5 of your Cost Estimate Worksheet. We'll use 26 pieces of 3/8 -inch 4 × 8 A-C exterior plywood in our Example Cost Estimate Worksheet.*

Porch Column Posts

Two column posts and bases (see Figure 13-5) are called for on the blueprints and specifications. In our estimate, we'll specify two colonial-style wood posts, 8 inches round by 8 feet tall.

> *Enter your estimate for **column posts** on line 13.7 of your Cost Estimate Worksheet. We'll use two 8-inch × 8-foot column posts in our Example Cost Estimate Worksheet.*

Molding

The molding shown between the soffit and siding in Figure 13-2, between the soffit and frieze board (Figures 13-3 and 13-4), and for the porch ceiling and frieze boards shown in Figure 13-5, is 9/16- × 1¾-inch bed molding. Figure 13-7 shows an illustration of bed molding. We need 360 linear feet of bed molding. Adding 10 percent for waste, the amount of material we need is 396 linear feet [360 + 36 (10 percent) = 396]. Rounding up to the next multiple of 10, we'll estimate 400 linear feet of bed molding.

> *Enter your estimate for **bed molding** on line 13.6.1 of your Cost Estimate Worksheet. We'll use 400 linear feet of 9/16- × 1¾-inch bed molding in our Example Cost Estimate Worksheet.*

We also need a total of 96 linear feet of drip cap for the horizontal siding joint at the gables, see Figure 13-6. No allowance for waste is necessary. Rounding up to the next multiple of 10 gives us an estimate of 100 linear feet of 1 1/16- × 1 5/8-inch drip cap mold (shown in Figure 13-8).

> *Enter your estimate for **drip cap mold** on line 13.6.2 of your Cost Estimate Worksheet. We'll use 100 linear feet of 1 1/16- × 1 5/8 -inch drip cap mold in our Example Cost Estimate Worksheet.*

We need 136 linear feet of ¾-inch quarter round rake mold, as shown in Figure 13-4. Rounding up, the next multiple of 10 is 140 linear feet (no allowance is made for waste).

> *Enter your estimate for **rake mold** on line 13.6.3 of your Cost Estimate Worksheet. We'll use 140 linear feet of ¾-inch quarter round for the rake mold in our Example Cost Estimate Worksheet.*

Garage Doors and Openers

Two 9- by 7-foot raised-panel steel rollup garage doors are listed on the blueprints and specs. They come equipped with automatic door openers and the necessary hardware and locks, as well as two remote-controlled door openers. They'll be installed by the local garage door supplier. His installation is included in our estimate of the cost of the doors and openers.

> *Enter your estimate for **garage doors and openers** on line 13.8 of your Cost Estimate Worksheet. We'll use two 9- × 7-foot garage doors with openers in our Example Cost Estimate Worksheet.*

Flashing

We need 91 linear feet of 7-inch flashing for the openings and intersections (Figure 13-6) shown on the blueprints and specs. We'll use 14-inch-wide × 0.0175-inch-thick aluminum, supplied in 10-foot rolls. The 14-inch strip will be cut in half to make 7-inch flashing.

Divide 91 linear feet by 2 to determine the linear feet of 14-inch aluminum we need. Then, divide that number by 10 for the number of 10-foot rolls we need.

$$\frac{91 \text{ lf}}{2 \times 10'} = 4.55 \text{ (rounded up to 5) rolls}$$

We'll estimate 5 rolls of 14-inch × 0.0175-inch aluminum flashing.

> *Enter your estimate for the number of 14-inch-wide rolls of **flashing** on line 13.9 of your Cost Estimate Worksheet. We'll use 5 rolls of 14-inch flashing in our Example Cost Estimate Worksheet.*

Nails

We'll use rust-resistant nails for all exterior trim. We'll break down the quantities needed for each trim type as follows.

Horizontal Siding

We need 6 pounds of 6d aluminum nails per 1,000 square feet of siding. We estimated 1,176 square feet of horizontal panel siding. Set up the following proportion:

$$6 \text{ lbs nails} : 1{,}000 \text{ sf} = X \text{ lbs nails} : 1{,}176 \text{ sf}$$

$$\text{Product of extremes} = \text{Product of means}$$

$$6 \text{ lbs nails} \times 1{,}176 \text{ sf} = 1{,}000 \text{ sf} \times X \text{ lbs nails}$$

$$X = \frac{6 \text{ lbs nails} \times 1{,}176 \text{ sf}}{1{,}000 \text{ sf}}$$

$$X = 7.06 \text{ (rounded up to 8) pounds of 6d aluminum nails}$$

Cornice

We estimated 580 linear feet of fascia, frieze, and rake boards. We'll need 1 pound of 8d galvanized nails for each 100 linear feet of board. Set up the following proportion:

$$1 \text{ lb nails} : 100 \text{ lf} = X \text{ lbs nails} : 580 \text{ lf}$$

$$\text{Product of extremes} = \text{Product of means}$$

$$1 \text{ lb nails} \times 580 \text{ lf} = 100 \text{ lf} \times X \text{ lbs nails}$$

$$X = \frac{1 \text{ lbs nails} \times 580 \text{ lf}}{100 \text{ lf}}$$

$$X = 5.8 \text{ (rounded to 6) pounds of 8d galvanized nails}$$

Soffit and Porch Ceiling

At 1½ pounds per 100 square feet, calculate how many 8d galvanized nails we'll need to install 832 square feet of soffit and porch ceiling material using the following proportion:

$$1\tfrac{1}{2} \text{ lbs nails} : 100 \text{ sf} = X \text{ lbs nails} : 832 \text{ sf}$$

$$\text{Product of extremes} = \text{Product of means}$$

$$1\tfrac{1}{2} \text{ lbs nails} \times 832 \text{ sf} = 100 \text{ sf} \times X \text{ lbs nails}$$

$$X = \frac{1\tfrac{1}{2} \text{ lbs nails} \times 832 \text{ sf}}{100 \text{ sf}}$$

$$X = 12.48 \text{ (rounded to 13) pounds of 8d galvanized nails}$$

Nail Totals

The total amount of nails estimated for the exterior trim are:

Horizontal siding	8 pounds of 6d aluminum nails
Cornice	6 pounds of 8d galvanized nails
Soffit and porch ceiling	13 pounds of 8d galvanized nails

We'll need a total of 8 pounds of 6d aluminum nails and 19 pounds of 8d galvanized nails.

> *Enter your estimate of the amount of **nails** required in Section 13.10 of your Cost Estimate Worksheet. We'll use the above amounts in our Example Cost Estimate Worksheet.*

Labor

Our labor records show a manhour factor of 0.140 for installing exterior trim on the first floor (with little or no scaffolding), and a manhour factor of 0.187 for the second floor (with scaffolding). Multiply by the total square feet of siding plus the total square feet of soffit and porch ceilings to get a reasonably accurate labor estimate. The manhour factor for the second floor includes building and dismantling the scaffolding.

We're using 2,008 square feet of material (1,176 square feet of siding and 832 square feet of soffit and porch ceiling). Since the first floor was finished with brickwork, most of the work is on the second floor, which will require scaffolding. At a manhour factor of 0.187, installing 2,008 square feet of material would take 375 manhours.

> **Note:** Cold weather can increase manhours by 10 percent or more.

Two carpenters and one laborer will be assigned to install the exterior trim. We'll estimate the manhours for each workman as follows:

$$\frac{\text{Total manhours}}{\text{Number of workmen}} = \frac{375}{3} = 125 \text{ manhours each}$$

> *Enter your estimate of the manhours of labor required in section 13.12 of your Cost Estimate Worksheet. We'll use 125 hours for each of the 3 workmen in our Example Cost Estimate Worksheet.*

COST ESTIMATE WORKSHEET FOR EXTERIOR TRIM

#	Qty	Item	Size		Cost	Per			Subtotal
13.1	**Windows**								
	1	horizontal sliding, Low E	108" × 62"		$447.00	each	=	$447.00	
	1	horizontal sliding, Low E	84" × 36"		$336.00	each	=	$336.00	
	1	horizontal sliding, Low E	72" × 48"		$268.00	each	=	$268.00	
	1	horizontal sliding, Low E	60" × 48"		$173.00	each	=	$173.00	
	5	double-hung, Low E	36" × 50"		$163.00	each	=	$815.00	
	3	horizontal sliding, Low E	36" × 24"		$105.00	each	=	$315.00	
								$2,354.00	$2,354.00
13.2	**Exterior doors**								
	1	mahogany, 6-panel	64" × 80"		$648.00	each	=	$648.00	
	3	mahogany, 6-panel	32" × 80"		$349.00	each	=	$1,047.00	
								$1,695.00	$1,695.00
13.3	**Siding**								
	12	squares 0.040" vinyl		@	$59.40	square	=	$712.80	
								$712.80	$712.80
13.4	**Fascia, frieze and rake boards (1" x 6")**								
13.4.1	220	lf fascia		@	$0.83	lf	=	$182.60	
13.4.2	200	lf frieze		@	$0.83	lf	=	$166.00	
13.4.3	160	lf rake		@	$0.83	lf	=	$132.80	
								$481.40	$481.40
13.5	**Soffit and porch ceiling**								
	26	pieces exterior plywood	⅜" × 4' × 8'	@	$15.80	piece	=	$410.80	
								$410.80	$410.80
13.6	**Molding**								
13.6.1	400	lf bed	9/16" × 1¾"	@	$0.71	lf	=	$284.00	
13.6.2	100	lf drip cap	1 1/16" × 1 5/8"	@	$2.12	lf	=	$212.00	
13.6.3	140	lf rake	¾" quarter round	@	$1.05	lf	=	$147.00	
								$643.00	$643.00
13.7	**Porch column posts and bases, wood**								
	2	column posts, colonial	8" × 8'	@	$134.00	piece	=	$268.00	
								$268.00	$268.00

COST ESTIMATE WORKSHEET FOR EXTERIOR TRIM, cont.

#	Qty	Item	Size		Cost	Per			Subtotal
13.8	**Garage doors and openers**								
	2	doors, raised-panel steel rollup	9' × 7'	@	$517.70	each	=	$1,035.40	
	2	openers, remote, screw-drive	½ HP	@	$300.20	each	=	$600.40	
								$1,635.80	$1,635.80
13.9	**Flashing, aluminum 0.0175"**								
	5	10-foot rolls	14" wide	@	$11.50	roll	=	$57.50	
								$57.50	$57.50
13.10	**Nails**								
	8	lbs	6d aluminum	@	$9.05	lb	=	$72.40	
	19	lbs	8d galvanized	@	$1.76	lb	=	$33.44	
								$105.84	$105.84
13.11	**Other material**								
	(List on separate sheet and enter cost here)							$0.00	$0.00
			Subtotal						$8,364.14
			Sales tax	@	7.75	%			$648.22
			Cost of material						$9,012.36
13.12	**Labor**								
	125	manhours - carpenter, BC		@	$32.09	hr	=	$4,011.25	
	125	manhours - carpenter, BC		@	$32.09	hr	=	$4,011.25	
	125	manhours - laborer, BL		@	$26.64	hr	=	$3,330.00	
								$11,352.50	$11,352.50
		Cost of exterior trim (entered on line 13 of Form 100)							**$20,364.86**

Chapter 14

Concrete Floors, Walkways and Terraces

CEMENT IS AS VERSATILE as it is durable. Cementitious materials were first put to use around 5600 BC. Aqueducts and other cement-bonded structures built by the Romans in the 8th century BC are still in good condition. The earth itself produced the first natural cement 12 million years ago, but little was known about the chemistry of cement until the mid-eighteenth century. Portland cement was introduced in 1824, when a patent was taken out for its manufacture. This material hardened into a yellowish-gray mass resembling the stone found in various quarries on the Isle of Portland, England, hence its name.

Concrete

Concrete is a mixture of fine and coarse aggregates surrounded and held together by hardened portland cement paste. These materials are inorganic, so concrete is resistant to decay, fire, termites, and rodents.

There's a direct link between the strength of concrete and the proportion of water used in the mix. Too much water and the paste will become thin and be weak and porous when it hardens. Cement made with the correct amount of water has strong binding qualities that hold the particles of aggregate firmly together, making a strong, dense concrete. Six gallons of water per sack of cement is the recommended proportion.

Concrete's compression strength is measured in pounds per square inch (psi). A six-bag mix (six bags of cement per cubic yard) is rated at 3,000 psi. A five-bag mix has the compression strength of around 2,500 psi. Any concrete exposed to traffic and weather needs to withstand pressure of at least 3,000 psi.

Estimating Concrete Quantities

Concrete pours in residential construction need to be coordinated with the carpentry work. For example, the basement floor must be poured before the basement stairs are built. Concrete is usually ordered by the cubic yard and delivered to the jobsite in a ready-mix truck. A cubic yard of concrete weighs approximately 4,000 pounds, or 2 tons.

To determine the cubic yards of concrete you need, calculate the cubic area of the pour (length × width × depth), and divide by 27 for cubic yards. For example, let's estimate the concrete you need for a garage floor that measures 24- by 26- feet and will have a 4½-inch concrete slab base. Multiply the garage length by the width and the desired depth of concrete to get the volume in cubic feet. Then divide by 27 to determine cubic yards.

$$24' \times 26' \times 0.375' \ (4\tfrac{1}{2}") = 234 \text{ cf}$$

$$\frac{234 \text{ cf}}{27} = 8.67 \text{ cy}$$

To be precise when estimating concrete quantities, use estimating tables. However, for a quick, reliable estimate (but not as precise), you can use the table in Figure 14-1. Multiply the length by the width for the area to be filled. Find the next highest number in the left hand column of the table, and follow that row to the right until it intersects the column for concrete thickness you need.

The garage area in our example is 624 square feet (24 × 26). Using the table in Figure 14-1, the next highest number in the left column is 650 (outlined). It intersects the 4½ inch or 0.375 foot column (shaded) at 9.03 cubic yards (outlined). This is reasonably close to the 8.67 cubic yards we calculated initially.

You can also use the concrete factors found in Figure 14-2 to determine concrete quantities. Using the appropriate concrete factor for 4½ inches from Figure 14-2, we can determine how many cubic yards of concrete we should estimate for the same garage area. Multiply the square footage by 0.01389 (4½ inch concrete factor):

$$624 \text{ sf} \times 0.01389 = 8.67 \text{ cy}$$

Concrete Accessories

There's more to concrete work than meets the eye. Environmental factors can make it necessary to use additional materials, reinforcing, or supplemental treatments in order to ensure maximum concrete strength.

Crushed Stone

When concrete is poured in areas prone to dampness (like floors, basements, or garages), you should place 4 inches of stone under the slab

Area	Thickness in Inches and Decimal Equivalents of a Foot				
Square Feet	3" 0.250'	3½" 0.292'	4" 0.333'	4½" 0.375'	5" 0.417'
5	0.05	0.05	0.06	0.07	0.08
10	0.09	0.11	0.12	0.14	0.15
20	0.19	0.22	0.25	0.28	0.31
30	0.28	0.32	0.37	0.42	0.46
40	0.37	0.43	0.49	0.56	0.62
50	0.46	0.54	0.62	0.69	0.77
60	0.56	0.65	0.74	0.83	0.93
70	0.65	0.76	0.86	0.97	1.08
80	0.74	0.87	0.99	1.11	1.24
90	0.83	0.97	1.11	1.25	1.39
100	0.93	1.08	1.23	1.39	1.54
150	1.39	1.62	1.85	2.08	2.32
200	1.85	2.16	2.47	2.78	3.09
250	2.31	2.70	3.08	3.47	3.86
300	2.78	3.24	3.70	4.17	4.63
350	3.24	3.79	4.32	4.86	5.41
400	3.70	4.33	4.93	5.56	6.18
450	4.17	4.87	5.55	6.25	6.95
500	4.63	5.41	6.17	6.94	7.72
550	5.09	5.95	6.78	7.64	8.49
600	5.56	6.49	7.40	8.33	9.27
650	6.02	7.03	8.02	9.03	10.04
700	6.48	7.57	8.63	9.72	10.81
750	6.94	8.11	9.25	10.42	11.58
800	7.41	8.65	9.87	11.11	12.36
850	7.87	9.19	10.48	11.81	13.13
900	8.33	9.73	11.10	12.50	13.90
950	8.80	10.27	11.72	13.19	14.67
1000	9.26	10.81	12.33	13.89	15.44

Figure 14-1
Cubic yard content

(shown in Figure 14-3) to minimize the hydrostatic pressure. Crushed stone is sold by both the cubic yard and the ton. To estimate crushed stone, calculate the cubic yards required (length × width × depth in feet, then divide by 27). Convert to pounds by multiplying the cubic yards by 2,700 pounds and dividing by 2,000 to get tons (2,000 pounds per ton). Figure 14-2 provides some shortcut factors you can use for calculating crushed stone quantities. Let's look at an example.

Thickness in Inches						
3"	3½"	4"	4½"	5"	5½"	6"
0.00926	0.01080	0.01235	0.01389	0.01543	0.01698	0.01852

Concrete

Area in Square Feet × Factor for Thickness = Cubic Yards

Example 695 sf @ 4" thickness
695 sf × 0.01235 factor = 8.58 cy

Example 1,065 sf @ 4½" thickness
1,065 sf × 0.01389 factor = 14.79 cy

Crushed Stone

Cubic Yards × 1.35 = Tons of Crushed Stone

Example 8.58 cy × 1.35 = 11.58 tons of crushed stone

Example 14.79 cy × 1.35 = 19.97 tons of crushed stone

Figure 14-2
Multiplying factors for cubic yards

A basement floor (42- × 28-feet) needs 4 inches of crushed stone under the slab. How many cubic yards of material should we estimate? Calculate the area in square feet, and multiply by the factor for 4 inches (0.01235) from Figure 14-2:

42' × 28' = 1,176 sf

1,176 sf × 0.01235 = 14.52 cy

Once you know the cubic yards, multiply the cubic yards by the factor found in the bottom section of Figure 14-2 (1.35) to convert cubic yards of crushed stone to tons:

14.52 cy × 1.35 = 19.60 tons

Vapor Retarder

Polyethylene film (4 or 6 mil) acts as a vapor retarder under concrete. It comes in various widths up to 40 feet, in rolls usually 50 feet long. Place it on the ground or stone immediately under the concrete slab. Calculate the area to be covered, and divide by the width of the roll to determine the length you need. Allow 10 percent for overlapping, and round up to the next whole number of rolls.

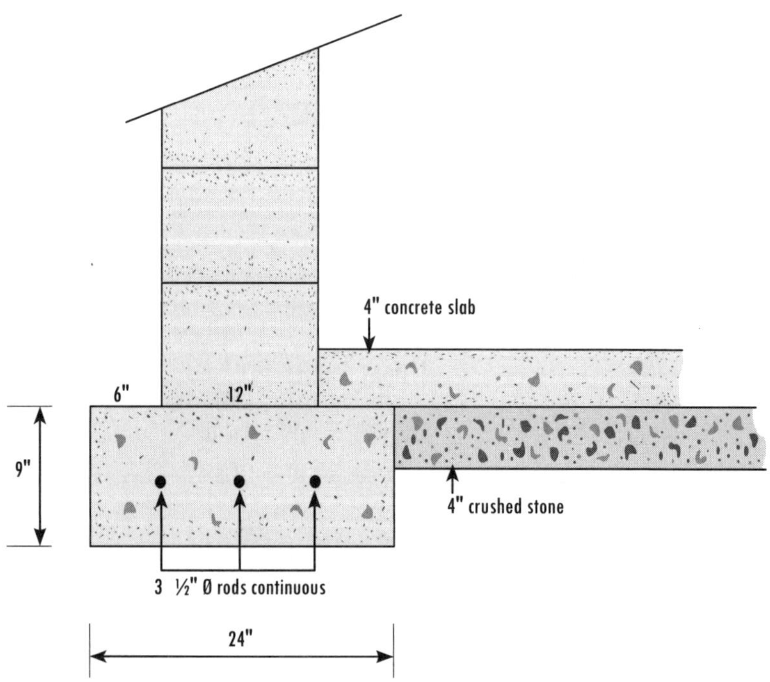

Figure14-3
Crushed stone under concrete slab

Reinforcing Rods		
Rod Size		Weight
Diameter in Inches	Rod Number	Pounds per Foot
¼	2	0.17
⅜	3	0.38
½	4	0.67
⅝	5	1.04
¾	6	1.50
⅞	7	2.04
1	8	2.67

Note: If the reinforcing rods are sold by weight, compute the total weight and cost as follows:
1. Multiply the total number of linear feet by the weight per foot

Example: The weight of 940 linear feet of ½" diameter rod is 629.80 lbs (940 x 0.67 = 629.80 lbs)
2. Total weight (rounded up) multiplied by the rate = cost

Figure 14-4
Reinforcing rods

Expansion Joints

We know that concrete shrinks as it cures, but the materials used in the production of concrete can affect its shrinkage in one way or another, too. The water content is by far the most important factor contributing to the strength or weakness of concrete. Restricted movement can cause cracking. If the concrete can move as it shrinks, cracking is less likely to occur. But if there's no room to move at the edges of a slab, you'll get a crack. Concrete will also crack when stress exceeds the tensile strength of the slab.

Cracking can be controlled by careful placement of *expansion joints*, areas where the concrete is free to expand and contract. The most common materials for expansion joints are asphalt, fiber, and asphalt-impregnated fiber. Various thicknesses (¼, ⅜, ½, ¾ and 1 inch) and widths (from 2 to 8 inches) of expansion joint material are available. They're estimated by the linear foot. Allow 1 manhour per 100 linear feet for installation.

Concrete Reinforcement

Welded wire fabric or reinforcing rods are used in slabs on grade to control cracking, but not to prevent it. If a reinforced slab cracks, the steel holds the cracks together so that loads can be transferred across the crack via the interlocked aggregate of the concrete.

Welded Wire Fabric Estimate welded wire fabric and wire mesh used for concrete reinforcement by the square foot. These materials are sold in rolls, typically 5 feet wide by 150 feet long (750 square feet), and are sized by the spacing and gauge of the wire.

Wire mesh labeled *6 × 6 : #10 × #10* or *6 × 6 : 10/10* means both the lengthwise and crosswise wires are spaced 6 inches on center, and both wires are 10 gauge. The mesh can be square or rectangular (4 × 8 : 8/12, etc). The 4 × 8 means the lengthwise spacing is 4 inches and crosswise spacing is 8 inches. The 8/12 means the gauge of the long wire is #8, with #12 gauge used for the cross wire.

Reinforcing Rods Reinforcing rods are designated by the number of eighths (⅛) in the rod diameter. Figure 14-4 shows rod sizes and weights. Reinforcing rods are used where heavy loading on the slab is expected, like

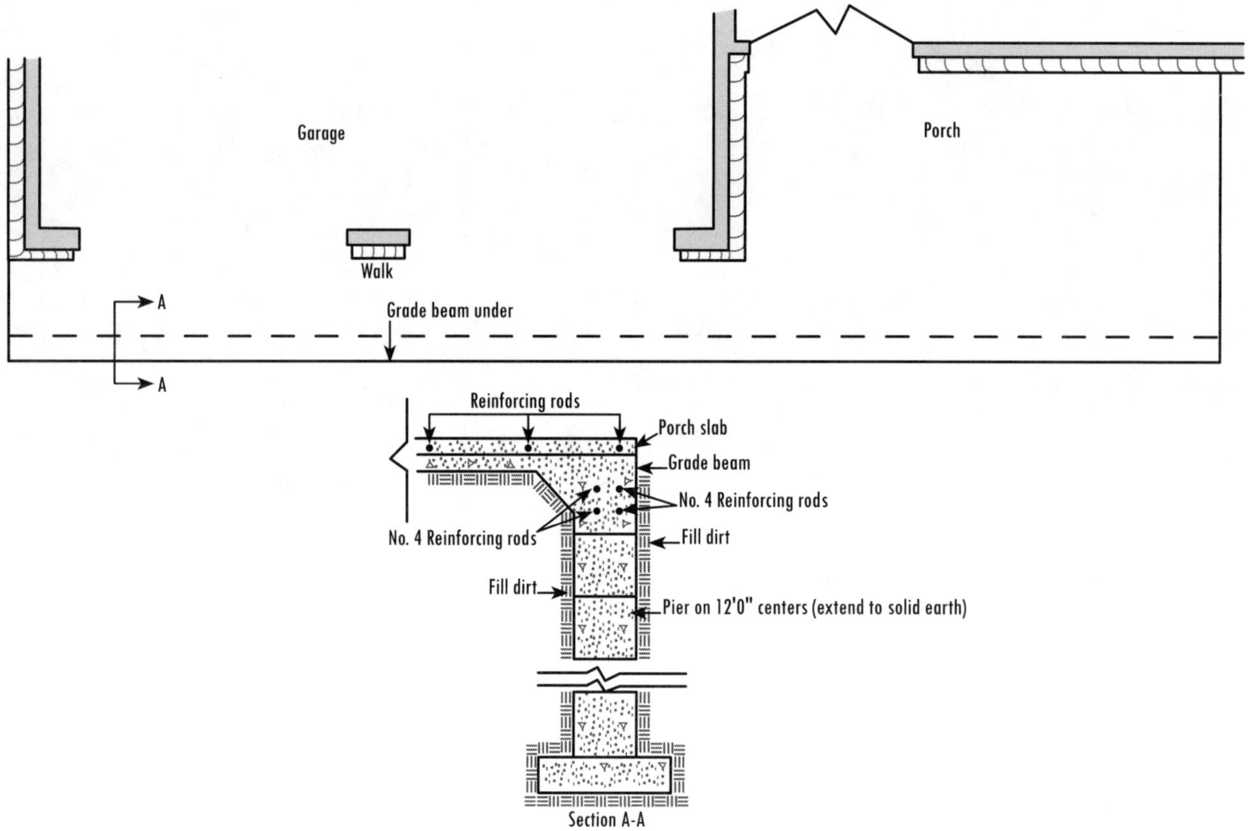

Figure 14-5
Grade beam

in a garage. Rods are also recommended when pouring over dirt fill because they're stronger than wire mesh, and they can be supported by masonry walls. In some instances, you may be required to install a grade beam extending down to solid earth to support the reinforcing rods, as shown in Figure 14-5. In this case, a walkway poured over fill that's adjacent to the structure needs reinforcing. Blueprints don't always show these grade beams, but don't forget to include them in your estimate.

A worksheet for estimating reinforcing rods for various slab dimensions is shown in Figure 14-6. A duplicate of this worksheet is included on the CD-ROM at the back of the book. There are calculations provided for reinforcing rods at both 16- × 16-inch and 12- × 12-inch on-center spacing. Calculate the reinforcing rods for your structure based on the dimensions noted on the blueprints.

Horizontal Rods and Traverse Rods @ 16" × 16"				
Width ____	× 0.75 less 1 =	0.00 × length 0.00	=	0.00 lf
Length ____	× 0.75 less 1 =	0.00 × width 0.00	=	0.00 lf
				0.00 lf
		Allow __% for overlapping		0.00
		Total		0.00 lf

Horizontal Rods and Traverse Rods @ 12" × 12"				
Width ____	less 1	= 0.00 × length 0.00	=	0.00 lf
Length ____	less 1	= 0.00 × width 0.00	=	0.00 lf
				0.00 lf
		Allow __% for overlapping		0.00
		Total		0.00 lf

Figure 14-6
Estimating reinforcing rod for slabs

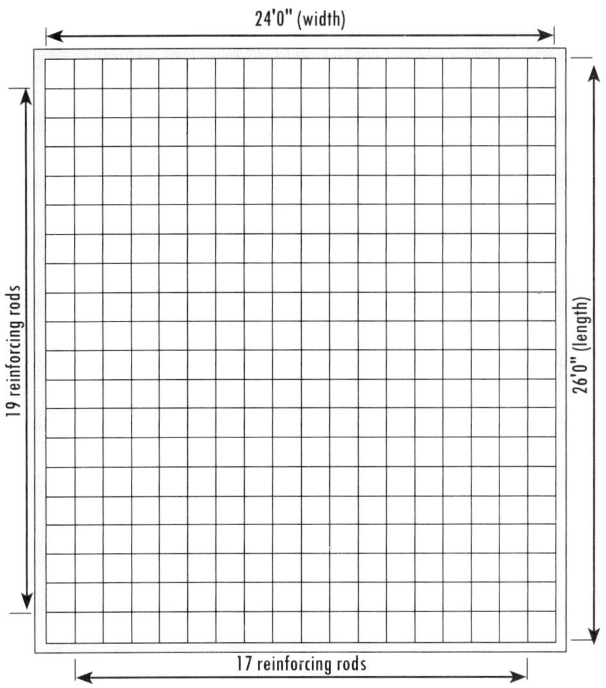

Figure 14-7
Reinforcing rods (16" × 16")

Use Figure 7-29 in Chapter 7 to convert inches and fractions to decimals of a foot. Enter the width and length of the slab, and the percentage of overlap specified for the reinforcing rods, which is typically 10 percent.

Figure 14-7 shows the number of reinforcing rods required for a 24- × 26-foot slab, with the rods at 16- × 16-inch on center spacing.

Tie wire for reinforcing rods is usually included in the cost of the rods. If not, allow 1 pound per 400 linear feet of rod.

Forms and Screeds

Forms protect and support concrete against hydrostatic pressure until it's sufficiently cured. Forms are used for porches, garages, walkways, terraces, and patio slabs. They should be braced with 2- × 2- × 18-inch stakes to prevent displacement, be rigid enough to support the concrete, and be tight enough to prevent concrete leakage. When curves are required, use plywood or hardboard that will bend to the proper radius and brace them as well.

Screeds, made of 1- × 3-inch material, act as thickness and leveling guides. Set them in place using a transit or level, and secure them with stakes. Allow 2 manhours per 100 square feet to build and remove forms and screeds.

Concrete Additives

You may need to include some of the following additives in your concrete estimate:

- Calcium chloride to accelerate the setting of concrete
- Air-entraining agents to improve the workability and durability of concrete, and increase its resistance to damage caused by frost
- Coloring agents

Cold Weather Pours

Pouring concrete during the winter in areas subject to freezing presents special problems. Concrete that freezes soon after placement gains very little strength and some permanent damage is done. This damage won't always be visible immediately. Surface scaling is easy to spot, but scaling an inch or more deep can take time to work its way to the surface. Frost action can also cause cracking, crumbling, or powdering.

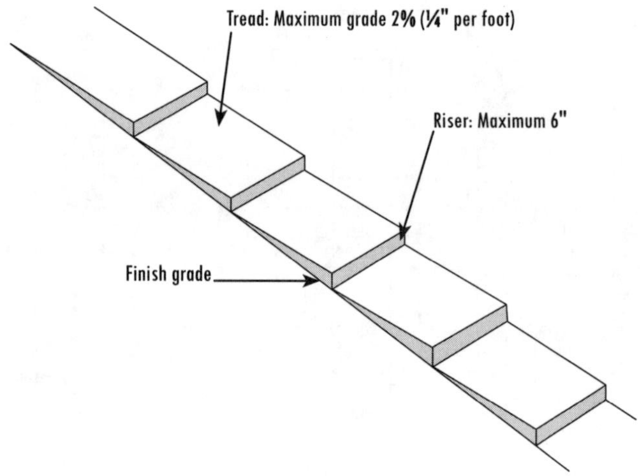

Figure 14-8
Stepped ramp with single risers

Before pouring concrete, frozen ground must be thawed. There are various methods you can use, from propane heaters to electric blankets to circulating hot liquid through pipes. A 2 percent calcium chloride additive in conjunction with air-entrainment additives in the concrete also provides cold weather protection. After pouring, cover the concrete with straw or Styrofoam to keep the new surface from freezing. You can also use temporary heaters along with polyethylene film to protect the concrete. Make sure there's adequate ventilation when using temporary fuel-burning heaters. These heaters produce carbon dioxide which, if allowed to mix with the calcium hydroxide present in fresh concrete, forms a thin, poorly-bonded layer of lime (calcium carbonate) on the surface that will "dust" under traffic.

Concrete will become stiff and difficult to finish if the ambient temperature is below 30 degrees Fahrenheit. If possible, delay exterior work such as sidewalks, driveways, patios and garage floors until the weather warms. In areas subject to freezing, walkways on a grade exceeding 5 percent (⅝ inch per foot) should be built with treads and risers (Figure 14-8). Ice on a sloping walk is an accident waiting to happen. You can expect concrete labor costs to increase about 5 to 15 percent during cold weather.

Finishing Concrete

Pouring and finishing concrete is precision work. Those who specialize in pouring concrete are faster and usually more economical to hire than workmen who only do it occasionally. Take that into consideration when you plan your pour. Also remember that weather and other conditions can delay a ready-mix concrete truck from unloading in the allotted time. Then, no matter how efficient your concrete crew is, you'll face a penalty charge.

The Material Estimate

Let's do a concrete materials estimate for our sample house. We'll figure each of the materials needed for our concrete pour separately by area: the basement, garage and utility area, front porch and walkway, and the terrace.

Estimating Concrete

We'll need to figure the square footage for each of the pour areas to determine how much concrete to estimate. Each area will require a 4-inch-thick concrete pour.

Basement

Our basement floor slab will cover 928.08 square feet, and be 4 inches thick. To determine how many cubic yard of concrete we should estimate for the basement, multiply 928.08 square feet by the thickness factor for 4 inches, from Figure 14-2.

$$928.08 \text{ sf} \times 0.01235 \text{ factor} = 11.46 \text{ cy}$$

Garage and Utility Area

The combined area of the garage and utility area is 669.90 square feet. The 4-inch concrete slab for this area will be poured over fill. To calculate the cubic yards of concrete, multiply 669.90 square feet by the thickness factor for 4 inches (0.01235) found in Figure 14-2.

$$669.90 \text{ sf} \times 0.01235 \text{ factor} = 8.27 \text{ cy}$$

Front Porch and Walkway

The front porch and walkway total 217.32 square feet (the porch is 12 feet × 10 feet and the walkway is 24.33 feet × 4 feet). The concrete for these areas will also be 4 inches thick, poured over fill. A grade beam (shown in Figure 14-5) is required for the walkway, which increases the quantity of concrete we'll need.

At 4 inches thick (using the 4-inch factor from Figure 14-2), we'll need to estimate the following amount of concrete for the 217.32-square-foot front porch and walkway area:

$$217.32 \text{ sf} \times 0.01235 = 2.68 \text{ cy}$$

The walkway grade beam will be 8 inches (0.67 feet) × 12 inches (1 foot) and run the length of the walkway and porch (24.33 feet + 12 feet). Calculate the area as follows, dividing by 27 to find the cubic yards:

$$\frac{0.67' \times 1.00' \times 36.33'}{27} = 0.90 \text{ cy}$$

We'll need 2.68 cubic yards of concrete for the front porch and walkway and an additional 0.90 cubic yards for the grade beam, giving us a combined total of 3.58 cubic yards of concrete.

Terrace

The dimensions of the proposed back terrace are 12 feet × 16.67 feet, which equal 200.04 square feet. The terrace will also be poured 4 inches thick.

Using the factor for 4 inches (0.01235) from Figure 14-2, we'll need the following amount of concrete:

$$200.04 \text{ sf} \times 0.01235 \text{ factor} = 2.47 \text{ cy}$$

Total Cubic Yards of Concrete

We'll need the following amount of concrete (at 3,000 psi) for the basement, garage and utility area, front porch and walk, and terrace:

Basement	11.46
Garage and utility area	8.27
Front porch and walk	3.58
Terrace	2.47
Total:	25.78 (rounded to 26) cubic yards

Enter your estimate for ***concrete*** *on line 14.1 of your Cost Estimate Worksheet. We'll use 26 cubic yards of concrete in our Example Cost Estimate Worksheet.*

Crushed Stone

The crushed stone for the basement was estimated back in Chapter 4. It was placed after the footings were poured and before the foundation walls were started. The front porch, walkway, and terrace will be poured over fill, and no crushed stone will be needed there.

The 4-inch slab in the garage and utility area will be poured over fill. No crushed stone is required for that area, either, so no more crushed stone will be estimated here.

Enter your estimate for ***crushed stone*** *on line 14.2 of your Cost Estimate Worksheet. We'll use 0 (zero) crushed stone in our Example Cost Estimate Worksheet.*

Wire Mesh

Wire mesh reinforcement usually comes in 750-square-foot rolls, and is estimated in whole rolls. We'll be using wire mesh for the basement and terrace. First we'll calculate how many rolls we need, then convert back to square feet by multiplying by the square feet per roll.

Basement

Our basement area is 928.08 square feet. We'll use 6 × 6 : 10/10 wire mesh for the concrete reinforcement. To find how many rolls of wire mesh we need, divide 928.08 square feet by the area of a roll of wire mesh (750 square feet).

$$\frac{928.08 \text{ sf}}{750 \text{ sf per roll}} = 1.24 \text{ rolls}$$

We'll need 1.24 rolls of 6 × 6 : 10/10 wire mesh for the basement.

Terrace

The terrace of our house is estimated at 200.04 square feet. How many 750-square-foot rolls of 6 × 6 : 10/10 wire mesh should we estimate for this area? Divide 200.04 square feet by the area of a roll of wire mesh (750 square feet):

$$\frac{200.04 \text{ sf}}{750 \text{ sf per roll}} = 0.27 \text{ roll}$$

We'll need 0.27 rolls of 6 × 6 : 10/10 wire mesh for the terrace.

Total Square Feet of Wire Mesh

Total the rolls for each area and add 15 percent for waste. Then multiply the number of rolls by 750 to find the total square feet of wire mesh we'll need.

Basement	1.24 rolls
Terrace	0.27 roll
Subtotal:	1.51 rolls
15 percent	0.23 roll
Total:	1.74 (rounded to 2) rolls

The total amount of 6 × 6 : 10/10 wire mesh that we need for the basement floor and terrace is two 750-square-foot rolls or 1,500 square feet.

> *Enter your estimate for **wire mesh** on line 14.3 of your Cost Estimate Worksheet. We'll use 1,500 square feet (2 rolls) of 6 × 6 : 10/10 wire mesh in our Example Cost Estimate Worksheet.*

Reinforcing Rod

We'll need #4 reinforcing rods for the garage and utility area, and the front porch and walkway, including the grade beam.

Garage and Utility Area

The combined garage and utility areas have been estimated at 669.90 square feet. We'll use #4 rods at 16 inches by 16 inches on center. They'll be supported by the foundation walls. We need to calculate how many linear feet we should estimate.

Using the top half of the worksheet in Figure 14-6, we can calculate the amount of rod required. Enter the width (22.33 feet) and the length (30.00

Horizontal Rods and Traverse Rods @ 16" × 16"	
Width 22.33 × 0.75 less 1 = 15.75 × length 30.00 =	472.50 lf
Length 30.00 × 0.75 less 1 = 21.50 × width 22.33 =	480.10 lf
	952.60 lf
Allow 10 % for overlapping	95.26
Total	1,047.86 lf

Figure 14-9
Estimating reinforcing rod for the garage/utility area

Horizontal Rods and Traverse Rods @ 16" × 16"	
Width 12.00 × 0.75 less 1 = 8.00 × length 10.00 =	80.00 lf
Length 10.00 × 0.75 less 1 = 6.50 × width 12.00 =	78.00 lf
	158.00 lf
Allow 10 % for overlapping	15.80
Total	173.80 lf

Figure 14-10
Estimating reinforcing rod for the front porch

feet) in decimal feet in the two underlined boxes, then do the calculations. The spreadsheet is also available on the accompanying CD-ROM. The correct result is shown in Figure 14-9.

We'll estimate 1,047.86 linear feet of reinforcing rod for the garage and utility area.

Horizontal Rods and Traverse Rods @ 16" × 16"	
Width 24.33 × 0.75 less 1 = 17.25 × length 4.00 =	69.00 lf
Length 4.00 × 0.75 less 1 = 2.00 × width 24.33 =	48.66 lf
	117.66 lf
Allow 10 % for overlapping	11.77
Total	129.43 lf

Figure 14-11
Estimating reinforcing rod for the walkway

Front Porch and Walkway

We previously estimated the area of the front porch (12 feet × 10 feet) and walkway (24.33 feet × 4 feet) at 217.32 square feet. How many linear feet of #4 reinforcing rod, at 16- × 16-inches on center (supported by the foundation wall and grade beam), should we estimate?

Again, use the worksheet in the top half of Figure 14-6 to calculate the linear feet of reinforcing rod required for each separate area. Using two separate spreadsheets, enter the widths (12.00 feet and 24.33 feet) and the lengths (10.00 feet and 4.00 feet) in decimals in the two underlined boxes, then Excel will do the math. The correct results are shown in Figures 14-10 and 14-11.

We'll estimate 303.23 (173.80 + 129.43) linear feet of reinforcing rod for the porch and walkway.

Now we need to figure how many linear feet of #4 reinforcing rod are required for the grade beam. Multiply the length of the grade beam by 4, which is the number of reinforcing rods needed (shown in Figure 14-5).

Grade beam length:
24.33′ + 12.00′ = 36.33′
Multiply length by 4:
36.33′ x 4 = 145.32 lf of reinforcing rod

Total Reinforcing Rod

The total linear feet of reinforcing rod we need is:

Garage and utility area	1,047.86
Front porch and walk	303.23
Grade beam	145.32
Total:	1,496.41 linear feet

Reinforcing bar comes in 20-foot lengths. Divide the total linear feet by 20 and round up to the next whole number to find how many bars you'll need. Then multiply that number of bars by 20 to find the linear feet to order.

$$\frac{1{,}496.41 \text{ lf}}{20 \text{ lf}} = 74.82 \text{ (rounded to 75) bars}$$

$$75 \times 20 \text{ lf} = 1{,}500.00 \text{ lf of reinforcing rod}$$

Enter your estimate for ***reinforcing rods*** *on line 14.4 of your Cost Estimate Worksheet. We'll use 1,500.00 linear feet of # 4 reinforcing rod in our Example Cost Estimate Worksheet.*

Expansion Joints

We need expansion joints in the concrete in the basement, garage and utility area, front porch, walkway, and terrace. The maximum spacing between these joints should be triple (in feet) the thickness of the concrete (in inches). For instance, a 4-inch slab needs an expansion joint at least every 12 feet.

We'll be using ½- × 4-inch expansion joint material for all of our expansion joints.

Basement

Our basement slab is 38.67 feet long and 24 feet wide. We'll need three expansion joints across the width, and one down the length. Calculate the linear feet of joint material as follows:

$$(24.00' \times 3) + 38.67' = 110.67 \text{ lf of expansion joint material}$$

Garage and Utility Area

The slab for the garage and utility area is 30 feet long and 22.33 feet wide. We'll need two expansion joints across the width, and one down the length. Calculate the linear feet of joint material for these areas as follows:

$$(2 \times 22.33') + 30.00' = 74.66 \text{ lf of expansion joint material}$$

Front Porch and Walkway

We won't need any expansion joints for the front porch because its dimensions (12 feet × 10 feet) are within the maximum requirements for a 4-inch-thick slab, which is one expansion joint every 12 feet.

The walkway, however, will need expansion joints at two points along its 24.33-foot length. The walkway is 4 feet wide, so we'll figure the joint material as follows:

$$2 \times 4.00' = 8 \text{ lf of expansion joint material}$$

Terrace

The back terrace is 12.00 feet × 16.67 feet. We'll need one expansion joint, requiring 12 linear feet of expansion joint material.

Total Expansion Joint Material

The total expansion joint material needed for the basement, garage and utility area, front porch and walkway, and the terrace is:

Basement	110.67
Garage and utility area	74.66
Porch and walkway	8.00
Terrace	12.00
Total:	205.33 (rounded to 206) linear feet

> *Enter your estimate for the linear feet of* ***expansion joint*** *material on line 14.5 of your Cost Estimate Worksheet. We'll use 206 linear feet of expansion joint material in our Example Cost Estimate Worksheet.*

Vapor Retarder

We'll use 4 mil polyethylene film as a vapor retarder under the concrete. We've already calculated these areas, so all we need to do is total our square feet.

Basement (38.67′ × 24.00′)	928.08
Garage and utility area (22.33′ × 30.00′)	669.90
Front porch (12.00′ × 10.00′)	120.00
Walkway (24.33′ × 4.00′)	97.32
Terrace (12.00′ × 16.67′)	200.04
Subtotal:	2,015.34
Add 15 percent for waste and overlaps	302.30
Total:	2,317.64 square feet

We'll use 14-foot-wide film in 50-foot rolls. That's 700 square feet per roll. Why not use 12-foot-wide rolls, since many of the widths we'll be covering are divisible by 12? Because that wouldn't allow for the overlap we need to keep moisture from leaking through the vapor retarder. For good coverage, it's better to go with the next width up — the 14-foot film.

Now, divide our 2,317.64 square feet by 700 square feet (the number of square feet in each roll), to find the number of rolls we need.

$$\frac{2{,}317.64 \text{ sf}}{700 \text{ sf per roll}} = 3.31 \text{ (rounded to 4) rolls}$$

We'll be ordering by the square foot, so we need to find how many square feet are in the 4 rolls:

4 x 700 sf = 2,800 sf of 4 mil polyethylene film

> *Enter your estimate for the square footage of **vapor retarder** on line 14.6 of your Cost Estimate Worksheet. We'll use 2,800 square feet of 4 mil polyethylene film in our Example Cost Estimate Worksheet.*

Forms and Screeds

All material for the forms and screeds, including stakes and nails, will come from surplus supplies, so no extra material needs to be estimated here. However, we'll still need to enter the cost of the labor required in the Labor Estimate section.

Optional Items

- Additives for concrete — none required
- Winter protection — none required

Labor Estimate for the Concrete

Concrete is typically delivered to the site by ready-mix truck, and placed directly into the forms by chute. Labor is based on the total cubic yards of concrete to be placed. This total was calculated for line 14.1 of the Cost Estimate Worksheet, and automatically transferred to line 14.11.1. You can use the labor rate from your records, or use the current *National Construction Estimator* rate.

> *Enter your **labor rate** per cubic yard for pumping concrete on line 14.11.1 of your Cost Estimate Worksheet. We'll use the rate from the National Construction Estimator in our Example Cost Estimate Worksheet.*

Concrete Finishing

The surface must be finished after the concrete is poured. Labor is based on the square footage of concrete to be finished. Since the concrete is uniformly 4 inches thick, we can divide the cubic yards by 0.33 feet (4 inches) and multiply by 27 to get the square footage of concrete to be finished. The number of cubic feet is automatically calculated on line 14.11.2 of the Cost Estimate Worksheet. Enter the slab thickness in feet (0.33 feet, in this case), along with your labor rate.

> *Enter your slab thickness and **labor rate** per square foot for **finishing concrete** on line 14.11.2 of your Cost Estimate Worksheet. We'll use 0.33' (4") and the National Construction Estimator rate in our Example Cost Estimate Worksheet.*

Wire Mesh

The labor for laying the wire mesh is based on the square feet of mesh to be laid. The total square footage was calculated for line 14.3 of the Cost Estimate Worksheet, and automatically transferred to line 14.11.3.

> *Enter your **labor rate** per square foot for laying **wire mesh** on line 14.11.3 of your Cost Estimate Worksheet. We'll use the National Construction Estimator rate in our Example Cost Estimate Worksheet.*

Reinforcing Rods and Tie Wire

The labor for laying and tying reinforcing rod is based on the linear feet of rod laid. That was calculated for line 14.4 of the Cost Estimate Worksheet and automatically transferred to line 14.11.4.

> *Enter your **labor rate** per linear foot for laying and tying **reinforcing rod** on line 14.11.4 of your Cost Estimate Worksheet. We'll use the National Construction Estimator rate in our Example Cost Estimate Worksheet.*

Expansion Joints

Labor for laying expansion joints is based on the quantity of material to be laid. This total was calculated for line 14.5 of the Cost Estimate Worksheet and automatically transferred to line 14.11.5.

> *Enter your **labor rate** per linear foot for laying **expansion joints** on line 14.11.5 of your Cost Estimate Worksheet. We'll use the National Construction Estimator rate in our Example Cost Estimate Worksheet.*

Vapor Retarder

Labor for laying vapor retarder film is based on the amount of polyethylene film material to be laid. This was calculated for line 14.6 of the Cost Estimate Worksheet and automatically transferred to line 14.11.6.

> *Enter your **labor rate** per square foot for laying vapor **retarder** on line 14.11.6 of your Cost Estimate Worksheet. We'll use the National Construction Estimator rate in our Example Cost Estimate Worksheet.*

Forms and Screeds

Labor to build and remove the forms and screeds is estimated at 4 manhours per 250 square feet of concrete slab. We have calculated the following square feet:

Basement (38.67′ × 24.00′)	928.08
Garage and utility area (22.33′ × 30.00′)	669.90
Front porch (12.00′ × 10.00′)	120.00
Walkway (24.33′ × 4.00′)	97.32
Terrace (12.00′ × 16.67′)	200.04
Total:	2,015.34 square feet

Our total area is 2,015.34 square feet. At 4 manhours per 250 square feet, it will take the following manhours to build and remove forms and screeds:

$$\frac{2{,}015.34 \text{ sf}}{250 \text{ sf}} \times 4 \text{ hrs} = 32.25 \text{ (rounded to 33) manhours}$$

One carpenter and one laborer will do this work. The manhours per workman will be:

$$\frac{33 \text{ manhours}}{2 \text{ workmen}} = 16.50 \text{ manhours each}$$

> *Enter your estimate of the **manhours** for **forms and screeds** on line 14.11.7 of your Cost Estimate Worksheet. We'll use 16.50 manhours for each of the two workmen in our Example Cost Estimate Worksheet.*

COST ESTIMATE WORKSHEET FOR CONCRETE FLOORS, FRONT PORCH, WALKWAYS AND TERRACES

#	Qty	Size	Cost Per		Amount	Subtotal
14.1	**Concrete**					
	26.00 cy	test 3,000 psi	@ $105.00 cy	=	$2,730.00	
14.2	**Crushed stone**					
	0.00 cy		@ $0.00 cy	=	$0.00	
14.3	**Wire mesh**					
	1,500.00 sf	6 × 6 : 10/10	@ $0.17 sf	=	$255.00	
14.4	**Reinforcing rods**					
	1,500.00 lf	# 4	@ $0.40 lf	=	$600.00	
14.5	**Expansion joints**					
	206.00 lf	½" × 4"	@ $0.69 lf	=	$142.14	
14.6	**Vapor retarder**					
	2,800.00 sf	4 mil	@ $0.08 sf	=	$224.00	
14.7	**Additives for concrete** (List on separate sheet and enter cost here)					
14.8	**Forms and screeds** (List on separate sheet and enter cost here)					
14.9	**Winter protection** (List any materials needed or cost of temporary heat to protect the concrete from freezing, and enter total cost here)					
14.10	**Other material** (List on separate sheet and enter cost here					
		Subtotal			$3,951.14	$3,951.14
		Sales tax	@ 7.75 %			$306.21
		Cost of material				$4,257.35
14.11	**Labor:**					
14.11.1	**Concrete** - pumping					
	26.00 cy		$15.59 cy	=	$405.34	
14.11.2	**Concrete** - finishing					
	702.00 cf	0.33 ft thick 2,127.27 sf	@ $0.43 sf	=	$914.73	
14.11.3	**Wire mesh**					
	1,500.00 sf		@ $0.15 sf	=	$225.00	
14.11.4	**Reinforcing rods**					
	1,500.00 lf		@ $0.46 lf	=	$690.00	
14.11.5	**Expansion joints**					
	206.00 lf		@ $0.88 lf	=	$181.28	
14.11.6	**Vapor retarder**					
	2,800.00 sf		@ $0.05 sf	=	$140.00	
14.11.7	**Forms and screeds**					
	16.50 manhours - carpenter, BC		@ $32.09 hour	=	$529.49	
	16.50 manhours - laborer, BL		@ $26.64 hour	=	$439.56	
		Total cost of labor			$3,525.40	$3,525.40
	Cost of concrete (entered on line 14 of Form 100)					**$7,782.75**

Chapter 15

Interior Trim

AFTER THE DRYWALL OR plaster is installed and the walls have had time to dry, interior trim work can begin. The floor underlayment and flooring must be in place before interior doors and baseboards are installed. The interior trim discussed in this chapter includes:

- Floor underlayment
- Flooring
- Interior doors with trim and hardware
- Window trim, finished
- Baseboard
- Base shoe
- Wall molding
- Paneling and molding
- Kitchen cabinets and countertops
- Vanities, bars and tops
- Closet shelves
- Permanent stairs
- Mirrors and medicine cabinets
- Tub and shower doors
- Bathroom accessories
- Miscellaneous materials

	Length in Feet																		
Width in Feet	2'	3'	4'	5'	6'	7'	8'	9'	10'	11'	12'	13'	14'	15'	16'	17'	18'	19'	20'
2'	4	6	8	10	12	14	16	18	20	22	24	26	28	30	32	34	36	38	40
3'	6	9	12	15	18	21	24	27	30	33	36	39	42	45	48	51	54	57	60
4'	8	12	16	20	24	28	32	36	40	44	48	52	56	60	64	68	72	76	80
5'	10	15	20	25	30	35	40	45	50	55	60	65	70	75	80	85	90	95	100
6'	12	18	24	30	36	42	48	54	60	66	72	78	84	90	96	102	108	114	120
7'	14	21	28	35	42	49	56	63	70	77	84	91	98	105	112	119	126	133	140
8'	16	24	32	40	48	56	64	72	80	88	96	104	112	120	128	136	144	152	160
9'	18	27	36	45	54	63	72	81	90	99	108	117	126	135	144	153	162	171	180
10'	20	30	40	50	60	70	80	90	100	110	120	130	140	150	160	170	180	190	200
11'	22	33	44	55	66	77	88	99	110	121	132	143	154	165	176	187	198	209	220
12'	24	36	48	60	72	84	96	108	120	132	144	156	168	180	192	204	216	228	240
13'	26	39	52	65	78	91	104	117	130	143	156	169	182	195	208	221	234	247	260
14'	28	42	56	70	84	98	112	126	140	154	168	182	196	210	224	238	252	266	280
15'	30	45	60	75	90	105	120	135	150	165	180	195	210	225	240	255	270	285	300
16'	32	48	64	80	96	112	128	144	160	176	192	208	224	240	256	272	288	304	320
17'	34	51	68	85	102	119	136	153	170	187	204	221	238	255	272	289	306	323	340
18'	36	54	72	90	108	126	144	162	180	198	216	234	252	270	288	306	324	342	360
19'	38	57	76	95	114	133	152	171	190	209	228	247	266	285	304	323	342	361	380
20'	40	60	80	100	120	140	160	180	200	220	240	260	280	300	320	340	360	380	400

Figure 15-1
Square feet in room (use for floor and ceiling)

Floor Underlayment

If an area is to be carpeted, tiled, or covered in resilient flooring (sheet vinyl), you may need to install an underlayment over the subfloor to give the flooring a smooth surface. This can be made of particleboard, plywood, or OSB. Attach the underlayment using waterproof glue anywhere moisture may be a possible problem. The most common underlayment in new construction is ⅝ -inch-thick sheets of 4 × 8s. Divide the square feet of area to be covered by the area of one sheet of underlayment (32 square feet for 4 × 8s) and round up to the next whole number to find the number of sheets you need. Figure 15-1 gives the square footage for various sized rooms.

Let's estimate how many pieces of ⅝-inch-thick 4 × 8 plywood underlayment are required for a 120 square foot kitchen. Divide the area by 32 and round up to the next whole number of sheets:

$$\frac{120 \text{ sf}}{32 \text{ sf per sheet}} = 3.75 \text{ (rounded to 4) sheets}$$

Nail floor underlayment 6 inches on center through intermediate supports, and 3 inches on center around the edges. Use 6d (2-inch) ring-grooved nails, allowing 2 pounds per 100 square feet. Labor to install the underlayment is estimated at 2 manhours per 100 square feet.

Wood Flooring

Wood flooring applied over a subfloor is manufactured from both hardwood and softwood trees. Generally, broad leaves are characteristic of hardwood trees and needles of softwoods. But those distinctions have nothing to do with the degree of hardness of the wood; they simply set the two types apart. Oak and maple are most frequently used for hardwood flooring, and southern pine and Douglas fir are the commonly used softwoods. Wood flooring comes in three forms: strip, plank, or block.

Strip Flooring

Strip flooring is available in 1½- and 2-inch widths, and in thicknesses of ⅜, ½, or 25/32 inch. For the 25/32-inch-thick strips, the widths are 1½, 2, 2¼ and 3¼ inches. The most commonly used strip flooring is 25/32-inch-thick × 2¼-inch-wide tongue and groove. Lengths of strip flooring vary from 1 foot to 16 feet.

To estimate 25/32 -inch × 2¼-inch tongue-and-groove flooring, multiply the square feet of area to be covered by a factor of 1.38, to allow for side and end matching and 5 percent for waste. This gives you the number of board feet required. Figure 15-1 shows the square footage for various room lengths and widths.

We'll be adding a factor (like 1.38 just mentioned) to cover the excess material needed to match ends and sides throughout the wood flooring section. The factors will vary, each pertaining to a specific material.

Wood flooring should be nailed 10 to 12 inches on center with 7d spiral flooring nails, or be machine-nailed with 2-inch power cleats. When installing 25/32 -inch × 2¼-inch flooring, allow 20 pounds of 7d spiral flooring nails, or 1 box of 2-inch power cleats (at 5,000 per box), per 1,000 square feet.

Plank Flooring

Plank flooring comes in random widths and has wood plugs of a contrasting color. This is meant to mimic the flooring used by settlers in the early American colonies. Once extremely popular, it's not often seen in today's new homes.

Block Flooring

Block or parquet flooring is strip flooring bonded together edgewise to form a square unit, most commonly 9 inches × 9 inches. The units usually have tongue-and-groove edges, and can be nailed to the subfloor or laid in mastic over concrete or wood.

Wood Flooring Finishing and Installation

Wood flooring comes either unfinished or with a factory finish. If unfinished, it must be sanded, filler applied, stained, and then have the finish applied. Finishing raw wood is usually done by specialists in woodworking. They price their labor and material by the square foot.

The labor to install wood flooring can vary from 4 manhours per 100 square feet (for $^{25}/_{32}$-inch × 2¼-inch strip flooring) to more than 10 manhours per 100 square feet (for genuine block or parquet flooring).

Interior Doors

When estimating interior doors, the floor plan or the door schedule show the number and size required, and the plans specify the type of doors to install.

Most common interior doors used today are 1⅜-inch-thick hollowcore flush or paneled doors. A complete door unit includes the frame, door, door butts, stops, casing on both sides, and a lock. Prehung doors are popular and widely used because they reduce labor costs. They come completely assembled in the frame, including door butts, stops, and casings. The unit may not include a lock, but the cross hole and bolt hole will be predrilled for a standard cylindrical lockset.

Install the interior door casing using 8d finish nails. Allow ½ pound of nails per opening. Setting one prehung door (installing and sanding the joints and the door lock) takes roughly 3 manhours. For other types of interior doors, like sliding or bi-fold, estimate 4 to 5 manhours per door to assemble and set the door frame, install the casing on both sides, and install the door and hardware.

Window Trim

The wood species and design for window side and head casings should match the door trim. Window trim typically consists of an apron, a casing, and a stool, with the apron and casing made of the same material. Metal windows normally need sills only; the drywall or plaster makes a return to the metal window, eliminating the need for casing and apron.

Window trim can be purchased by the linear foot, or may be included as part of the window set. If you purchase trim by the linear foot, add 20 percent for cutting and waste. Many manufacturers of wood windows make matching trim. If that's the case, it's usually packaged along with each window, and since all the parts fit together, no allowance for waste is needed.

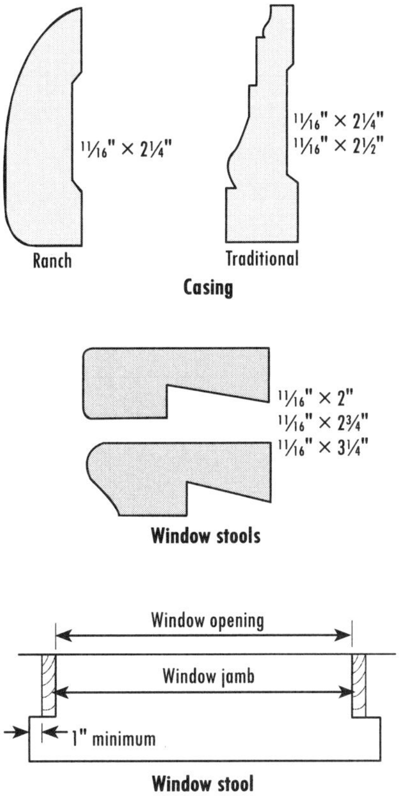

Figure 15-2
Window and door trim

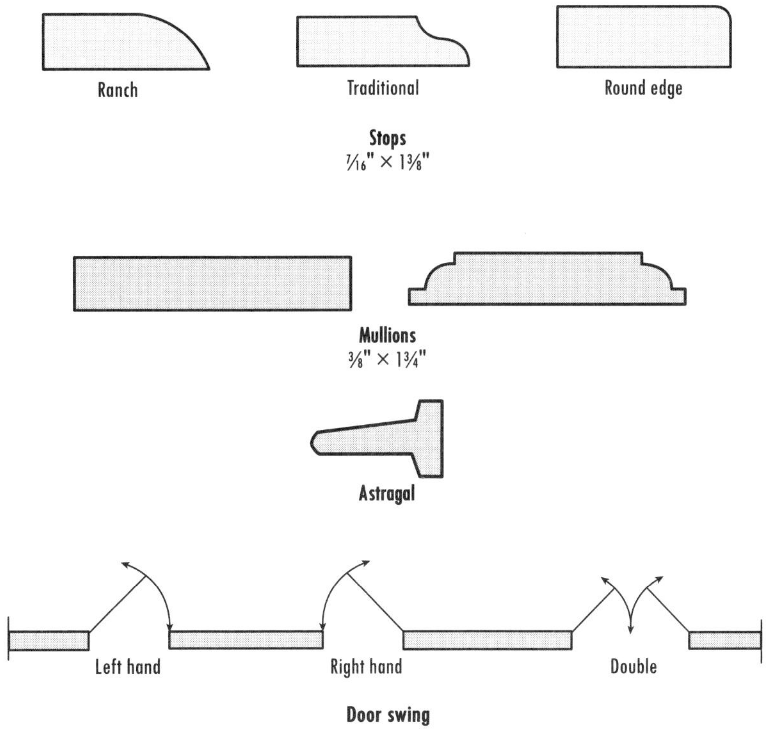

Figure 15-3
Window and door trim

The number and type of windows and the trim are on the plans, window schedule, and specs. Figures 15-2 and 15-3 show common window and door trim details.

Use 6d or 8d finishing nails for the window trim. All the joints should be glued, nailed, and sanded before being finished. Allow ¼ pound of nails for each side.

Installation labor for window trim can vary, depending on the size and type of window. A 3 × 3 single window unit takes about 50 percent less labor to trim than a large picture window. Your own cost records will provide the best estimate. In general, allow 3 manhours per window, regardless of size.

Baseboard and Base Shoe

A baseboard runs continuously around the floor perimeter of each room, hall, and closet, hiding the gap between the finished floor and the wall. Baseboard is usually the same wood as the window and door trim. Baseboard joints should be made over studs, which provide a secure nailing base. Add 10 percent for waste.

Width in Feet \ Length in Feet	2'	3'	4'	5'	6'	7'	8'	9'	10'	11'	12'	13'	14'	15'	16'	17'	18'	19'	20'
2'	8	10	12	14	16	18	20	22	24	26	28	30	32	34	36	38	40	42	44
3'	10	12	14	16	18	20	22	24	26	28	30	32	34	36	38	40	42	44	46
4'	12	14	16	18	20	22	24	26	28	30	32	34	36	38	40	42	44	46	48
5'	14	16	18	20	22	24	26	28	30	32	34	36	38	40	42	44	46	48	50
6'	16	18	20	22	24	26	28	30	32	34	36	38	40	42	44	46	48	50	52
7'	18	20	22	24	26	28	30	32	34	36	38	40	42	44	46	48	50	52	54
8'	20	22	24	26	28	30	32	34	36	38	40	42	44	46	48	50	52	54	56
9'	22	24	26	28	30	32	34	36	38	40	42	44	46	48	50	52	54	56	58
10'	24	26	28	30	32	34	36	38	40	42	44	46	48	50	52	54	56	58	60
11'	26	28	30	32	34	36	38	40	42	44	46	48	50	52	54	56	58	60	62
12'	28	30	32	34	36	38	40	42	44	46	48	50	52	54	56	58	60	62	64
13'	30	32	34	36	38	40	42	44	46	48	50	52	54	56	58	60	62	64	66
14'	32	34	36	38	40	42	44	46	48	50	52	54	56	58	60	62	64	66	68
15'	34	36	38	40	42	44	46	48	50	52	54	56	58	60	62	64	66	68	70
16'	36	38	40	42	44	46	48	50	52	54	56	58	60	62	64	66	68	70	72
17'	38	40	42	44	46	48	50	52	54	56	58	60	62	64	66	68	70	72	74
18'	40	42	44	46	48	50	52	54	56	58	60	62	64	66	68	70	72	74	76
19'	42	44	46	48	50	52	54	56	58	60	62	64	66	68	70	72	74	76	78
20'	44	46	48	50	52	54	56	58	60	62	64	66	68	70	72	74	76	78	80

*Note: No door openings are deducted

Figure 15-4
Perimeter of room in linial feet* (use for base and shoe)

Estimate baseboard in linear feet. Total the linear feet required for all areas; don't deduct door openings, to allow for waste. Figure 15-4 shows linear feet for various room lengths and widths.

Use base shoe to trim wood or resilient flooring. You can also use it to cover the perimeter of paneled walls, with or without baseboard. Base shoe isn't needed when you install wall-to-wall carpet. Estimate base shoe the same as baseboard. Figure 15-5 shows different types and sizes of baseboard and base shoe.

Use 8d finish nails to secure baseboard, allowing 1 pound per 100 linear feet. For base shoe, use 6d finish nails, estimating ½ pound per 100 linear feet. It takes roughly 4 manhours per 100 linear feet to install baseboard, including finishing the joints. For base shoe, estimate 2 manhours per 100 linear feet.

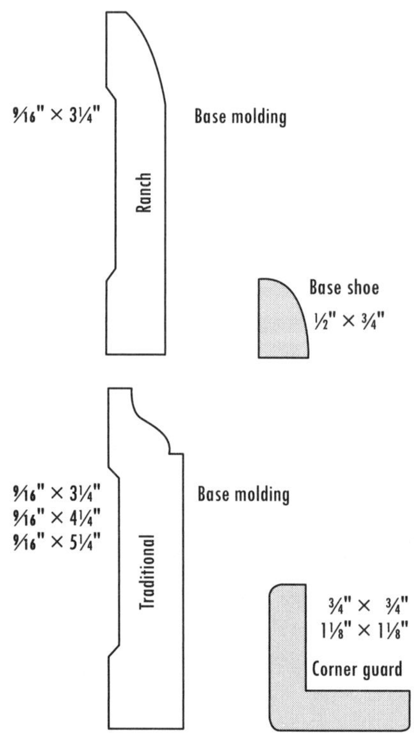

Figure 15-5
Base molding and outside corner guard

Wall Molding

Interior moldings are both decorative and functional. Baseboard, casing and chair rail serve to protect the walls and surfaces from damage. Panel molding and ceiling or crown

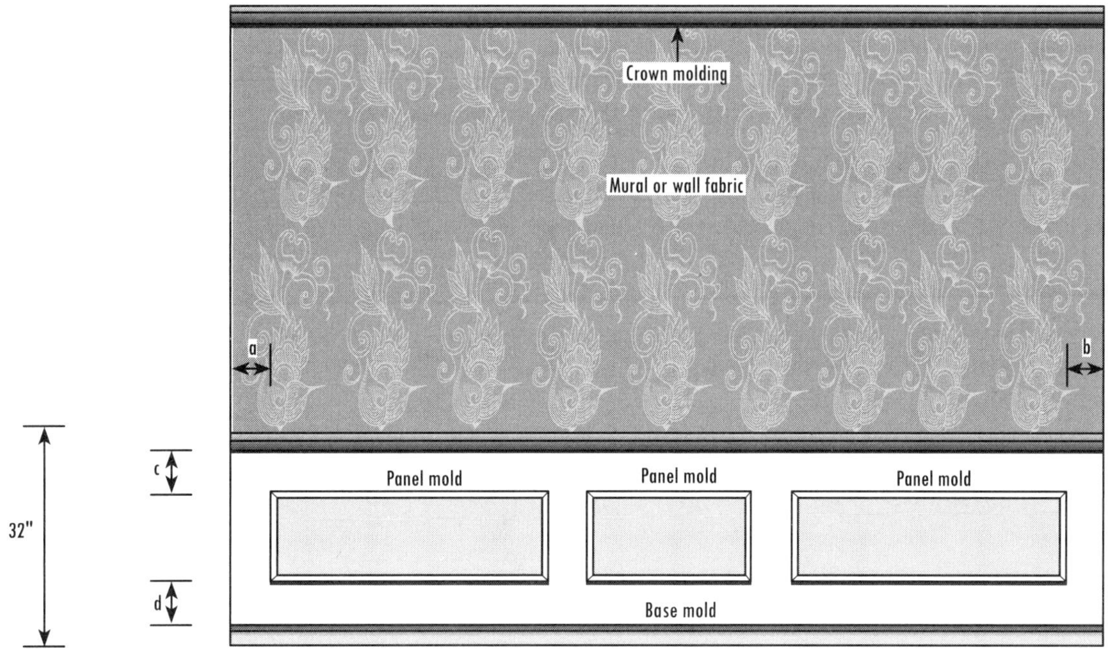

Legend:

(a) and (b)	Distance from inside wall corner to panel mold. *(Distance (a) and (b) should be equal.)*
(c)	Distance from bottom of chair rail to top of panel mold.
(d)	Distance from top of base mold to bottom of panel mold. *(Distance (c) and (d) should be equal.)*

Figure 15-6
Wall molding

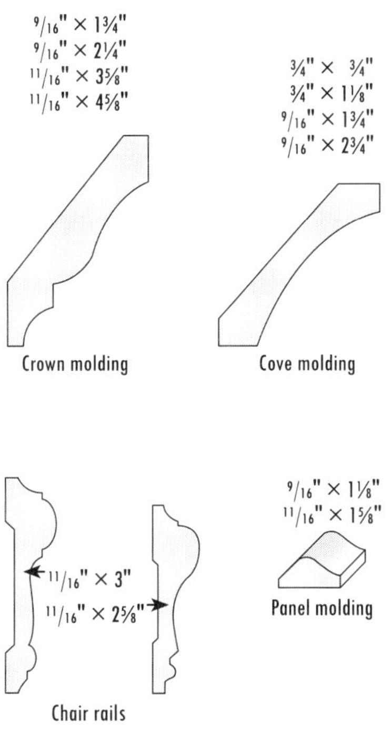

Figure 15-7
Wall molding shapes and sizes

molding add decorative detail and elegance to a room. Molding can be installed anywhere in a house, but it's most commonly found in living rooms, dining rooms, family rooms and foyers.

Figure 15-6 shows various molding applications. Nail chair rail (Figure 15-7) at studs, and use a good adhesive to complete the installation. Adhesive prevents the chair rail from separating from the wall between studs. The recommended height from the floor to the top of a chair rail is 32 inches. You can use chair rail on one wall or all the walls in a room; it's just a matter of preference. A decorative wallcovering between the chair rail and crown molding can be very effective.

Panel mold is generally installed between the chair rail and baseboard. Like chair rail, it should be attached with a good grade of adhesive and nailed at the studs. Panel mold is flexible, so designs can be customized to suit any room. Follow the guidelines in Figure 15-6 to give the design a professional look. As shown in the Figure, the distance from the inside corners of the room, (a) and (b), to the panel mold should be equal; it's also ususally equal between panels. If the chair rail meets a window or door casing between two opposing walls, the distance from this casing to the panel mold must also be equal to that spacing established from the wall corners to the panel mold (4 to 6 inches). The distance from the bottom of the chair rail to the top of the panel mold (c) should equal the distance from the top of the base mold to the bottom of the panel mold (d). This distance should also be the same as (a) and (b) for an overall balanced design. After the panel mold is installed, finish the wall inside the molding with paint, wallpaper, or decorative fabric.

Both chair rail and panel mold are estimated by the linear foot. It takes roughly 6 manhours per 100 linear feet to install chair rail and panel mold. Calculate the linear feet for the area where chair rail is to be installed, as well as the linear feet of design patterns for the panel mold, if they're indicated, and add 10 percent for waste.

Use 8d finish nails and adhesive for the chair rail, allowing 1 pound of nails per 100 linear feet. Use 6d finish nails along with adhesive for panel mold, allowing ½ pound of nails per 100 linear feet.

Paneling

Wood-paneled walls add warmth and beauty to a home. They also increase the strength of the wall. Paneling is most commonly made of hardboard, plywood, or solid wood.

Hardboard Paneling

Hardboard paneling comes in a variety of simulated wood grains. It's a smooth, hard, and dense material composed of wood fibers. The panels are available in various lengths, and come in thicknesses of 1/8 inch or ¼ inch. The most popular sizes are 4 × 8 and 4 × 10 feet.

Plywood Paneling

Plywood paneling is available in a wide variety of wood species, including aromatic cedar, ash, birch, cherry, maple, mahogany, red and white oak, and walnut. The standard panel is 4 × 8 feet, ¼ inch thick, with V-grooves at random widths, and a factory finish. They're also available in ⅛, ½, and ¾ inch thicknesses in some types of wood.

To estimate paneling in 4-foot widths, divide the perimeter of the room by 4 and round up. Use the table in Figure 15-4 to easily find the perimeter.

Let's estimate how many pieces of 4-foot-wide plywood paneling pieces we need for a 12- by 15-foot room, with a standard 8-foot-high ceiling. The total perimeter of this room is 54 linear feet. Divide by 4 and round up:

$$\frac{\text{54 lf}}{\text{4 ft per panel}} = \text{13.50 (rounded up to 14) pieces}$$

Paneling is typically nailed and glued to the studs. Use nails colored to match the paneling, which come in ¼-pound boxes, or 4d (1½-inch) finish nails. Estimate one box of colored nails or one pound of finish nails per four 4 × 8 panels. Estimate adhesive at one 10.5-ounce tube per 100 square feet of paneling, based on a ¼-inch bead and studs at 16 inches on center. Allow 6 percent for waste. Other bead sizes and stud spacing will have different costs, so be sure you have a reliable pricing source from which to make your estimates.

Inside corners are normally scribed. Prefinished inside corner molding is available to match the paneling. Prefinished outside corner molding or *corner guards*, shown in Figure 15-5, complete those particular areas. Corner molding is typically supplied in 8-foot lengths. To estimate the corner molding required, count the number of corners where they'll be used and multiply by the length of each piece.

Prefinished cove molding (used where the walls meet the ceiling) comes in precut lengths, but is sold by the linear foot. Determine the perimeter of the room, divide by the length of one piece, and round up. Then multiply the number of pieces by the piece length to find the number of linear feet you need.

Let's find how many linear feet of cove molding (in 10-foot lengths) we need to estimate for a 12- × 15-foot room. The perimeter is 54 linear feet. Divide by 10 (the length of 1 piece) and round up:

$$\frac{\text{54 lf}}{\text{10 ft per piece}} = \text{5.40 (rounded to 6) pieces}$$

Multiply the length of one piece by 6 to find the total linear feet to estimate:

$$6\ (\text{pieces}) \times 10'\ (\text{per piece}) = 60'$$

We need 60 linear feet of ceiling molding.

Solid Wood Paneling

Solid wood paneling is made from either hardwood or softwood. It comes in random widths and normally has tongue-and-groove joints. It's graded either *knotty* or *clear*. Knotty wood paneling has a rustic appearance, while clear paneling has a more formal look.

When estimating solid wood paneling, you need to use a factor of 1.25 to allow for waste and for tongue-and-groove installation. Multiply the wall area (in square feet) by 1.25. For the same 12- × 15-foot room, calculate the area like this:

$$(12' \times 8' \times 2) + (15' \times 8' \times 2) = 432\ \text{sf}$$
$$432\ \text{sf} \times 1.25 = 540\ \text{sf of wood paneling}$$

Install paneling using 6d finish or casing nails; allow 2 pounds per 100 square feet.

If the type of paneling material hasn't been selected by the time the estimate is made, allow a reasonable cost for the paneling. If the owner then selects a more expensive paneling, charge the difference to the owner at the time it's selected. If the paneling chosen is less expensive, you can give the owner a credit.

Labor to Install Paneling

To install hardboard or plywood paneling, labor is estimated at 3 manhours per 100 square feet. Estimate the labor for solid wood paneling at 6 manhours per 100 square feet. For molding installation, estimate 5 manhours per 100 linear feet.

Kitchen Cabinets

The kitchen detail section on the floor plan gives cabinetry information, including the material to be used and the type of countertop. If cabinets are to be installed in other areas — a wet bar, for instance — they'll be shown in a special detail section and described in the specs. If the owner doesn't know what specific cabinets he wants at the time you prepare the estimate, include a cabinetry allowance until he makes his selection.

There are two ways to estimate kitchen cabinetry: per running foot or by counting individual cabinets. The running foot method is easy and gives a fairly accurate estimate, unless you'll be installing specialty cabinets. The costs listed here are based on mill-fabricated and assembled kitchen cabinets, including a sink base cabinet, one three-drawer base cabinet, and six base cabinets with doors. Estimate base and wall cabinets separately. Costs are per linear feet of the front or back edge, whichever is longer.

Countertops can be made of a variety of materials. Laminated plastic (Formica, Textolite or Wilsonart) or ceramic tile on particleboard, engineered stone (quartz particles with acrylic or epoxy binder, like Crystalite, Silestone, or Cambria) or granite, as well as treated concrete, are all popular today. Estimate laminated plastic countertops with square edges and a 4-inch square inside-corner backsplash (the least expensive option), and give the owner the option to upgrade at additional cost.

The labor to install factory-built cabinets, including the countertops and molding, can be estimated at 7 manhours per 100 square feet of face area.

Vanities

The bathroom detail section on the floor plan gives the linear feet, type, and size of cabinets required. The specifications indicate the cabinets

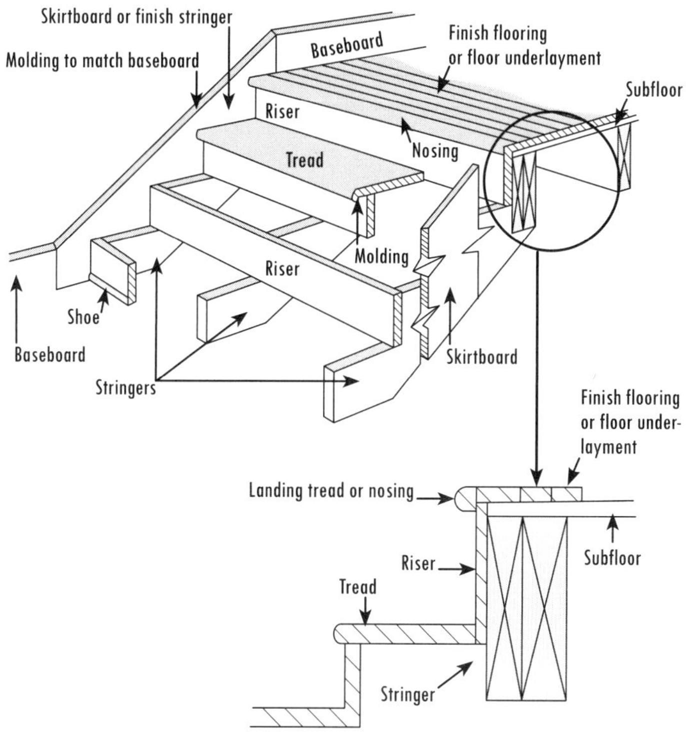

Figure 15-8
Stair components

and countertops to use (material and type). There's a range of quality available (standard, good, or premium), and a choice of sizes (24, 30, 36, 48 or 60 inches wide, and 18 or 21 inches deep). Standard vanities are 31½ inches high. Countertops with built-in sinks are estimated separately. The usual practice is to estimate standard quality, with the option to upgrade at additional cost.

Estimate labor to install factory-built vanities with countertops at 7 manhours per 100 square feet of face area.

Closet Shelves

The blueprints show the size and location of closets. A dotted line indicates a rod under a shelf in a clothes closet. For linen closets, the number of shelves will be specified.

Clothes closets generally consist of 10-, 12- or 18-inch shelving material, 1- × 4-inch cleats or ribbons, clothes rods, and rod supports every 3 feet. The shelf is normally installed 66 inches up from floor level. Closet dimensions are also shown on the floor plan. Use the information provided to calculate the closet shelving required in running feet.

Linen closets have shelves and cleats built in. Determine the width of the shelves for each closet, and multiply by 5 (for 5 linen closet shelves) to get the total running feet required. Estimate labor at 3 manhours per 100 linear feet of shelving.

Stairs

Stair components are shown in Figure 15-8. They are: the stringers (or *carriage*), treads, risers, skirtboard (*finish stringer),* and molding. A balustrade — balusters, handrail, turnout and newel posts — is used on the open side of a main staircase (see Figure 15-9). If the stair opening has a wall on each side, you only need a handrail and handrail brackets.

Risers are usually optional for basement stairs. The handrail is secured with brackets at the top of the stairs, and by a post (normally 4 × 4 inches) at the bottom.

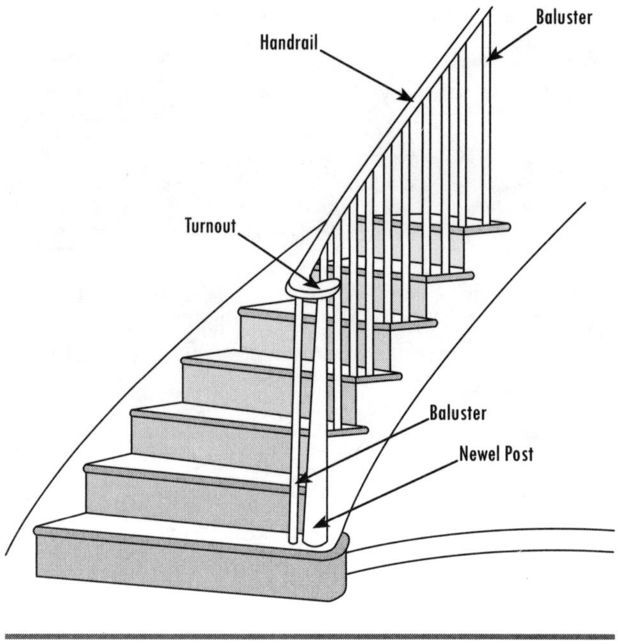

Figure 15-9
Circular stair balustrade

The main stair stringers are usually built during the framing of the house. Use temporary treads until the interior trim is installed. Any stairs 30 inches wide or more need three stringers, one on each side and an additional stringer for center support. The length of the stringer varies with the number and dimension of the risers and treads. Look back at Figure 7-58 (in Chapter 7) for help determining stringer length. Stringers are usually included in the estimate for the superstructure (covered in Chapter 7).

When estimating for stairs, refer to the blueprints for floor-to-floor rise. If the number of risers isn't on the plans, convert the floor-to-floor rise into inches (refer to Figure 7-58 in Chapter 7), divide by 7.75 (riser height), and round up.

Let's figure how many risers we need for stairs with a floor-to-floor rise of 8′10⅜″. First, convert the floor-to-floor rise to decimal inches, then divide by 7.75 to determine the number of risers needed:

$$(8' \times 12") + 10.375" = 106.38"$$

$$\frac{106.38"}{7.75"} = 13.73 \text{ (rounded to 14) risers}$$

If we chose to use 13 risers instead of 14, each riser would be 8$\frac{3}{16}$ inches — greater than the 7¾ inches that code allows. We need to install 14 risers (7⅝ inches each) to meet code. Keep in mind that there's always one less tread than riser (because the top step is the floor of the second story), so we need 14 risers and 13 treads.

You need to know the numbers of risers and treads as well as the material you'll use in order to estimate your stairway cost. Include skirtboard, molding, handrails, handrail brackets, posts (for basement stairs), and all materials for the balustrade. If a landing is required, you'll find that information on the plans. Estimate landing material as part of the stairs.

Labor costs vary with the type of stairs to be installed and the efficiency of the craftsmen installing them. Using craftsmen experienced in building stairs is recommended. Labor for basement stairs with open risers and two handrails could be as low as 10 manhours per set of stairs. Circular or ornate main stairs with elaborate balustrades could require 80 or more manhours.

Bathroom Trim

Although mirrors are found in most bathrooms, installing custom mirrors can add glamour to function. You can use mirrors to accent architectural features or to make the room appear larger.

Mirrors and Medicine Cabinets

Mirrors and medicine cabinets are shown on the floor plan, in the detail section. You'll find the number of mirrors and their size, either here or on the construction specs.

Labor to install mirrors and medicine cabinets is generally estimated at 8 manhours per 55 square feet of mirror area. Use a factor of 0.145 manhour per square foot of mirror area (8 hours divided by 55 square feet). For example, how much labor should we estimate to install a 3- × 6½-foot (19.5-square-foot) mirror? Multiply the square footage of the mirror by the 0.145 factor:

$$19.5 \text{ sf} \times 0.145 = 2.83 \text{ (rounded to 3) manhours}$$

We should estimate 3 manhours for this installation.

Tub and Shower Doors

All tub and shower door information is shown on the floor plans, detail sections, and specs. Allow 2 manhours per door for installation.

Bathroom Accessories

Bathroom accessories are towel bars, toilet paper holders, soap dishes, grab bars, and shower curtain rods (if a shower door isn't to be installed). These accessories are usually selected by the owner. Allow 0.25 manhour per unit to install bathroom accessories.

Miscellaneous Items

This category includes any special items the owner has chosen that aren't on the plans, such as valances, room dividers, or artificial beams. Also include here any necessary items, such as nails, glue, adhesive, wood filler, sandpaper, and tub and tile caulking, that aren't listed elsewhere.

Estimating Trim

We'll do an estimate of the interior trim materials needed for our sample house (Figures 7-1 and 7-2) and follow up with an estimate of the labor for installation.

Flooring Underlayment

We'll put wall-to-wall carpet in all areas of the house except the foyer, kitchen, bathrooms and utility room. We'll put down a particleboard underlayment for the carpeting. The foyer, kitchen, master and family bathrooms will have vinyl flooring with a plywood underlayment. (Remember, it's best to estimate the least expensive material and let the owner upgrade to a higher quality, at an additional cost to him.) The utility area and downstairs bathroom will have a concrete floor. We'll estimate the cost of these floor covering materials in Chapter 16. Right now, we'll just estimate the underlayment.

Particleboard

The combined floor area of the rooms to be carpeted, the living room, family room, dining room and bedrooms, is 1,760 square feet. How much ⅝-inch-thick, 4 × 8 particleboard will we need for the underlayment? Divide the square footage by 32 (4 × 8 sheets), and round up:

$$\frac{\text{1,760 sf}}{\text{32 sf per piece}} = \text{55 pieces}$$

We'll need 55 pieces of particleboard.

> *Enter your estimate for **particleboard** on line 15.1.1 of your Cost Estimate Worksheet. We'll use 55 pieces of ⅝-inch particleboard in our Example Cost Estimate Worksheet.*

Plywood

Now let's calculate how many pieces of ⅝-inch-thick, 4 × 8 plywood underlayment we should estimate for the foyer, kitchen, and upstairs bathrooms. The total floor area is 296 square feet. Divide the square footage by 32 (4 × 8 sheets), and round up:

$$\frac{\text{296 sf}}{\text{32 sf per piece}} = \text{9.25 (rounded to 10) pieces}$$

We'll need 10 pieces of plywood for underlayment.

> *Enter your estimate for the **plywood** on line 15.1.2 of your Cost Estimate Worksheet. We'll use 10 pieces of ⅝-inch plywood in our Example Cost Estimate Worksheet.*

Number	Size	Opening
1	18 inch × 80 inch	LH
2	24 inch × 80 inch	LH
3	28 inch × 80 inch	RH
1	28 inch × 80 inch	LH
6	30 inch × 80 inch	RH
2	30 inch × 80 inch	LH
2	60 inch × 80 inch	Double
2	30 inch × 80 inch	Bi-fold
2	36 inch × 80 inch	Bi-fold
2	48 inch × 80 inch	Bi-fold
2	60 inch × 80 inch	Bi-fold
LH (left hand), RH (right hand)		

Figure 15-10
Interior door schedule

Nails

We'll install the underlayment using 6d ring-grooved nails. We need 2 pounds of nails per 100 square feet of underlayment. The total area to be covered is 2,056 (1,760 + 296) square feet. Divide the total square footage to be covered by 100 square feet, then multiply by 2 pounds, and round up:

$$\frac{2{,}056 \text{ sf}}{100 \text{ sf}} \times 2 \text{ lbs} = 41.12 \text{ (rounded up to 42) lbs}$$

We'll need 42 pounds of 6d ring-grooved nails to install the underlayment.

*Enter your estimate for the number of pounds of **6d ring-grooved nails** on line 15.17.1 of your Cost Estimate Worksheet. We'll use 42 pounds of 6d ring-grooved nails in our Example Cost Estimate Worksheet.*

Interior Doors

We'll use prehung, 6-panel Colonial hollowcore doors inside the house. These include the door, jamb, hinges and trim. The double doors specified are 10-lite French doors, and the bi-fold doors are birch 4-panel doors that come with the hardware and tracks. The other doors have been bored for the lockset, but the hardware isn't included. We'll estimate those separately. The doors we need are listed in Figure 15-10.

*Enter your estimate for the **number and type of interior doors** in section 15.2.1 of your Cost Estimate Worksheet. We'll use the number and types of interior doors listed in Figure 15-10 in our Example Cost Estimate Worksheet.*

Locksets

We'll use passage latch locksets on all interior doors, except the 3 bathrooms, which need privacy latch locksets, and the bi-fold doors which don't need locksets at all. There are a total of 17 doors, 14 passage doors and 3 privacy doors that need locksets (see Figure 15-10).

*Enter your estimate for the number of **locksets** on line 15.2.2 of your Cost Estimate Worksheet. We'll use 14 passage and 3 privacy locksets in our Example Cost Estimate Worksheet.*

Nails

We'll use 8d finish nails for the door openings. There are 25 openings listed, and we need 0.5 pound of nails per opening. Multiply the total openings by 0.5 and round up:

$$25 \times 0.5 = 12.50 \text{ (rounded to 13) pounds of 8d finish nails}$$

> *Enter your estimate for the number of pounds of **8d finish nails** on line 15.17.2 of your Cost Estimate Worksheet. We'll use 13 pounds of 8d finish nails in our Example Cost Estimate Worksheet.*

Window Trim

Estimate the interior window trim components (casing, stool, and apron) separately. Use the dimensions given in Chapter 10, Figure 10-11 to calculate the linear feet of casing, stool, and apron. We'll assume the linear feet of stool and apron are the same.

Casing

We'll need casing at the top and the right and left sides of each window. Multiply the height of each window by 2 (for the left and right sides), and add in the width. If you use Figure 10-11, you'll come up with a total of 1,644 inches (137 linear feet). Add 20 percent for the mitered corners and waste, and round up.

$$137 \times 0.20 = 27.40$$

$$137 + 27.40 = 164.40 \text{ (rounded to 165) linear feet}$$

We'll need 165 linear feet of ½- × 1⅝-inch window casing.

> *Enter your estimate for linear feet of **casing** on line 15.3.1 of your Cost Estimate Worksheet. We'll use 165 linear feet of ½" × 1⅝" casing in our Example Cost Estimate Worksheet.*

Stool

We need stool for the bottom of each window. Measure the width of each window. If you use Figure 10-11, you'll get a total of 612 inches (51 linear feet). Add 20 percent for waste, and round up.

$$51 \times 0.20 = 10.20$$

$$51 + 10.20 = 61.20 \text{ (rounded to 62) linear feet}$$

We'll need 62 linear feet of 1- × 3½-inch beveled stool material.

> *Enter your estimate for linear feet of **stool** on line 15.3.2 of your Cost Estimate Worksheet. We'll use 62 linear feet of 1-inch × 3½-inch beveled **stool** in our Example Cost Estimate Worksheet.*

Apron

We also need apron for the bottom of each window. We'll need the same number of linear feet of apron as we needed for the stool. The apron will be made from the same ½- × 1⅝-inch material that we use for the casing.

We'll need 62 linear feet of ½- × 1⅝-inch casing material for the apron.

> *Enter your estimate for linear feet of **apron** on line 15.3.3 of your Cost Estimate Worksheet. We'll use 62 linear feet of ½-inch × 1⅝-inch casing for the apron in our Example Cost Estimate Worksheet.*

Nails

We'll use 8d finish nails for the window trim, and we'll need ¼ pound of nails per window face. From Figure 10-11, we know there are 12 windows. Multiply 12 by 2 for the 2 faces of each window (inside and out), and then multiply by 0.25 (for the ¼ pound of nails), and round up:

$$12 \times 2 \times 0.25 = 6 \text{ pounds of 8d finish nails}$$

> *Enter your estimate for the number of pounds of **8d finish nails** on line 15.17.3 of your Cost Estimate Worksheet. We'll use 6 lbs of 8d finish nails in our Example Cost Estimate Worksheet.*

Baseboard

We'll install ½- × 3½-inch baseboard throughout the entire house except in the family room, which will be paneled.

To estimate the number of linear feet of baseboard, list all the areas where baseboard is needed. Use the table in Figure 15-4 to calculate how much material we'll need for each area. If the room dimension isn't a whole number, use the next larger number of feet in the table. This will allow for waste. The total for our sample house comes to 850 linear feet.

> *Enter your estimate for the number of linear feet of **baseboard** on line 15.4 of your Cost Estimate Worksheet. We'll use 850 linear feet of ½"× 3½" baseboard in our Example Cost Estimate Worksheet.*

Nails

We'll need 1 pound of 8d finish nails per 100 linear feet of baseboard. We're installing 850 linear feet of baseboard. Multiply 850 by 1 pound, divide by 100 linear feet, and round up.

$$\frac{850 \text{ lf} \times 1 \text{ lb}}{100 \text{ lf}} = 8.5 \text{ (rounded to 9) lbs}$$

We need 9 pounds of 8d finish nails for the baseboards.

> *Enter your estimate for the number of pounds of* ***8d finish nails*** *on line 15.17.4 of your Cost Estimate Worksheet. We'll use 9 pounds of 8d finish nails in our Example Cost Estimate Worksheet.*

Base Shoe

We need base shoe in addition to the baseboard in the rooms where we're going to install vinyl flooring: the foyer, kitchen, master and family bathrooms.

We'll use the table in Figure 15-4 to calculate the linear feet of base shoe we need for each area. Our total is 250 linear feet of base shoe.

> *Enter your estimate for the number of linear feet of* ***base shoe*** *on line 15.5 of your Cost Estimate Worksheet. We'll use a figure of 250 linear feet in our Example Cost Estimate Worksheet.*

Nails

For the base shoe, we'll use 6d finish nails. We need ½ pound per 100 linear feet. Multiply 250 linear feet by 0.50 (½ pound), divide by 100 linear feet, and round up.

$$\frac{250\text{ lf} \times .50\text{ lb}}{100\text{ lf}} = 1.25\text{ (rounded to 2) lbs}$$

We need 2 pounds of 6d finish nails for the base shoe.

> *Enter your estimate for the number of pounds of* ***6d finish nails*** *on line 15.17.5 of your Cost Estimate Worksheet. We'll use 2 pounds of 6d finish nails in our Example Cost Estimate Worksheet.*

Wall Molding

We won't be installing any wall molding in our sample house.

Paneling

We're installing prefinished oak plywood paneling in the family room, with prefinished oak cove molding at the ceiling. We'll use ¼-inch-thick, 4- × 8-foot panels.

The family room dimensions are 15'10" × 18'9", with a standard 8-foot ceiling. We'll use Figure 15-4 to get the perimeter, rounding up to the next whole number of feet for the room dimensions, because they aren't whole numbers. The fireplace, doorways, and windows obviously don't require paneling. But unless you use paneling that's very expensive, it's best to over estimate and have a surplus to allow for waste, than to do a precise estimate.

Using Figure 15-4, a 16- × 19-foot room has a perimeter of 70 linear feet. We'll divide the linear feet by 4 feet (the width of each panel) to find the number of panels we need:

$$\frac{70 \text{ lf}}{4 \text{ ft per panel}} = 17.5 \text{ (rounded up to 18) pieces}$$

We'll estimate 18 pieces of ¼-inch-thick, 4- × 8-foot oak plywood panels.

*Enter your estimate for the number of **panels** on line 15.7.1 of your Cost Estimate Worksheet. We'll use 18 pieces of ¼-inch-thick, 4 × 8 oak plywood panels in our Example Cost Estimate Worksheet.*

Cove Molding

We'll use ½- × ½-inch oak cove molding at the ceiling in our family room. Cove molding is estimated in linear feet. We calculated the perimeter of the family room at 70 feet, which includes an allowance for waste (the perimeter was overestimated). We'll need 70 linear feet of cove molding.

*Enter your estimate for the number of linear feet of **cove molding** on line 15.7.2 of your Cost Estimate Worksheet. We'll use 70 linear feet of ½- × ½-inch oak cove molding in our Example Cost Estimate Worksheet.*

Nails

We'll use 1⅝-inch colored nails for the paneling and molding in the family room. Estimate 1 pound per four 4 × 8 panels. Divide the number of panels by 4 and round up:

$$\frac{18 \text{ panels}}{4 \text{ panels}} \times 1 \text{ lb} = 4.50 \text{ (rounded to 5) lbs}$$

We'll need 5 pounds of 1⅝-inch colored nails.

*Enter your estimate for the number of pounds of **nails** on line 15.17.6 of your Cost Estimate Worksheet. We'll use 5 pounds of 1⅝-inch colored nails in our Example Cost Estimate Worksheet.*

Adhesive

Estimate adhesive per 100 square feet of paneling. Figure 1 tube of gun-applied adhesive per 100 square feet of paneling, using a ¼-inch bead on studs 16 inches on center, with an allowance of 6 percent for waste. Calculate the square feet to install, and round to the nearest 100 square feet:

18 panels × 32 sf (4 × 8) = 576 (rounded to 600) square feet of panel area

We'll estimate we need adhesive for 600 square feet of paneling (6 tubes). Apply the cost per square foot from the *National Construction Estimator*, which includes 6 percent for waste.

> *Enter your estimate for the number of square feet of **panel adhesive** on line 15.18.1 of your Cost Estimate Worksheet. We'll use 600 square feet in our Example Cost Estimate Worksheet.*

Kitchen Cabinets

Use the running (linear) feet of cabinets to be installed as the basis for the estimate. You'll find the dimensions on the plans.

We'll install a set of mill-fabricated and assembled kitchen cabinets, including a sink base cabinet, one three-drawer base cabinet, and six base cabinets with doors. Estimate the base and wall cabinets separately. The costs are per linear feet of the front or back edge, whichever is longer.

Base Cabinets

Total the running (linear) feet of the base cabinets we're installing.

Sink cabinet	3.00
Rear wall cabinet	4.25
Stove cabinet	4.50
Cabinet next to refrigerator	3.50
Total:	15.25 running (linear) feet

> *Enter your estimate for the linear feet of **base cabinets** on line 15.8.1 of your Cost Estimate Worksheet. We'll use 15.25 linear feet of base cabinets in our Example Cost Estimate Worksheet.*

Wall Cabinets

Total the running (linear) feet of wall cabinets we're installing.

Stove cabinet	4.50
Cabinet next to refrigerator	3.50
Total:	8.00 running (linear) feet

> *Enter your estimate for the linear feet of **wall cabinets** on line 15.8.2 of your Cost Estimate Worksheet. We'll use 8 linear feet of wall cabinets in our Example Cost Estimate Worksheet.*

Countertops

The running feet of countertops are the same as the base cabinets. In our kitchen, it's 15.25 linear feet. We'll estimate laminated plastic countertops with a 4-inch backsplash (the minimum option), and let the owner upgrade at an appropriate additional cost if he wants a higher quality.

> *Enter your estimate for the linear feet of **countertop** on line 15.8.3 of your Cost Estimate Worksheet. We'll use 15.25 linear feet of plastic laminate countertop in our Example Cost Estimate Worksheet.*

Vanities

First estimate vanities by the sink base alone, and then add the vanity tops, which include a lavatory bowl. You'll find the dimensions on the plans. You could also have countertops without bases (indicated as *knee spaces* on the drawings) that need to be estimated.

Vanity Bases

Vanity bases typically have ¾-inch furniture-grade sides with veneer interior, overlay doors and drawers, and concealed hinges. The usual height is 31½ inches. Costs are generally based on standard, good, and premium quality, and various dimensions (24, 30, 36, 48 and 60 inches wide and 18 or 21 inches deep). We'll estimate for premium quality.

Here's a list of the bases we need:

Downstairs bathroom	36-inch base
Master bathroom	60-inch base
Dressing area	30-inch base
Family bathroom	60-inch base

> *Enter your estimate for the **vanity bases** required in section 15.9.1 of your Cost Estimate Worksheet. We'll use the above list of vanity bases in our Example Cost Estimate Worksheet.*

Vanity Tops

We're going to estimate vanity tops with integrated sinks. Vanity tops are typically 1 inch wider and 1 inch deeper than the vanity base. Here's a listing of vanity tops with integrated sinks that we'll need to match our list of vanity bases:

Downstairs bathroom	37-inch top
Master bathroom	61-inch top
Dressing area	31-inch top
Family bathroom	61-inch top

> *Enter your estimate for the **vanity tops with sinks** required in section 15.9.2 of your Cost Estimate Worksheet. We'll use this list of vanity tops in our Example Cost Estimate Worksheet.*

Countertop Knee Spaces

Estimating for the bathroom countertops (knee spaces) in the dressing area and second bathroom requires measuring the running feet of the counter area. There are 3 running feet (36 inches) of countertop in the dressing area, and 4 running feet (48 inches) in the family bathroom, for a total of 7 feet (84 inches). We'll estimate laminated plastic countertops with a 4-inch backsplash (the least expensive option) and allow the owner to upgrade at additional cost.

> *Enter your estimate for the running feet of **countertop** on line 15.9.3 of your Cost Estimate Worksheet. We'll use 7 running (linear) feet of plastic laminate countertop in our Example Cost Estimate Worksheet.*

Closet Shelves

The closets will have 1-inch-diameter clothes poles with support brackets every 3 feet on center. You can estimate either by the number of 6-foot sections, or by the total running feet of closet shelving to be installed. In new housing, it's generally easier to use running feet.

We'll need the following 18-inch-deep closet shelving:

Foyer	6.00 feet
Master bedroom	6.00 feet
Dressing area	7.75 feet
	10.75 feet
	4.75 feet
Bedroom #2	4.25 feet
	3.50 feet
Bedroom #3	5.00 feet
	3.50 feet
Total:	51.50 running feet of closet shelving

> *Enter your estimate for the running feet of **closet shelving** required on line 15.11.1 of your Cost Estimate Worksheet. We'll use 51.50 feet of closet shelving in our Example Cost Estimate Worksheet.*

Linen Closet Shelves

You'll also need an estimate for the linen closets, which usually have 5 shelves, either 18- or 24-inches deep. You can estimate either by the number of 4-foot-wide shelves needed, or by the running feet required. Again, using running feet is usually easier, because there are fewer calculations.

There are two linen closets in our house, one in the dressing area and one in the upstairs hallway. Each will have five 18-inch-deep shelves. To find the linear feet of shelving we need, add the widths of the shelves in each linen closet, and multiply the total by 5:

Dressing area closet	4.50 feet wide
Upstairs hallway closet	2.50 feet wide
Total:	7.00 feet wide

7.00 × 5 shelves = 35 linear feet of shelving

*Enter your estimate for the linear feet of **linen closet shelving** required on line 15.11.2 of your Cost Estimate Worksheet. We'll use 35 feet of 18-inch-deep linen closet shelving in our Example Cost Estimate Worksheet.*

Stairs

The stair stringers were estimated in Chapter 7. We'll install temporary treads during the superstructure phase of building. However, we now need to estimate the permanent treads and risers, newel post, stair rail, and balusters, using the information from Chapter 7, Figures 7-1, 7-2, and 7-60. We'll be using oak for all of these stair components.

The stringer length will be 13 feet 7½ inches, so we'll need 13 treads 4 feet wide and 10 inches deep, and 14 risers, each 7⅝ inches tall. We'll have one post at the bottom of the stairs, with 14 feet of handrail (rounded up from 13 feet 7½ inches, which is the length of the stringers).

The balusters are 2¼ inches wide and spaced so there's no more than 4 inches between them, for safety. Because of this safety issue, they can't be spaced more than 6¼ inches on center (the gap plus 2 half-widths of a baluster). To determine how many balusters we need, calculate the horizontal run of the stairs. Multiply the stringer length by 0.8:

$$13.63'\ (13'7\tfrac{1}{2}'') \times 0.8 = 10.90'\ \text{of horizontal run}$$

Divide the horizontal run in inches by the 6¼ inches (6.25 inches) on-center spacing:

$$\frac{130.80''\ (10.90')\ \text{horizontal run}}{6.25''\ \text{oc}} = 20.93\ \text{(rounded to 21) balusters}$$

There's no need to add a starter baluster, since the newel post takes the place of a starter.

We need the following oak material for the main stairs:

13 treads	10" × 4'0" each
14 risers	7⅝" × 4'0" each
1 newel post	3" × 3" × 4'0"
1 handrail	2⅜" × 2¼" × 14'0"
21 balusters	1¼" × 2¼" × 3'0"

*Enter your estimate for the **main stairs** material in section 15.12.1 of your Cost Estimate Worksheet. We'll use the materials and quantities listed above in our Example Cost Estimate Worksheet.*

Basement Stairs

The basement stairs have a floor-to-floor rise of 8 feet 5½ inches (or 101.50 inches), and we'll estimate them separately.

Basement stairs can be purchased preassembled with 12 treads and 13 risers, and three choices of riser height (7¹³⁄₁₆ inches, 8 inches and 8³⁄₁₆ inches), depending on the floor-to-floor rise. To determine what riser height will be best for our 101.50-inch floor-to-floor rise, divide the total rise (101.50 inches) by the number of risers (13).

$$\frac{101.50''}{13 \text{ risers}} = 7.81''(\text{very close to } 7\tfrac{13}{16}'')$$

We'll estimate a preassembled set of basement stairs with 13 risers 7¹³⁄₁₆ inches in height, and 12 treads, 2 feet 11¾ inches wide.

*Enter your estimate for the **basement stairs** material on line 15.12.2 of your Cost Estimate Worksheet. We'll use one set of 7¹³⁄₁₆-inch riser stairs in our Example Cost Estimate Worksheet.*

Bathroom Mirrors

Our plans and specs show that a 4- × 3-foot mirror will be installed over the vanity in the master bathroom, and a 7- × 3-foot mirror over the vanity in the family bathroom (bathroom 2, upstairs). No mirror is specified for the downstairs bathroom (bathroom 3). Beveled-edge, distortion-free float glass mirrors are typically used in bathrooms, and are estimated by the square foot. We have 12 square feet of mirror in the master bath, and 21 square feet in the family bathroom.

*Enter your estimate for the **mirror** areas needed in section 15.13 of your Cost Estimate Worksheet. We'll use these mirror dimensions in our Example Cost Estimate Worksheet.*

Medicine Cabinets

No medicine cabinets are called for in the master bathroom or in the family bathroom (2). There will be a medicine cabinet in the downstairs bathroom (3). We'll estimate a 25- × 26-inch better-quality cabinet, with two adjustable shelves and an etched and beveled mirrored door with polished brass knobs.

> *Enter your estimate for the **medicine cabinets** needed on line 15.14 of your Cost Estimate Worksheet. We'll use the medicine cabinet we just described in our Example Cost Estimate Worksheet.*

Tub and Shower Doors

The tubs in the master bath and family bathroom (#2), and the shower in the downstairs bathroom will have doors installed. They weren't included in the plumbing fixture costs. The master bath and bathroom #2 will each have 2-panel sliding shower/tub doors made of 5/32-inch opaque tempered glass with an anodized aluminum frame, 60 inches wide and 54 inches high, with an outside towel bar. For the downstairs bathroom, we'll estimate a swinging shower door with tempered safety glass in a gold anodized aluminum frame, 32 inches wide and 64 inches high.

> *Enter your estimate for **tub and shower doors** in section 15.15 of your Cost Estimate Worksheet. We'll use the tub and shower doors just described in our Example Cost Estimate Worksheet.*

Bathroom Accessories

We'll have an allowance for all the bathroom accessories not included in the plumbing contract, such as towel bars and rings, toilet roll holders, soap holders, etc.

> *Enter your estimate for **bathroom accessories** on line 15.16 of your Cost Estimate Worksheet. We'll estimate $500 for accessories in our Example Cost Estimate Worksheet.*

Miscellaneous Items

There are no valances, room dividers, artificial beams or other additional cost items shown on the blueprints. We'll estimate the following miscellaneous materials:

8 pounds of 8d finish nails
5 pounds of 6d finish nails
2 pounds of 4d finish nails
1 gallon of carpenter's glue
4 tubes (10.5 ounce) of tub and tile caulking

Enter the extra nails in nail section 15.17. There's a miscellaneous section where we'll enter paneling adhesive. We'll enter carpenter's glue and caulking there as well.

> *Enter your estimate for **nails** in section 15.17, and **miscellaneous** items in section 15.18 of your Cost Estimate Worksheet. We'll use the above materials and quantities in our Example Cost Estimate Worksheet.*

Labor

You can estimate labor for installing the interior trim either by using the manhour rates we've shown in the Cost Estimate Worksheet (Option 1, lines 15.19.1 through 15.19.15), or by using your own records for similar work, Option 2 on the worksheet. I recommend Option 2, but if you have no records, you'll need to use Option 1.

If you decide to use Option 1, complete the material quantity entries in the Cost Estimate Worksheet using the amounts that you calculate, just as we did earlier in the chapter. The spreadsheet will use the recommended manhour rates to give you a total manhour figure. Then complete the entries on line 15.19.16 (the number of carpenters and laborers you plan to use), and the spreadsheet will give you the manhours per worker estimate. Enter this number on lines 15.20.1 through 15.20.4 plus the labor rates for a carpenter and a laborer.

We'll use Option 2 in our Example Cost Estimate Worksheet.

From our records for a similar quality 2,500-square-foot two-story house, the labor for interior trim was 520 manhours. This included floor underlayment, interior doors, window trim, trim for the basement door, all stairs, installing tub and shower doors, and bathroom accessories. Our sample house has 2,238 square feet. We can set up the following proportion to calculate the manhours based on our previous job.

$$520 \text{ manhours} : 2{,}500 \text{ sf} = \text{X manhours} : 2{,}238 \text{ sf}$$

$$\text{Product of extremes} = \text{Product of means}$$

$$520 \text{ manhours} \times 2{,}238 \text{ sf} = 2{,}500 \text{ sf} \times \text{X manhours}$$

$$X = \frac{520 \text{ manhours} \times 2{,}238 \text{ sf}}{2{,}500 \text{ sf}}$$

$$X = 465.50 \text{ manhours}$$

We will have two carpenters and two laborers installing the interior trim. The manhours per workman will be:

$$\frac{465.50 \text{ manhours}}{4 \text{ workmen}} = 116.25 \text{ (rounded to 117) manhours/workman}$$

Our estimate will be 465.50 manhours, or 117 hours per workman.

> *Enter your estimate for the **labor** (manhours) for the interior trim in section 15.20 of your Cost Estimate Worksheet. We'll use 117 hours for the 4 workmen in our Example Cost Estimate Worksheet.*

COST ESTIMATE WORKSHEET FOR INTERIOR TRIM

#	Item	Qty			Size		Cost	Per			Subtotal
15.1	**Floor underlayment**										
15.1.1		55	pieces	⅝" particleboard	4' × 8'	@	$16.90	piece	=	$929.50	
15.1.2		10	pieces	⅝" plywood	4' × 8'	@	$21.70	piece	=	$217.00	
										$1,146.50	$1,146.50
15.2	**Interior doors**										
15.2.1		1	pieces	LH	18" × 80"	@	$63.80	piece	=	$63.80	
		2	pieces	LH	24" × 80"	@	$64.50	piece	=	$129.00	
		3	pieces	RH	28" × 80"	@	$66.20	piece	=	$198.60	
		1	pieces	LH	28" × 80"	@	$66.20	piece	=	$66.20	
		6	pieces	RH	30" × 80"	@	$68.30	piece	=	$409.80	
		2	pieces	LH	30" × 80"	@	$68.30	piece	=	$136.60	
		2	pieces	Double	60" × 80"	@	$241.00	piece	=	$482.00	
		2	pieces	Bi-fold	30" × 80"	@	$40.50	piece	=	$81.00	
		2	pieces	Bi-fold	36" × 80"	@	$44.30	piece	=	$88.60	
		2	pieces	Bi-fold	48" × 80"	@	$63.70	piece	=	$127.40	
		2	pieces	Bi-fold	60" × 80"	@	$81.00	piece	=	$162.00	
15.2.2	**Locksets**										
		14	sets		passage	@	$40.00	piece	=	$560.00	
		3	sets		privacy	@	$48.00	piece	=	$144.00	
										$2,649.00	$2,649.00
15.3	**Window trim**										
15.3.1	**Casing**										
		165	lf		½" × 1⅝"	@	$1.07	lf	=	$176.55	
15.3.2	**Stool (beveled)**										
		62	lf		1" × 3½"	@	$3.64	lf	=	$225.68	
15.3.3	**Apron**										
		62	lf		½" × 1⅝"	@	$1.07	lf	=	$66.34	
										$468.57	$468.57
15.4	**Baseboard**										
		850	lf		½" × 3½"	@	$1.36	lf	=	$1,156.00	$1,156.00
15.5	**Base shoe**										
		250	lf		7/16" × ¾"	@	$0.50	lf	=	$125.00	$125.00
15.6	**Wall molding**										
			lf			@		lf	=	$0.00	$0.00
15.7	**Paneling**										
15.7.1		18	pieces	¼" oak plywood	4' × 8'	@	$21.76	piece	=	$391.68	
15.7.2		70	lf	oak cove molding	½" × ½"	@	$0.79	lf	=	$55.30	
										446.98	$446.98

COST ESTIMATE WORKSHEET FOR INTERIOR TRIM, cont.

#	Item	Qty		Quality/Type	Size	Cost	Per	Amount	Subtotal
15.8	**Kitchen cabinets and tops**								
15.8.1	**Base cabinets**			Quality	Standard				
		15.25	lf		34½" × 24"	@ $132.00	lf	= $2,013.00	
15.8.2	**Wall cabinets**			Quality	Standard				
		8	lf		30" × 12"	@ $68.70	lf	= $549.60	
15.8.3	**Countertops**			Quality	Laminate				
		15.25	lf		25" deep	@ $14.50	lf	= $221.13	
								$2,783.73	$2,783.73
15.9	**Vanities and tops**								
15.9.1	**Bases**			Quality	Premium				
		1	30" wide		21" deep	@ $283.00	piece	= $283.00	
		1	36" wide		21" deep	@ $311.00	piece	= $311.00	
		2	60" wide		21" deep	@ $529.00	piece	= $1,058.00	
15.9.2	**Tops with sinks**			Quality	Corian				
		1	31" wide		22" deep	@ $502.00	piece	= $502.00	
		1	37" wide		22" deep	@ $586.00	piece	= $586.00	
		2	61" wide		22" deep	@ $805.00	piece	= $1,610.00	
15.9.3	**Countertops**			Quality	Laminate				
		7	lf		25" deep	@ $14.50	lf	= $101.50	
								$4,451.50	$4,451.50
15.10	**Other cabinets**								$0.00
	(List on separate sheet and enter cost here)								
15.11	**Closet shelves and linen closet shelving**								
15.11.1	**Closet shelves, 1" thick, 18" deep**								
		51.50	lf	Quality	Wood	@ $1.56	lf	= $80.34	
15.11.2	**Linen closet shelving, 1" thick, 18" deep**								
		35	lf	Quality	Wood	@ $1.19	lf	= $41.65	
								$121.99	$121.99
15.12	**Stairs**								
15.12.1	**Main Stairs**			Type					
	Treads	13	pieces	Oak	10" × 4'0"	@ $16.26	piece	= $211.38	
	Risers	14	pieces	Oak	7⅝" × 4'0"	@ $17.35	piece	= $242.90	
	Post	1	pieces	Oak	3" × 3" × 4'0"	@ $37.70	piece	= $37.70	
	Handrail	14	lf	Oak	2⅜" × 2¼"	@ $5.37	lf	= $75.18	
	Balusters	21	pieces	Oak	1¼" × 2¼"	@ $5.73	piece	= $120.33	
15.12.2	**Basement Stairs**				Riser height				
		1	set		$7\frac{13}{16}$"	@ $242.00	set	= $242.00	
								$929.49	$929.49
15.13	**Mirrors**								
		12	sf		4' × 3'	@ $6.75	sf	= $81.00	
		21	sf		7' × 3'	@ $6.75	sf	= $141.75	
								$222.75	$222.75

COST ESTIMATE WORKSHEET FOR INTERIOR TRIM, cont.

#	Item	Qty	Size		Cost	Per			Subtotal
15.14	**Medicine cabinets**								
		1 unit	25" × 26"	@	$111.00	piece	=	$111.00	$111.00
15.15	**Tub and shower doors**								
		2 sets	60" × 54"	@	$216.00	set	=	$432.00	
		1 set	32" × 64"	@	$161.00	set	=	$161.00	
								$704.00	$704.00
15.16	**Bathroom accessories**							$500.00	$500.00
15.17	**Nails**								
15.17.1		42 lbs	6d ring grooved	@	$4.25	lb	=	$178.50	
15.17.2		13 lbs	8d finish	@	$2.04	lb	=	$26.52	
15.17.3		6 lbs	8d finish	@	$2.04	lb	=	$12.24	
15.17.4		9 lbs	8d finish	@	$2.04	lb	=	$18.36	
15.17.5		2 lbs	6d finish	@	$2.06	lb	=	$4.12	
15.17.6		5 lbs	1⅝" colored nails	@	$2.83	lb	=	$14.15	
15.17.7		8 lbs	8d finish	@	$2.04	lb	=	$16.32	
15.17.8		5 lbs	6d finish	@	$2.06	lb	=	$10.30	
15.17.9		2 lbs	4d finish	@	$2.04	lb	=	$4.08	
								$284.59	$284.59
15.18	**Miscellaneous**								
15.18.1		600 sf	adhesive for paneling	@	$0.07	sf	=	$42.00	
15.18.2		1 gal	carpentry glue	@	$12.40	gal	=	$12.40	
15.18.3		4 tubes	caulking	@	$4.73	tube	=	$18.92	
								$73.32	$73.32
			Subtotal						$16,174.42
			Sales tax	@	7.75	%			$1,253.52
			Cost of material						$17,427.94

#	Item	Qty		# Manhours	Per		Manhours	Subtotal
15.19	**Labor (Option 1)**							
15.19.1	Floor underlayment		sf	2	100.00	sf	$0.00	
15.19.2	Doors		doors	3	1.00	door	$0.00	
15.19.3	Window trim		windows	3	1.00	window	$0.00	
15.19.4	Baseboard		lf	4	100.00	lf	$0.00	
15.19.5	Base shoe		lf	2	100.00	lf	$0.00	
15.19.6	Chair rail and panel mold		lf	6	100.00	lf	$0.00	
15.19.7	Hardboard/plywood paneling		sf	3	100.00	sf	$0.00	
15.19.8	Solid wood paneling		sf	6	100.00	sf	$0.00	
15.19.9	Factory-built kitchen cabinets		sf	7	100.00	sf	$0.00	
15.19.10	Factory-built vanities		sf	7	100.00	sf	$0.00	
15.19.11	Closet shelves		lf	3	100.00	lf	$0.00	
15.19.12	Stairs		set	20	1.00	set	$0.00	
15.19.13	Bath mirrors & medicine cab		sf	8	55.00	sf	$0.00	
15.19.14	Tub and shower doors		doors	2	1.00	door	$0.00	
15.19.15	Bathroom accessories		units	0.25	1.00	unit	$0.00	
					Total Manhours		$0.00	

15.19.16 **Divide Total Manhours by** 2 **carpenters and** 2 **laborers**

0.00 **manhours per workman**

Enter hours per workman on Lines 15.20.1 through 15.20.4

#	Item		Cost	Per			Subtotal
15.20	**Labor (Option 2)**						
15.20.1	117 manhours - carpenter, BC	@	$32.09	hour	=	$3,754.53	
15.20.2	117 manhours - carpenter, BC	@	$32.09	hour	=	$3,754.53	
15.20.3	117 manhours - laborer, BL	@	$26.64	hour	=	$3,116.88	
15.20.4	117 manhours - laborer, BL	@	$26.64	hour	=	$3,116.88	
						$13,742.82	$13,742.82
	Cost of interior trim (entered on line 15 of Form 100)						**$31,170.76**

Chapter 16

Painting, Floor Covering and Appliances

THE FINAL STAGES OF construction include the exterior and interior finish work: painting or other wall finishes, the floor coverings, and the installation of the built-in appliances.

Painting

Paint not only provides an attractive finish to both interior and exterior surfaces, it adds weather-resistance to the exterior materials. But that's the case only if you use the paint recommended for the job. You shouldn't use masonry paint on a metal surface, or indoor paint on a fence post. Every paint is labeled with information about exposure (exterior or interior), the type of surface it's recommended to cover (wood, metal, concrete, plaster), and the type of wear it's designed to take.

Exterior wood surfaces should be primed when the trim is in place and given a minimum 4.0 mil (0.004-inch) finish coat. Normally, two coats of finish will provide that level of coverage. All exterior wood doors and window frames should get at least two coats of paint or sealer.

Interior walls made of plaster or gypsum wallboard should get a finish coat of paint, unless you're installing a wall covering. Painting is by far the easiest way to finish walls. The most frequently used interior paints are flat, semigloss, or gloss formulas. You can apply paint by brush, roller, or spray gun. Latex paint is popular because it dries fast, cleans up easily, and it's the most economical finish to apply.

Use one coat of sealer and a minimum of one coat of paint on gypsum or plaster walls. Semigloss or gloss paint in kitchens and bathrooms provide the durable, washable surface needed for those areas.

Apply paint or a natural finish to all interior wood surfaces, using at least three coats including the primer. Sand between coats to ensure a smooth surface. Be sure to paint or seal the top and bottom of interior doors.

Wall coverings come in a wide variety of patterns and materials, from simple paper to ornate fabric. If you're installing a wall covering in the kitchen or bathroom, make sure the material is waterproof and mildew-resistant, to withstand the moisture common to those areas.

Paint costs vary depending on the size and shape of the building, the type of paint, the number of coats needed, how many window and door openings you must cut around, and the amount of prep work, like sanding and caulking, that's required. Painting is usually done by paint contractors. They'll estimate their costs by the square foot, calculating for both labor and materials or labor only. Obviously, the cost will be different from job to job. Most paint contractors prefer to look over and assess the areas to be painted before submitting a firm bid, but they can usually make a ballpark estimate based on square footage and wall height.

Explicit instructions and specifications should be listed on every painting contract. The specs usually include the type of paint, number of coats to be applied, locations where caulking and sanding are needed, and other pertinent information. The agreement should be signed by all parties involved. Covering these points in detail will prevent any misunderstanding about the level of quality you expect from the job.

Wallpaper or wall fabric is usually chosen by the owner. If the material isn't selected at the time you prepare the estimate, provide an allowance that covers both labor and material, based on a specific cost per roll. Any additional cost or refund can be made at the time of final selection.

Floor Covering

There's a wide range of floor coverings available today, including carpet, wood, ceramic tile, natural stone and resilient flooring. The selection usually depends on the area's intended use and the quality expected by the buyers or owners of the property.

Carpet

Wall-to-wall carpet is common in residential living areas. Carpet is generally made of wool, nylon, polyester, Olefin, or a blend of synthetic fibers. Carpet and padding can be laid directly on a concrete slab or plywood subfloor, without any additional material. Carpet comes in widths of 9, 12, 15, and 18 feet. The 9- and 12-foot widths are most commonly used in residential construction.

Carpet padding is made of natural or synthetic fibers, rubber, or foam. Foam is used in about 90 percent of all installations. Padding increases the resilience and durability of the carpet, as well as providing insulation and soundproofing. Padding is graded in ounces per square yard — the heavier (thicker) the padding, the better the quality. Labor and material costs for the carpet and the padding are generally estimated separately, although carpet with padding bonded to the underside is available for some types of installations and estimated at a single unit price.

Both carpet and padding are estimated by the square yard. To estimate carpet for a room that's 11 feet by 17 feet (187 square feet or 20.78 square yards), an experienced installer would figure 204 square feet or 22.67 (rounded to 23) square yards of carpet. He would choose to cut 17 feet of carpet from a 12-foot-wide carpet roll to eliminate any extra seams. Even though the extra 2 square yards may seem like a waste of material, eliminating the seam will give a more professional result. When estimating carpet, you want to keep waste to a minimum, but you must always consider the end product and your customer's satisfaction. With most types of carpet, the fewer the seams, the better the result.

If the carpet isn't selected at the time you prepare the estimate, specify either a dollar amount per square yard or a dollar figure to install all the carpet and padding in the house.

Resilient Flooring

Resilient floor covering is frequently used in kitchens, bathrooms and laundry areas. Sheet vinyl and linoleum are sold by the square yard, and available in rolls 6 or 12 feet wide. Discuss the room layout with the installer to determine the minimum number of seams and the least amount of waste before preparing an estimate for resilient floors.

Resilient tile comes in asphalt, vinyl-asbestos, vinyl, rubber, cork, and linoleum. The tile, available in 9- or 12-inch squares, is priced by the square foot. Most suppliers don't sell partial boxes of tile, so be prepared to round your estimate up to the next whole number of boxes. Always make an allowance of 10 to 20 percent for waste.

Resilient flooring provides reasonable durability and economy of maintenance, and can be installed with adhesive on concrete or over a wood underlayment. Resilient flooring is commonly estimated with a provision allowing the customer to upgrade to a higher quality of floor covering. The customer would then pay the additional material costs and any extra labor when the upgrade is selected.

Masonry Flooring

Masonry floor materials are usually installed in more expensive homes. Ceramic tile is generally used in bathrooms, while porcelain, slate, travertine, or other natural stone materials are mostly seen in entry, family room and kitchen areas.

Laying stone or tile floors is normally done by contractors who specialize in that work. As with carpet, the customer may not have made a final decision on the type and grade of floor covering when you make your original estimate. If that's the case, estimate for a specific material, and inform the customer that any additional costs for the selected material (and labor) will be included in your final estimate.

Appliances

Built-in appliances — the oven, cooktop, range, range hood, dishwasher and garbage disposal — are usually included in the house estimate. If other built-ins are planned (a microwave oven or trash compactor, for instance), include them in the estimate, as well. Appliances like the refrigerator, washer and dryer aren't usually built in, and so aren't usually included in the estimate.

The carpenters' work should be coordinated with that of the carpet layers. For example, carpenters can normally trim the bottom of interior doors after flooring material is installed. Carpenters may also be responsible for installing built-in appliances; however, the electrician and plumber make the electrical and plumbing connections. Don't forget to add these labor costs to your estimate.

A Sample Estimate

Let's estimate the painting, floor covering, and appliances for our sample house, shown in Figures 7-1 and 7-2.

Painting

The interior and exterior painting, including all caulking and sanding, is generally performed by a subcontractor. A small to medium builder can rarely afford to have a permanent team of painters on staff. If that describes your company, you'll be including the painting subcontractor cost in your estimate.

We'll need two finish coats of exterior latex paint over the prime coat on the outside of the house. All interior wallboard will get two coats of latex flat paint, except the kitchen, bathrooms, and utility area, which will be finished with semigloss. The interior wood trim will need three coats of latex paint, finished with semigloss. All woodwork will have a smooth finish. The doors and trim in the family room will have a natural wood finish.

Estimate the cost of painting from your records for similar work. Our records show a cost of $8,500 for a similar, 2,500-square-foot two-story house, so we'll use that as a basis for our sample house estimate. Our house is 2,238 square feet. We'll show the cost of painting as X, and set up the following proportion:

$$\$8{,}500 : 2{,}500 \text{ sf} = X : 2{,}238 \text{ sf}$$

$$\text{Product of extremes} = \text{Product of means}$$

$$\$8{,}500 \times 2{,}238 \text{ sf} = 2{,}500 \text{ sf} \times X$$

$$X = \frac{\$8{,}500 \times 2{,}238 \text{ sf}}{2{,}500}$$

$$X = \$7{,}609.20$$

We'll round up and estimate $7,610 to paint our 2,238 square-foot house.

*Enter your estimate for the cost of **painting** on line 16.1 of your Cost Estimate Worksheet. We'll use $7,610 in our Example Cost Estimate Worksheet.*

Floor Covering

We will install wall-to-wall carpet throughout the house (including the main stairs), except in the entry, kitchen, bathrooms and utility area. The entry floor will be tiled in slate, and the bathrooms and utility area will be floored with ceramic tile. The kitchen floor covering will be vinyl tile. We'll estimate each of these areas separately.

Entry

We'll calculate the slate for the entry on a per-square-foot basis. The entry is 12′4″ × 7′3″. We'll round up each dimension before calculating the area to be estimated, which gives us dimensions of 13 × 8 feet or 104 square feet. We're going to install 12-inch slate squares, so the installer will have to cut some pieces from full squares. Since we rounded up our dimensions, we shouldn't need any other allowance for waste. The installer should be able to cut many of the smaller pieces from one tile.

We'll estimate 104 square feet of 12-inch square slate tiles for the entry.

*Enter your estimate of the **square footage** and **floor covering type (slate)** on line 16.2.2 of your Cost Estimate Worksheet. We'll use 104 square feet of slate in our Example Cost Estimate Worksheet.*

Bathrooms and Utility Area

The tile we've chosen comes in 8-inch squares, and we plan to lay them on thinset mortar. Let's calculate the floor area of the three bathrooms and the utility area and estimate on a per-square-foot basis.

Master bath (5′6″ × 10′9″)	5.50′ × 10.75′ = 59.13
Family bath (5′6″ × 10′9″)	5.50′ × 10.75′ = 59.13
Downstairs bath (6′9″ × 10′0″)	6.75′ × 10.00′ = 67.50
Utility area (9′9″ × 10′0″)	9.75′ × 10.00′ = 97.50
Total floor area:	283.26 (rounded to 284) sf

Ceramic tile is sold by the box, and suppliers rarely split boxes. If you're building a single property, determine the square footage a single box of tiles will cover, and round up to the next number of boxes. If you're building several homes, the surplus from one house can be used in another.

There's no need to allow for waste; there are floor areas included in the square footage that won't be tiled — under the tub, for instance. These areas will give us our allowance for waste.

> *Enter your estimate for* ***square footage*** *and* ***floor covering type (ceramic tile)*** *on line 16.2.3 of your Cost Estimate Worksheet. We'll use 284 square feet of ceramic tile in our Example Cost Estimate Worksheet.*

Kitchen

Our homeowner is undecided about kitchen flooring. But he wants to move into his home as soon as possible. Finished floors and countertops must be installed before a house can pass inspection for final occupancy. We'll install vinyl flooring tile so the house will pass final inspection, and he can move in.

The kitchen floor area is 151.71 square feet (11′8″ × 13′00″ or 11.67′ × 13.00′). We'll round that up to an even 152 square feet and estimate on a per-square-foot basis, using 12-inch square vinyl tiles, set on adhesive.

Vinyl tiles are sold by the whole box. So, we may have to round up again to the next whole box of tiles. Again, there's no need to allow for waste. There's enough area counted in our square footage — under the cabinets, for example — that won't need floor covering to account for waste.

> *Enter your estimate for* ***square footage*** *and* ***floor covering type (vinyl tile)*** *on line 16.2.4 of your Cost Estimate Worksheet. We'll use 152 square feet of Harbour vinyl 12-inch-square tile in our Example Cost Estimate Worksheet.*

Carpeted Living Areas

We'll install medium-quality (25- to 35-ounce face weight) nylon wall-to-wall carpet with 5-pound, ½-inch-thick rebond foam padding throughout the rest of the house (living room, dining room, family room, bedrooms, dressing area, hallways and the main stairs). Our house has a total of 2,238 square feet, but we've already calculated other floor coverings for the following square footage:

Entry	104 sf
Bathrooms and utility area	284 sf
Kitchen	152 sf
Total:	540 sf

That leaves 1,698 square feet or 188.66 square yards of area to be carpeted. We'll round up to 190 square yards.

You have three choices to help you estimate the carpet and pad — get a price from a carpet installer, calculate the area and make your own estimate, or make a proportional estimate based on a previous job. The last choice is the quickest and gives a good ballpark number, especially if you plan to give the owner a carpet allowance.

We'll base our estimate on cost records for a similar project. The cost of laying 225 square yards of the same quality carpeting in another two-story house was $8,000. Set up the following proportion, with the cost of our 190 square yards as X:

$$\$8{,}000 : 225 \text{ sy} = X : 190 \text{ sy}$$

$$\text{Product of extremes} = \text{Product of means}$$

$$\$8{,}000 \times 190 = 225 \times X$$

$$X = \frac{\$8{,}000 \times 190 \text{ sf}}{225 \text{ sy}}$$

$$X = \$6{,}755 \text{ (rounded up to \$6,800)}$$

We'll estimate laying 190 square yards of medium-quality carpet and padding at $6,800.

*Enter your estimate for the cost of **carpeting** on line 16.2.1 of your Cost Estimate Worksheet. We'll use $6,800 in our Example Cost Estimate Worksheet.*

Appliances

The plans and specs call for a double oven, counter-mounted cooktop, range hood, dishwasher, and garbage disposal. The owner is supplying the refrigerator. If you give your customer the option to select the appliances before installation, make an allowance for them, with the owner paying any difference in price when the final purchase is made. A reasonable guesstimate for appliances is $3,800. If you're building a spec house, appliances have to be installed before the house can be sold, so you'll need an accurate estimate. We'll estimate our kitchen appliances now, since the client has already chosen what he wants. We'll install:

1 double oven: wall-mounted, SS finish, 30 inches wide
1 cooktop: electric, 4-element, smooth ceramic
1 range hood: under-cabinet mount, 400 CFM, 30 inches wide, SS finish
1 dishwasher: standard, under-counter installation
1 garbage disposal: standard, ¾ HP

*Enter your estimate for **appliances** on line 16.3 of your Cost Estimate Worksheet. We'll use the above appliances in our Example Cost Estimate Worksheet.*

Labor

We'll estimate 10 manhours for a carpenter to trim the interior doors after the carpet is installed, and install the appliances. The cost of the appliances includes the electrical and plumbing connections. Our electrical contract includes the cost of electrical hook-up for the appliances.

*Enter your estimate for carpenter **manhours** on line 16.4 of your Cost Estimate Worksheet. We'll use 10 hours in our Example Cost Estimate Worksheet.*

COST ESTIMATE WORKSHEET FOR PAINTING, FLOOR COVERING, AND APPLIANCES

#	Item	Qty		Size		Cost	Per			Subtotal
16.1	**Painting, labor and material and tax**								$7,610.00	
	Wallpaper, labor and material								$0.00	
									$7,610.00	$7,610.00
16.2	**Floor covering**									
16.2.1	**Carpet, installed, incl tax**								$6,800.00	
16.2.2	**Slate tile, installed, incl tax**									
		104 sf	natural slate	12" x 12"	@	$11.96	sf	=	$1,243.84	
16.2.3	**Ceramic tile, installed, incl tax**									
		284 sf	glazed ceramic	8" x 8"	@	$9.20	sf	=	$2,612.80	
16.2.4	**Vinyl tile, installed, incl tax**									
		152 sf	Harbour, 0.08"	12" x 12"	@	$2.31	sf	=	$351.12	
									$11,007.76	$11,007.76
16.3	**Appliances, incl tax**									
	Cooktop		Electric, smooth ceramic 4-element						$465.05	
	Oven		Double wall oven, SS, 30" wide						$1,637.43	
	Range hood		Under-cabinet mount 400 CFM, 30" wide						$540.64	
	Dishwasher		Standard, under-counter installation						$715.60	
	Garbage disposal		Standard ¾ HP kitchen disposal						$278.00	
									$3,636.72	$3,636.72
16.4	**Labor for carpenters***									
		10	manhours - carpenter, BC		@	$32.09	hour	=	$320.90	$320.90
			Total (entered on line 16 of Form 100)							**$22,575.38**

* Note: Coordinate work with the subcontractors.

Chapter 17

Gutters, On-site Improvements and Miscellaneous

IN THIS CHAPTER WE'LL look at items that we haven't specifically covered in previous chapters, but that need to be included in your final estimate.

Gutters

Gutters control runoff from roofs, and are designed to prevent damage by directing water away from the structure's foundation. They're sometimes omitted when there's a roof overhang exceeding 12 inches for a single-story house or 24 inches for a two-story home.

Gutters can be made of galvanized steel, aluminum, copper, plastic, or wood. Seamless aluminum gutters are popular because they're prefinished and don't need painting. They're made from rolled aluminum and can be constructed without seams, except at corners and downspouts. For a large housing project, set up gutter-making equipment on-site if gutters are needed on all the structures. If only one house is being constructed, you can purchase gutters in 10-foot lengths, and join them with section connection kits. The simplest way to estimate the number of connection kits you need is to divide the total length of gutter by 10.

As a rule, downspouts, or *leaders*, should have 1 square inch of cross section for each 100 square feet of roof surface. Using that as a guide, you can get a rough estimate of the number and size of downspouts needed. Typically, gutters slope toward each corner of a roof, where downspouts are installed.

Installing gutters is specialty work, and the cost of labor and materials is normally based on the linear feet of material installed. To estimate, determine the linear feet of the eaves, including overhangs. Round to the next increment of 10, if you use 10-foot sections. You'll also need to estimate for downspout material and the number of connection kits required. Each downspout typically includes the two elbows that connect to the gutter and downspout, and the one that connects either to the drain or diverts the water to a splash block.

Add in the cost of trenching if a drain pipe is run from the downspouts to a point of discharge. If you don't need drain pipe, you'll have to supply splash blocks. Splash blocks are usually made of plastic or concrete, at least 12 inches wide and 24 inches long. Make sure they're heavy and embedded securely enough in the ground to prevent movement.

On-Site Improvements

The plans and specs provide dimensions for the driveway and the type of finished surface it's to have. Compact the soil thoroughly before spreading asphalt or concrete if the driveway is over filled ground. The fill dirt may have been compacted when you placed it, but the weight of construction vehicles and the effects of the elements (freezing and thawing) will cause additional settling for several months.

Other improvements on your estimate could include retaining walls, fences, patios, and accessory buildings. All work outside the basic estimate can be done on a cost-plus basis, or for a contract price. Any required extra work should be detailed in writing, including the method of payment.

Miscellaneous

Some items that may not be listed on the specification sheet but must be included in the cost estimate are:

- Additional lumber for scaffolding
- Polyethylene for covering materials
- Equipment rental
- Additional nails, lag screws, and bolts
- Weatherstripping materials, if not included elsewhere
- Access doors to crawl spaces, attics, etc.
- Folding attic stairs
- Gravel for a temporary driveway during construction
- Hardware, such as screen wire (for vents, where needed), corner braces for shelves, magnetic catches and hinges

- Molding not included elsewhere
- Additional caulking
- Light bulbs needed during construction
- Railings not included elsewhere
- Undereave soffit vents, if not included elsewhere
- Special equipment, such as attic fans, not included elsewhere
- Cleaning materials and cleanup expense

These are just a few of the things that can get overlooked. Some items may be listed elsewhere in the bid, but they're worth noting here as a precaution. Their combined cost can be considerable.

Example Estimate

Let's estimate the costs of gutters, on-site improvements, and the miscellaneous items needed for our sample house.

Gutters, Downspouts and Splash Blocks

We'll calculate the gutters and downspouts we plan to install in linear feet. The water discharge will be directed onto splash blocks through elbow joints at the bottom of each downspout. All connecting pieces are included in the cost of the downspouts.

Gutters

Figure the length of the eaves at the front and rear of the house and the garage front to determine the length of gutter we'll need. Allow 10 percent for waste, and round up to the next 10-foot length.

Two-story section, 2 × 42′ lengths =	84.00
Single-story section, 2 × 25′ lengths =	50.00
Garage front, 1 × 26′4″ length =	26.33
Subtotal:	160.33
Add 10 percent for waste:	16.03
Total:	176.36 (rounded to 180) linear feet

We'll need 180 linear feet, or 18 10-foot pieces, of gutter.

Enter your estimate for linear feet of ***gutter*** *on line 17.1.1 of your Cost Estimate Worksheet. We'll use 180 linear feet of aluminum gutter in our Example Cost Estimate Worksheet.*

Gutter Kits

To find the number of gutter connection kits we need, divide the total length of gutter by 10 (for 10-foot lengths). This will give a little extra connection kit material.

$$\frac{180 \text{ lf}}{10} = 18$$

We'll need 18 gutter connection kits.

> *Enter your estimate for the number of **connection kits** on line 17.1.2 of your Cost Estimate Worksheet. We'll use 18 connection kits in our Example Cost Estimate Worksheet.*

Downspouts

We need downspouts at each corner of the two main roof sections and at the two front corners of the garage. At the rear of the two-story section, the downspouts will be the full two stories (20-foot lengths) long. At the front of the two-story section, the downspouts will lead to the garage offset (10-foot lengths). At the front and rear of the one-story section, and at the front of the garage, the downspouts will be 10-foot lengths. We need the following lengths:

Two-story, rear	2 × 20′
Total:	2 – 20′ lengths
Single-story, front and rear	4 × 10′
Two-story, front	2 × 10′
Garage front	2 × 10′
Total:	8 – 10′ lengths

We'll need two 20-foot lengths and eight 10-foot lengths of downspout.

> *Enter your estimate for the number and lengths of **downspouts** on line 17.1.3 of your Cost Estimate Worksheet. We'll use these totals in our Example Cost Estimate Worksheet.*

Splash Blocks

We'll need 8 splash blocks for the downspouts that terminate at grade; the two other downspouts terminate at the garage roof offset, so they don't need splash blocks.

> *Enter your estimate for the number of **splash blocks** on line 17.1.4 of your Cost Estimate Worksheet. We'll use 8 splash blocks in our Example Cost Estimate Worksheet.*

On-Site Improvements

The only on-site improvements in the specifications for our sample house are paving the driveway and an allowance for landscaping. The driveway needs a base of crushed stone and recompaction before it can be paved.

Crushed Stone

The driveway dimension from the garage to the street is 16 feet × 77 feet. Since the driveway is on fill, we need to lay a final 4-inch-thick layer of crushed stone before the permanent driveway is placed. Sometimes crushed stone is laid before construction of the house begins, to provide an adequate working surface. However, by the end of construction, the surface has usually been eroded and needs to be compacted again prior to final surfacing. Estimate stone as follows:

$$\frac{16.00' \times 77.00' \times 0.33'\ (4")}{27\ (\text{cf to cy})} = 15.06\ \text{cy}$$

$$\text{Cubic yards} \times 1.35\ \text{factor} = \text{tons}$$

$$15.06\ \text{cy} \times 1.35\ \text{factor} = 20.33\ (\text{rounded to } 21)\ \text{tons}$$

We'll need 21 tons of crushed stone for the driveway.

> *Enter your estimate for **crushed stone** on line 17.2.1 of your Cost Estimate Worksheet. We'll use 21 tons of crushed stone in our Example Cost Estimate Worksheet.*

Paving

After the fill is compacted, and the crushed stone placed, the driveway can be finished with asphalt paving or reinforced concrete. If you opt for reinforced concrete, use the information from Chapter 14 to estimate for concrete, reinforcing bar, ties, and expansion joints. If your plans call for asphalt paving, lay a 3-inch layer of asphalt over the consolidated crushed stone, and top it with an oil seal finish coat. Typically, the installation of residential driveways is subcontracted, with the cost based on square footage, including material, labor, equipment, and the subcontractor's overhead and profit.

For our sample house, we'll have a subcontractor lay a 3-inch mat of asphalt paving on the 16- × 77-foot driveway. Let's estimate how many square feet of asphalt we'll need by calculating the driveway area:

$$16.00' \times 77.00' = 1{,}232\ (\text{rounded to } 1{,}250)\ \text{sf}$$

We'll need our subcontractor to put down 1,250 square feet of asphalt in a 3-inch layer.

> *Enter your estimate for **asphalt** on line 17.2.2 of your Cost Estimate Worksheet. We'll use 1,250 square feet of 3-inch asphalt in our Example Cost Estimate Worksheet.*

Landscaping

If you're asked to estimate or give an allowance for landscaping, the details of which have yet to be determined, make a cost allowance that can be credited toward landscaping. You'll most likely estimate that work separately at a later date.

> *Enter your estimate or allowance for **landscaping** on line 17.3 of your Cost Estimate Worksheet. We'll use an allowance of $3,000 in our Example Cost Estimate Worksheet.*

Miscellaneous

It's good to add a little "insurance" to cover items you might've missed in your estimate. Additional nails, hardware, cleaning materials and clean-up expenses can be entered here. Past job records help you determine costs for items to include that aren't covered elsewhere in the estimate. Always compare your actual costs with your estimates, once a job is completed. This helps you see how close or how far off your estimates are, and what you've forgotten to add in the past. The goal is for your estimated and actual costs to be as close as possible.

> *Enter your estimate for **miscellaneous** items on line 17.5 of your Cost Estimate Worksheet. We'll use an allowance of $500 in our Example Cost Estimate Worksheet.*

Labor

Labor for installing the gutters and downspouts and laying the asphalt driveway were included in the contractors' quotes for those jobs. The labor to spread the stone on the driveway can be estimated using previous job records. For example, in Chapter 5 we estimated that it took 12 manhours to spread 18 tons of crushed stone in the basement of our sample house.

Let's use the figures from Chapter 5 and set up a proportion to estimate the labor it will take to spread the 21 tons of crushed stone for our driveway. The number of manhours to spread the driveway stone will be X. Set up the proportion as follows:

$$\text{12 manhours : 18 tons of stone} = \text{X manhours : 21 tons of stone}$$

$$\text{Product of extremes} = \text{Product of means}$$

$$\text{12 manhours} \times \text{21 tons} = \text{18 tons} \times \text{X manhours}$$

$$X = \frac{\text{12 manhours} \times \text{21 tons}}{\text{18 tons}}$$

$$X = \text{14 manhours}$$

Our calculation shows that it will take 14 manhours to spread the 21 tons of crushed stone for the driveway.

> *Enter your estimate for the number of* ***manhours*** *on line 17.6.1 of your Cost Estimate Worksheet. We'll use 14 manhours in our Example Cost Estimate Worksheet.*

That completes the cost estimate for our sample house shown in Figures 7-1 and 7-2. In Chapter 18, we'll discuss Overhead, Contingency, and Profit and show you how to add it into your estimate.

COST ESTIMATE WORKSHEET FOR GUTTERS, ON-SITE IMPROVEMENTS, AND MISCELLANEOUS

#	Qty	Item	Size	Cost Per		Subtotal
17.1	**Gutters and downspouts, incl labor**					
17.1.1	**Gutters**					
	180 lf	aluminum	5" × 10'	@ $3.11 lf	= $559.80	
17.1.2	**Connection kits**					
	18	aluminum	5"	@ $3.91 piece	= $70.38	
17.1.3	**Downspouts (incl ells, etc)**					
	2	aluminum	20'	@ $67.76 piece	= $135.52	
	8	aluminum	10'	@ $46.96 piece	= $375.68	
17.1.4	**Splash blocks**					
	8	heavy plastic	11¼" × 23½"	@ $10.29 piece	= $82.32	
					$1,223.70	$1,223.70
17.2	**On-site improvements**					
	Driveway					
17.2.1	21 tons	crushed stone		@ $17.21 ton	= $361.41	
17.2.2	1,250 sf	asphalt		@ $2.25 sf	= $2,812.50	
17.2.3	cy	concrete		@ ____ cy	= $0.00	
17.2.4	lf	rebar		@ ____ lf	= $0.00	
17.2.5	lbs	tie wire		@ ____ lb	= $0.00	
17.2.6	lf	expansion joints		@ ____ lf	= $0.00	
17.2.7	lf	culvert pipe		@ ____ lf	= $0.00	
					$3,173.91	$3,173.91
17.3	**Landscaping, incl tax**				$3,000.00	$3,000.00
17.4	**Other on-site improvements** (List on separate sheet and enter cost here)				$0.00	$0.00
17.5	**Miscellaneous** (If necessary list on separate sheet and enter cost here)				$500.00	$500.00
17.6	**Labor**					
17.6.1	14	manhours - one laborer, BL		@ $26.64 hour	= $372.96	
					$372.96	$372.96
				Total (entered on line 17 of Form 100		**$8,270.57**

Chapter 18

Overhead, Contingency and Profit

HAVING GREAT BUILDING SKILLS won't guarantee your success as a contractor; you also need good business skills. You have to *plan your profit* into each job. That means knowing your overhead and estimating it as a percentage of every job; planning for the unexpected; and always adding a fair profit for your business into your estimates.

In this chapter, we'll go over how to approach each of these items so you'll come out with a profit on the jobs you estimate.

Overhead

Expenses that aren't shown on plans and specifications are usually combined and referred to as *overhead.* Some of your overhead expenses are:

- FICA (employer's share of Social Security taxes)
- FUTA (federal and state unemployment taxes paid by employer)
- Workers' compensation insurance
- Liability insurance
- Administrative costs (payroll and end-of-year statements)
- Car and truck expenses
- Cell phone cost
- Interest on borrowed money
- Office costs
- Employee sick leave, vacation, and holiday pay

If you're building on spec, you'll have to pay these costs, plus most of the following:

- Fire and liability insurance
- Real estate taxes
- Advertising expenses
- Sales commissions
- Legal fees

These expenses, which are the cost of doing business, are commonly separated into two categories: *direct* and *indirect overhead.*

Direct Overhead

Direct overhead, sometimes called *labor burden,* includes the costs involved in employing labor, other than actual wages or salaries. We'll be discussing those costs categories next.

Fringe Benefits

Fringe benefits include vacation pay, holiday pay and sick leave. Fringe benefits account for roughly 5.15 percent of base wages in the building industry. Remember, these benefits are *in addition to* the basic wage you're paying employees.

Insurance and Employer Taxes

Included under these categories are workers' compensation, contractor's casualty and liability insurance, and unemployment insurance. Rates vary considerably from state to state and by trade type (the higher the risk of accident, the higher the rate; i.e. roofers have a higher workers' comp rate than secretaries). Another important factor affecting the rate is the contractor's loss record. Insurance and taxes add between 21 (electricians) and 44 percent (roofers) to the wage, generally averaging 25 to 35 percent. Social Security and disability insurance are included under FICA, to which employees also contribute.

Medical and Pension Costs

These costs include the employer's contribution to medical insurance and tax-deferred pension plans, such as a 401k and profit sharing, adding another 4.55 percent of base wages to your expenses.

Let's see how all this adds up for one carpenter. Numbers vary for different trades, but detailed costs for all trades can be found in the *National Construction Estimator*, published by Craftsman Book Company.

Carpenter's hourly wage	$22.42
Fringe benefits (5.15 percent of wage)	1.15
Subtotal:	23.57
Insurance and employer taxes (31.83 percent of subtotal)	7.50
Medical and pension costs (4.55 percent of hourly wage)	1.02
Total hourly cost:	$32.09

None of these costs are paid by the employee. Employees *do* contribute to FICA, but that contribution isn't included here. The additional amounts are part of the rate you use to estimate your labor costs. Rates used in the Example Cost Estimates at the end of each chapter include these direct costs. They're the total hourly costs for the tradesman doing the work covered in the chapter.

Indirect Overhead

Indirect overhead may include two types of expenses. Those that you haven't assigned to a particular job, but should have, and your general business costs that ultimately should be divided among all your jobs.

The first group, indirect overhead costs associated with jobs, is often overlooked in estimates and so becomes a general business expense. These expenses are usually listed in the *General Conditions* section of a contract, and they're the responsibility of the general contractor. They have to be included somewhere, either in indirect overhead, or better, added separately to the specific job. It's up to you where you want to list them. But remember, they're real costs of the job and should be included as a part of your estimate. Make a list of these incidental costs and put it on the wall over your computer so you don't forget to include them when necessary.

This group of costs may not be associated with any particular trade or material. Instead, they may be incurred as a result of performing a particular job — setting up barricades, paying special bonds (maintenance, street repair), applying for building permits and fees, clean-up, design fees, rent for a job shack, 'port-a-potty', or job signs, public protection during construction, repairing damage, sewer connection fees, street closing fees, surveying, temporary fencing, temporary utilities (phone, water, electricity), hiring watchmen and paying water meter fees.

Most often, your indirect overhead falls into the second group. This group includes all the costs of doing business other than buying materials and paying wages and benefits for employees working on the job. Even if all construction stops and there are no jobs, you'll still have this indirect overhead to pay. These expenses exist as the day-to-day operating cost of the business, and include: office rent, office insurance, office phones, office staff salaries and benefits, office supplies, office utilities, accounting fees, legal fees, licenses, association dues and subscriptions, advertising, postage, equipment purchases, interest, automobiles, depreciation on vehicles and equipment, maintenance and repairs, small tools, taxes, travel expenses, donations, entertaining, and uncollectibles. You must assign a percentage of these costs to each job.

Indirect overhead is somewhat harder to estimate than direct overhead. The good news is that you generally only have to do it once annually, unless your business circumstances change dramatically during the year. In that case, you'd need to recalculate. The best way I've found to determine indirect overhead is to make two calculations, both based on the answers to these two questions:

1. What do you anticipate your total administrative or non-job-specific costs to be for the coming year? Either you or your accountant should be able to make a pretty good estimate using your records from the previous year. Total the indirect overhead costs, then adjust up or down to reflect the expected costs for the coming year.
2. What do you expect your gross volume of business to be in the coming year? Be realistic. It's better to underestimate here than overestimate.

Divide the total anticipated indirect overhead costs by your expected gross volume, and convert that to a percentage (multiply by 100). In most average-sized construction companies, the result will be about 15 percent. Subcontractors may run a little lower, while remodelers and insurance repair contractors may be higher. Enter that percentage on line 19.1, the *Overhead* line on Form 100, found at the end of this chapter, the last sheet in our Cost Estimate Workbook.

*Enter your indirect **overhead** figure on line 19.1 on Form 100. We'll use 15 percent in our example. That's 15 percent of the subtotal from line 18.*

Contingency

Contingency allows you to cover unforeseen expenses, from errors in your takeoff to unanticipated weather delays or labor strikes. Contingency can be a fixed dollar amount or a percentage of labor and material costs plus indirect overhead. It's fairly common to include a 2 to 5 percent contingency in an estimate. Be aware, though, it also increases your job bid by that amount, which could be a disadvantage if you're competing for a contract. It's a matter of judgment and experience. You decide whether you need to allow for contingencies and — if you do include them in your estimate — how much you can include without reducing your chances of getting the contract.

> *Enter your **contingency** figure on line 19.3 of Form 100. We'll use 3 percent in our example. That's 3 percent on top of the subtotal including overhead.*

Profit

You're in business to make a profit. And you want to make the most profit you can; not just get by. After all, building homes is your livelihood and you want to be successful.

Is it better to make a small profit margin on a high volume of business or a high profit margin on a small volume of business? The bottom line: the amount of actual profit you earn may be the same either way; it's just the route you take to get there that can be very different. Gauging your profit margin correctly depends on how well you can predict your volume of business each year. A business that flourishes with a small number of high-end sales per year, like a Ferrari dealership, needs to make a very high profit on each sale to succeed. A supermarket chain with huge retail numbers can do very well on low profit margins.

What profit margin percentage should you apply to your estimates? You'd probably like to use as high a figure as possible. But construction is a competitive business — you're almost always bidding against other companies. You have to be practical about what you can charge for your work. If your bid is too high, you won't get the contract. If you bid low, you could get every contract you bid on, but not make any profit. Your goal is to make a profitable living, not break even — and certainly not to lose money.

Typically, estimators or contractors begin by applying a 10 percent profit margin to their estimates, and see how it goes. If you give that percentage a try and continually get underbid, it could be for a number of reasons. Maybe your company isn't very efficient, or maybe your estimates aren't as accurate as they should be. The construction business could be in a slump, as it is periodically. Or there's always the possibility that the competitors are poor estimators, and *they're* underbidding and not making money. If you believe you're running an efficient operation and your estimates are accurate, maybe you need to lower your profit margin percentage a bit to capture some of the business that's getting away.

On the other hand, if you're winning more contracts than your company can effectively handle, you could be using too low a profit margin percentage. You should increase it. If you're using a fairly high percentage figure and still getting plenty of business, increase it just enough to win the contracts the company can comfortably handle.

It's clearly a juggling act. Your profit margin percentage shouldn't be etched in stone. Adjust it to correspond with current business trends and the number of competitors bidding for jobs. If you can keep your profit margin consistently above 10 percent, you should be on your way to success.

Contingency is a percentage of your labor and material costs plus indirect overhead. Profit is a percentage of labor and material costs plus indirect overhead *plus* contingency. Remember that, as you're finalizing your costs. If you only take your profit based on a percentage of the labor and material costs, your actual profit margin will be too low. Always apply profit margin to the indirect overhead *and* contingency costs.

*Enter your **profit** percentage figure on line 19.5 of Form 100. We'll use 10 percent in our Example Form 100. That's 10 percent on top of all the other amounts that have been added.*

All costs from the previous chapters have been entered on our Example Form 100, including the indirect overhead, contingency, and the profit margin percentage we've selected. The sales price for the property is calculated on line 20. That's the price you ideally think you can get for the house, including profit and expenses, always taking into consideration the current market. Remember to add enough to your sales price to allow for some buyer/seller negotiation, and for a real estate broker's commission, if one is involved in the sale. Brokers generally get 6 percent of the selling price, so make sure your asking price covers that additional amount.

If you're bidding on a project and don't anticipate any price negotiation, you can bid using your calculated sales price with some degree of confidence. If you really want the job and expect stiff competition, you may opt to reduce your bid somewhat. Again, it's a matter of judgment and business expertise — skills you'll acquire through experience.

FORM 100
Estimating Home Building Costs

Date ____________

1	**Building site**							$70,900.00
2	**Preliminary costs**							$10,320.00
3	**Site clearing, excavation and fill dirt**							$15,830.00
4	**Footings**							$10,965.75
5	**Foundation**							$17,678.41
6	**Floor system**							$9,783.15
7	**Superstructure**							$30,701.75
8	**Roofing**							$5,056.50
9	**Electrical, plumbing, heating and air conditioning**							$53,250.00
10	**Brickwork**							$24,485.57
11	**Energy-saving materials**							$3,824.66
12	**Interior wall and ceiling finish**							$14,685.23
13	**Exterior trim**							$20,364.86
14	**Concrete floors, walks and terrace**							$7,782.75
15	**Interior trim**							$31,170.76
16	**Painting, floor covering and appliances**							$22,575.38
17	**Gutters, on-site improvements and miscellaneous**							$8,270.57
18	**SUBTOTAL**							$357,645.34
19	**Overhead, contingency and profit**							
19.1	Indirect overhead	Line 18	$357,645.34	×	15 %	=		$53,646.80
19.2								$411,292.14
19.3	Contingency	Line 19.2	$411,292.14	×	3 %	=		$12,338.76
19.4								$423,630.90
19.5	Profit	Line 19.4	$423,630.90	×	10 %	=		$42,363.09
20	**SALES PRICE**							**$465,993.99**

How to Use the CD-ROM

Contained on the CD-ROM bound into the back of this book are all the estimate worksheets that appear at the end of each chapter, the summary sheet, and any extra worksheets used within the chapters. These are provided as both PDFs and as Excel spreadsheets. If you prefer blank forms that you'll fill out by hand, print the PDFs.

To install the files to your computer, put the CD in the drive and follow the instructions on your screen. When installation is complete, you'll notice a new folder under *Documents* (or *My Documents*). This folder will be labeled *Craftsman*. After opening this folder, you'll find another folder labeled *Estimating Home Building Costs.* Contained in this folder will be the Excel and PDF versions of the *Cost Estimate Workbook* files, along with four folders, one for each of the four chapters that reference Excel and PDF documents. The files copied off the CD will be placed into the *Excel* and *PDF* folders under each chapter folder.

The original unmodified forms will always be available on the CD. These forms can simply be copied off the CD and pasted to any location on your computer. Files copied off the CD in this manner are *Read Only.* You can remove this by *right-clicking* on the file (located on your computer, not on the CD), selecting **Properties**, then unchecking the **Read Only** attribute. Then click **OK**.

Many of the *Excel* forms already have the formulas necessary to do the arithmetic calculations. The cells that contain these formulas have been locked, or protected, so the user won't accidentally change or delete a formula. If you want to change any of these formulas to suit your needs, you must first unprotect the form. The forms aren't password protected.

To *un*lock cells in a protected *Excel* worksheet:

1. Click **Tools**
2. Click **Protection**
3. Click **Unprotect Sheet**

To lock/protect cells in an Excel worksheet:

1. Select **All** (Ctrl-A)
2. Click **Format**
3. Click **Cells**
4. Click on the **Protection** tab
5. Uncheck the **Locked** option
6. Click **OK**
7. Select the cells you wish to protect
8. Click **Format**
9. Click **Cells**
10. Click the **Protection** tab
11. Check the **Locked** option
12. Click **OK**
13. Click **Tools**
14. Click **Protection**
15. Click **Protect Sheet**
16. Leave *Password* blank = no password
17. Click **OK**

Note: To re-install any of these files, insert the CD into the CD drive and follow the instructions to install the files. When asked, choose the **Repair** process.

Glossary

A

Abrasive: Granular coating on sandpaper and grinding wheels used for smoothing.

Access door: A small door used for entry into attics and crawl spaces.

Accessory buildings: Small buildings used for storage and utility purposes.

Adhesive: A sticky or bonding substance used to hold materials together.

Aesthetics: Dealing with visual attractiveness.

Aggregate: Sand and stone added to cement to make concrete.

Air entraining agent: A chemical added to concrete which causes microscopic air bubbles for the purpose of resisting freezing.

Anchor bolts: Steel bolts embedded in concrete to tie the sill to the foundation.

Angle iron: A structural piece of steel shaped to form the letter "L" in a cross section.

Apron: Inside window trim placed against the wall under the stool.

Architect: One who is specially trained to design buildings.

Areaway: An enclosed space below grade adjacent to a basement window which allows ventilation and light to enter.

Arterial street: The principal street for through traffic.

Asphalt shingles: Shingles made from asphalt-impregnated felt and covered with mineral granules.

Astragal: A small molding shaped to form the letter "T" used between double doors.

Atrium: An open court within a building.

Attic: The space under a roof and above the ceiling of a house.

B

Backfill: Earth used to fill around foundation walls or as fill for excavation.

Baluster: One of a set of small vertical members that support the handrail of a balustrade for a stairway.

Balustrade: The complete set of balusters and handrail serving as an enclosure for stairways and balconies.

Base shoe: A small molding between the floor and baseboard.

Baseboard: The finish board lining the plaster or gypsum wallboard where it meets the floor.

Basement: The lowest floor area in a building. Normally it is partially or entirely below ground level.

Batten: A narrow strip of wood nailed vertically over the joints of vertical siding.

Batter boards: Horizontal boards set to a predetermined elevation constructed at the corners of a proposed building from which lines are stretched to locate the outline of the foundation.

Beam: A horizontal structural member used to support loads.

Bearing: In architecture, the wall that supports a load, such as a bearing wall. Also the portion of a piece of timber or steel which rests upon a wall.

Bench mark: A fixed point on some permanent object used as a reference in determining floor and grade elevations.

Bird's mouth: The cutout at the lower end of a rafter that rests on the wall plate.

Blueprints: An exact plan reproduced on sensitized paper in white lines on a blue background. Now more frequently done in blue or black lines on a white background.

Board foot (abbreviated bfm): A unit of measure for lumber. One bfm is 1 inch thick, 1 foot wide and 1 foot long.

Bottom chord: The lower horizontal member of a truss replacing the ceiling joist.

Brick: Masonry units made from baked or burned clay and molded into oblong blocks used in buildings and walls.

Bridging: Cross bracing between floor joists for reinforcement and distribution of floor loads.

BTU (British Thermal Unit): One BTU is the amount of heat required to raise the temperature of one pound of water 1°F.

Builder: One who specializes in one type of construction such as residential building, as opposed to a general contractor who engages in many types of construction.

Building line: The line of the outside of a foundation.

Building permit: An authorization from the local governing body to construct or remodel a building.

Building site: The location of an actual or planned structure.

Built-up beam: Several framing members nailed together to act as one beam. For example, a 6" × 10" built-up beam is made up of three 2″ × 10″ members nailed together.

Butts, door: Door hinges.

C

Calcium chloride: A white crystalline compound used in concrete to accelerate curing time and to retard freezing.

Carport: An open-sided shelter for an automobile.

Carriage (or stringer): The framing material that supports stair steps.

Casing: Finish trim around windows and doors.

Ceiling joists: Horizontal framing members resting on the wall plate to support the ceiling.

Cesspool: An underground catch basin to receive and retain sewage.

Chalking: Loose powder formed from a paint surface.

Chimney: Sometimes called a flue. A passage through which smoke or fumes escape from the furnace and/or fireplace.

Chimney saddle: A small sloping roof in back of a chimney, used to shed water.

Collar beam: A horizontal beam just below the ridge tying rafters together.

Collector street: A secondary street that serves as a feeder street.

Column post: A vertical shaft or pillar used to support loads.

Common rafter: A roof timber sloping up from the wall plate to the ridge.

Compaction: The process of applying pressure to loose material to remove air pockets and to make the mass denser.

Composition roofing: A term referring to roofing with asphalt materials.

Concrete: A hard compact substance made of sand, gravel, portland cement and water.

Condensation: The moisture produced when warm moist air comes in contact with a cold surface.

Conduit: A tube for electrical wires.

Contingency: Something likely to happen, such as uncertain conditions and unforeseen costs.

Contingency fund: A sum of money set aside for unforeseen costs.

Contractor: An individual or company who contracts to do certain work for a stipulated sum.

Corner bead: A metal bead used on outside corners for plaster and gypsum boards.

Corner brace: Diagonal braces at the corners of stud walls.

Cornice: The part of the roof overhang that projects from the wall.

Cost-Plus contract: Work done on actual cost plus a fixed fee or percent basis.

Counter flashing: A secondary and overlapping layer of flashing.

Crawl space: An unfinished area under a floor normally with little head room.

Cripple: A structural member that is cut less than full length.

Cripple rafter (or jack rafter): A rafter whose length is less than a common rafter. They run between the ridge and valley rafter, and between the wall plate and hip rafter.

Cripple stud: Studs used over door openings and over and under window openings.

Critical Path Method: A graphical method used to control the planning and scheduling of a construction job to minimize loss time.

Cubic foot: A unit of volume having three dimensions. One cubic foot equals one foot in length multiplied by one foot in width multiplied by one foot in depth.

Cubic yard: A unit of volume having three dimensions. One cubic yard equals three feet in length multiplied by three feet in width multiplied by three feet in depth, or 27 cubic feet.

Cul-de-sac: A dead-end street with a turnaround at the end.

Culvert: A pipe under a driveway or street used for drainage.

Cupola: A small dome or similar structure on a roof.

Custom-built: A house made to order and not factory-built.

D

Deck: Subfloors are often referred to as floor decks, and the roof sheathing is sometimes referred to as the roof deck.

Deed: A legal document showing transfer of property.

Detail: The separate items of a structure drawn for the purpose of clarity.

Dimension line: A line showing the distance between two points.

Disappearing stairs: Stairs that fold out of sight when not in use, usually into an attic.

Door buck: The rough framing around the door opening in masonry or concrete walls to which the finish door frame is attached.

Door stop: A small molding around the door jambs and head against which the door closes. Also metal door stops attached to baseboards, door butts, doors and walls to protect walls and furniture from damage.

Dormer: A structure projecting from a sloping roof.

Downspout (or leader): The vertical pipe connected to the gutter for the purpose of discharging the roof water.

Drain tile: Pipe used to carry off water.

Drip edge (or drip cap): A metal strip placed over the roof sheathing along the cornice to allow the roof water to drip free into the gutters.

Drywall: Interior wall covering other than plaster.

E

Easement: The legal right a person or corporation has to use a designated portion of the land owned by another for the purpose of installing and maintaining underground utilities, or for use as a road.

Eaves: The lower portion of a roof that extends beyond the wall.

Efflorescence: Crystalline compounds appearing on masonry walls that change to a whitish powder when the water that carried them to the surface evaporates.

Elevation: Drawings of a section of a building made as though the observer was looking straight at it. Also used to denote grade and floor heights above or below a reference point such as a bench mark.

Estimator: One who prepares a bid or cost estimate on a construction project.

Excavation: Earth removal, normally for the purpose of constructing a dwelling or other structure.

Existing grade elevation: The grade elevation before any excavation or fill is done.

Expansion joint: A flexible joint used in concrete construction to allow thermal expansion and contraction, and thus prevent cracking.

F

Face brick: Brick of better appearance and quality used on the face of walls.

Factor: A number multiplied by another number to form a product.

Factory-built house: A house built in component sections in a factory and transported to the job site for final assembly.

Fascia: The vertical board on the end of rafters that are part of the cornice.

Felt: An asphalt impregnated rag-fiber paper used under roofing.

Fenestration: The arrangement of windows in a building.

Fiberboard: Fibers of various substances pressed together to form sheets.

Fiberglass shingles: Shingles with an inorganic base composed of glass fiber mat and covered with mineral granules.

F.I.C.A. (Social Security): Federal Insurance Contribution Act. Known as FICA taxes paid by employer and employee.

Fill: Adding earth, etc. to an existing grade until a required elevation is reached.

Finish grade: The final grade elevation.

Fire brick: A brick used in fireplaces that is heat resistant.

Flashing: Sheet metal, or other material, over windows and doors and around chimneys to prevent water leakage into the building.

Flitch plate: A steel plate fastened with bolts joining wood timbers to form a beam.

Floor covering: Carpet, tile, sheet goods, etc. laid over a floor underlayment for the finish floor.

Floor elevation: The height of a floor above or below a designated point such as a bench mark.

Floor joists: Framing timbers that support the floor.

Floor plan: A drawing showing room arrangements, door and window locations and all needed dimensions.

Floor system: That part of the floor that includes the sill plate, girder, floor joists, bridging and subfloor.

Floor underlayment: Materials, normally with a smooth surface such as particleboard or plywood, that are laid over the subfloor to provide a smooth base for the carpet, tile, etc.

Flooring: A name normally applied to wood flooring laid over a subfloor.

Flue: A passage in a chimney to allow the escape of smoke and fumes into the outer air.

Footing: A concrete base for a foundation wall, wider than the wall, designed to distribute the load to the soil.

Foundation: The supporting part of a structure, including the footing and the base course.

Framing: The wood skeleton of a building.

Frieze: A flat vertical board fastened to the wall.

Frost line: The depth to which frost will penetrate into the ground.

Furring strips: Wood strips used to level and receive finish surface material.

F.U.T.A. (FUTA tax): Federal Unemployment Tax Act. A tax paid by the employer on his employees.

G

Gable roof: A triangle roof sloping up from two walls.

Gambrel roof: A roof with two roof pitches from the eave to the ridge sloping up from two walls. The lower pitch is steeper than the upper pitch.

Girder: A large horizontal structural member used to support the ends of joists.

Glazing: Setting glass in windows, etc.

Grade: The surface of the ground, normally referred to around buildings.

Grade beam: A reinforced concrete beam poured over unstable soil to support loads.

Grantee: A person to whom a grant is made.

Grantor: A person who makes a grant.

Grounds: Wood strips used to control the thickness of plaster.

Gutters: A horizontal trough used to collect and carry off water.

Gypsum board: A board with a gypsum core faced with paper used in drywall construction.

H

Handrail: A protective railing used on stairways.

Hasp: A locking device used on doors.

Head joint: The vertical joint between bricks or masonry blocks.

Header: A structural member which supports the ends of the joists. It is also the name used for beams over window and door openings.

Hearth: The masonry floor in front of a fireplace.

Heat pump: A unit used for both heating and cooling.

Hip rafter: A diagonal rafter at the junction of two sloping roofs running from the plate to the ridge to form a hip.

Hip roof: A roof sloping up from all walls of a building.

Hose bibb: A water faucet used for the attachment of a hose.

House package: A term used to indicate all items purchased in a factory-built house.

I

I-Beam: A steel beam shaped in the form of the letter "I".

Infiltration: Air seepage around windows and doors.

Insulation: Special materials used over ceilings, under floors and in walls to retard transfer of heat.

Invert: The lowest point in a pipe or ditch.

Isosceles triangle: A triangle with two equal length sides.

J

Jack rafter: A rafter whose length is less than a common rafter. It runs between the ridge and a valley rafter, and between the wall plate and a hip rafter.

Jamb: The vertical member of a door or window frame.

Joist: One of the framing members that supports the ceiling or floor.

Joist hanger: A metal strap or hanger to support one end of a joist.

K

Kick plate: A metal plate attached to the bottom of a door used to protect it.

Kiln-dried lumber: Lumber dried in artificial heat such as in a kiln.

Knee wall: A low wall normally required in 1½-story houses.

L

Lally column: A steel column pipe filled with concrete for added strength.

Laminated beam: A beam glued together in layers under pressure.

Leader (or downspout): See Downspout.

Ledger strip: A wood strip nailed to the bottom of a girder or beam on which notched floor joists are attached.

Lightweight concrete: Concrete made with an aggregate lighter in weight than sand. Vermiculite, perlite and pumice are lightweight aggregates.

Linear foot: One foot in length.

Lintel: A horizontal support of wood, steel or concrete across the head of a door or window opening.

Lookout: A short horizontal framing member that extends from the wall to the rafter that supports the finish soffit material.

Louver: An opening that provides ventilation while providing protection from rain and insects.

Lumber: Timber sawed in a sawmill, such as boards and framing members.

M

Manhole: An opening to allow access to a sewer line.

Mansard roof: A roof with two roof pitches sloping up from the eave to the ridge on all four sides. The lower pitch is steeper than the upper pitch.

Mantel: The facing and/or shelf about a fireplace.

Masonry: A term used for building material such as brick, concrete block, stone, etc. bonded with mortar.

Masonry cement: A mixture of cement and hydrated lime.

MBM: Thousand (feet) board measure.

Meeting rail: The horizontal center rails of a window.

Merchant builder: One who builds houses to sell on the open market.

Metes and bounds: A system used in land surveying that describes the direction and distances of property lines until the perimeter has been traced around to the starting point.

Mil: Used to measure the diameter of wire and the thickness of paint.
1 mil = 0.001 inch.

Miter: The beveled surface cut on the end of casing and molding.

Modular house: A house built and finished in a factory and transported to the job site in two or more sections to be erected on a foundation.

Molding: A strip of material used for decoration.

Mortar: Masonry cement, sand and water used to bond together brick, concrete blocks, stone, etc.

Mullion: A vertical bar between two windows.

Muntin: Small bars separating glass panes in a window.

Mural: A large picture on wall fabric or painted on a wall.

Muriatic acid: Dilute hydrochloric acid used to clean bricks.

N

Newel post: The post at the head or foot of stairs supporting the handrail.

Nominal size: The size of lumber before it is dressed.

Norman brick: Brick with dimensions of 2¼″ × 3¾″ × 11½″, or thereabouts.

Nosing: The rounded edges of a stair tread that extend beyond the riser.

O

On center (oc): Measurement from center-to-center of framing members.

OSB: Oriented strand board, an engineered wood product formed by layering strands (flakes) of wood in specific orientations.

Outlet: The point where a lighting fixture, heater, appliance or other current-consuming device is attached to a wiring system.

Overhang: The projection of one part of a structure over another, such as a roof overhang.

Overhead, direct: Expenses associated with employing labor, other than wages and salary, such as FICA, FUTA, holiday pay, vacation pay, sick leave, etc.

Overhead, indirect: Expenses charged to a construction project that are not directly a part of the construction costs, such as administrative expenses.

P

Paneling: A term used for finishing walls with wood panels.

Parapet: A low wall around the edge of a roof.

Parging: A cement coating applied to masonry walls.

Parquet floor: Flooring laid in geometric patterns.

Particleboard: Wood chips bonded together with heat and pressure to form sheets.

Penny (d): A term indicating the length of nails, such as 8d (2½ inches) or 10d (3 inches).

Percolation test: A test to determine the ability of soil to absorb water, normally used before a permit is issued to construct a septic tank.

Perpendicular: A line at right angles to another line.

Pier: A masonry pillar to support the floor framing.

Pilaster: A pier attached to a wall used for the purpose of strengthening the wall.

Piling: Wood or concrete posts driven down to a solid base in the earth to provide safe footing.

Plank: A long, broad, thick board at least 1½ inches thick.

Planning Commission: A commission appointed by the local governing body to prepare a comprehensive plan as a general guide for the development of the area, and as a basis for the preparation of zoning and other regulations. The commission has the power of enforcement as prescribed by law.

Plat: A drawing, map or plan of a piece of land.

Plate: A horizontal framing member such as sill plate, sole plate and top plate.

Plot plan: A drawing showing the description of the lot, location and dimensions of the house, garage, walks and driveway, easements, utility lines, and finish floor and grade elevations.

Plumb: A wall that is in true vertical alignment.

Plywood: Wood built up from veneer sheets glued together under pressure with their grains at right angles to one another.

Polyethylene film: A lightweight thermoplastic film that is resistant to chemicals and moisture. It is primarily used as a vapor retarder in construction work.

Porch: A covered entrance to a building that projects from the wall and has a separate roof.

Portland cement: A cement that derives its name from a patent that was taken out in 1824 in England for the manufacture of an improved cement, which, when hardened produced a yellowish-gray mass resembling the stone found in various quarries on the Isle of Portland, England. This cement is the most widely used in concrete construction.

Precast concrete: Reinforced concrete structural units manufactured in a plant and transported to the job site.

Prefab fireplace: A fireplace built in a factory and transported to the job site.

Prefab house: See Factory-built house.

Prime coat: The first coat of paint applied to wood or metal.

Proportions: The relation between things with respect to size, amount, quantity, etc.

Purlin: A structural member laid over trusses to support rafters.

PVC: Polyvinyl Chloride pipe used for water lines, waste lines, etc.

Q

Quarry: A place where stone is excavated.

Quoins: Exterior corners of a building distinguished from the adjoining surfaces by color, size or projection.

R

Radii: The plural of radius. A straight line from the center to the periphery of a circle.

Rafter: A diagonal framing member supporting a roof.

Rafter supports: Framing members inserted (usually diagonally) under rafters to strengthen the roof.

Railing: Horizontal top member of a balustrade.

Rake: The slanting edge of a gable roof.

Rake board: The vertical board along a rake.

Ratio: A proportion of fixed relation between two similar things.

Ready-mix concrete: Concrete mixed in a concrete plant and transported to the job site in specially designed trucks.

Realty: Real estate.

Reinforced concrete: Concrete with steel rods or wire mesh inserted to increase its tensile strength.

Reinforcing rods: Steel rods used to reinforce concrete.

Renovation: A term used to describe remodeling a building.

Retaining wall: A wall that retains earth, used to eliminate steep banks.

Ridge: The highest point on a roof where two slopes meet.

Rise: The vertical height of a roof or stairs.

Riser: The vertical member of stairs between treads.

Roman brick: Bricks with dimensions of 1½″ × 3¾″ × 11½″, or thereabouts.

Roof: The top covering of a building made of joists, trusses, or rafters, roof sheathing and roofing.

Roof pitch (or roof slope): The steepness of a roof measured in the number of inches of vertical rise per 12 inches of horizontal run. A roof with a pitch of 6 in 12 will rise 6 inches for every 12 inches of run.

Roof underlayment: See Felt.

Roof ventilating louvers: Louvers placed on top of a roof for ventilation.

Roofing: Waterproof roof covering.

Rough grade: Shaping of the ground to an approximate elevation and contour.

Rough opening: An unfinished opening in a building.

Rowlock: Brick laid on edge.

Run: One half the span of a roof. The horizontal distance of a flight of stairs.

R-Value: The resistance value of insulating material measured by its ability to resist heat transfer. R-Values are numbered: The higher the number, the more effective the insulation.

S

Sash: A frame for holding the glass of a window or door.

Scaffold: A temporary wooden or metal frame for supporting workmen and materials.

SCR brick: Brick with dimensions of 2¼″ × 5½″ × 11½″, or thereabouts.

Screed: A wooden or metal form used as a thickness and leveling guide for concrete.

Scribing: Marking and fitting a piece of lumber to an irregular surface.

Septic tank: A private sewerage disposal system where waste matter is collected in a tank, putrefied and decomposed through bacterial action, and the liquid discharged into a disposal bed where it is absorbed into the soil.

Sheathing: The rough covering over the frame of a house.

Shed roof: A roof with only one slope. Normally built against a higher wall.

Shim: A thin wedge-shaped piece of wood or metal used for filling and leveling.

Shingles: A term used for pieces of material used for covering roofs and the sides of houses, such as asphalt shingles.

Shoring: Timbers used for temporary support.

Shutters: A movable louvered cover for a window.

Siding: The boarding on the outside of a house.

Sill: A horizontal wood or masonry unit under a window or door. The horizontal member bolted to the foundation wall is called the mudsill or sill plate.

Sill plate (or mudsill): See Sill.

Slab: A term used for a concrete floor.

Soffit: The underside of an overhang.

Solar: Something coming from the sun, such as heat and energy.

Soldier course: Bricks set in a vertical position.

Sole plate: The horizontal framing member under a stud wall.

Span: Distance between two supports.

Specifications: Written instructions that describe the type and quality of material that is not shown or specified on the blueprints. The specifications become part of the blueprints and are referred to as the "plans and specifications" in a contract.

Speculative builder: See Merchant builder.

Splash block: A masonry unit placed under a downspout to disperse the water away from the foundation.

Square: The material needed to cover 100 square feet.

Standard brick: Brick with dimensions of 2¼″ × 3¾″ × 8″, or thereabouts.

Starter course: The roof shingles or felt laid around the eave and under the first course of roofing.

Stile: The vertical piece of a window or door frame.

Stool: A wood shelf across the bottom and inside of a window.

Stops: See Door stops.

Storm window: A window used with a regular window for added protection against the cold weather.

Stretcher course: Brick laid lengthwise so the side only appears on the face of a wall.

Stringer: See Carriage.

Struck joint: A finished mortar joint.

Struts: The vertical or inclined members between the top chord and bottom chord of a truss.

Stucco: A cement coating for walls.

Stud: The vertical framing member of a wall.

Subcontractor: One who contracts part of a job, such as electrical or masonry work, from the principal contractor.

Subfloor: Sheathing nailed to the floor joists to receive the finish floor.

Sump pump: A pump used to move water from a lower to a higher elevation.

Superstructure: That part of a building above the foundation.

Surfaced lumber: Lumber dressed in a planing mill designated as S4S, etc.

Symbol: A mark, sign or object used to represent another thing.

T

Terrace: An unroofed, paved area adjacent to a house.

Terrain: The natural contour of land.

Thermal glass: Two or more glass panes sealed together in a factory and separated by hermetically sealed air or other gas spaces.

Thimble: A horizontal pipe running through the chimney wall into the flue.

Tie wire: Wire used to tie reinforcing rods together.

Toe nail: Nailing at an angle.

Ton of refrigeration: 12,000 BTUs.

Top chord: The upper horizontal framing member of a truss which is the nailing base for the roof sheathing.

Top plate: The upper horizontal framing member of a stud wall.

Trapezoid: A quadrilateral (four-sided figure) of which only two sides are parallel.

Tread: The horizontal surface of a step.

Triangle: A three-sided figure.

Trim: The finish carpentry work in a house.

Trimmer: Joist or rafter around an opening in the floor or roof into which a header is framed.

Truss: A single framework consisting of top chord, bottom chord and struts joined together for supporting a roof.

Turnout: The lower end of a handrail on a balustrade that makes a spiral or twisting turn.

U

Underpinning: A support under a wall already built.

Undressed lumber: Rough lumber without a smooth finish.

Utilities: Water, sewer, electric, gas, telephone, TV cable, etc.

V

Valley: The bottom intersection of two roof slopes.

Valley jacks: Rafters between the ridge and valley rafter.

Valley rafter: The diagonal rafter under a valley.

Vapor barrier: Waterproof membrane such as polyethylene film used to retard passage of moisture.

Variance: A deviation from a zoning ordinance, such as noncompliance with the required setback regulations. A waiver may be granted by the local Planning Commission for some variances.

Vent: A small opening to permit the escape of gas and fumes.

Vertex: The highest point of two intersecting lines furthest from the base.

Vestibule: A small entrance hall or room.

Volume: The cubic contents of the space occupied by three dimensions.

Volute: See Turnout.

W

Walers: Supports for concrete forms.

Wall plate: See Top plate.

Wall ties: Corrugated metal strips used to tie brick veneer walls to the adjacent wall.

Wallpaper (or wall fabric): Decorative paper or fabric for covering walls and ceilings.

Wainscot: The lower part of a room with a different finish from the upper part, such as paneling on the lower wall.

Waterproofing: Preventing the entrance of water.

Weatherstripping: A strip of fabric or metal around doors and windows to prevent air leakage.

Weeping tile: See Drain tile.

Winders: A stair tread that is wider on one end than the other.

Window stool: See Stool.

Wire mesh: Wires welded together in rectangular or square grids used to reinforce concrete.

Workmanship: The skill of a worker.

XYZ

Zoning ordinance: An ordinance adopted by the local governing body to protect the health, safety, property and welfare of the public. All building construction is regulated by the zoning ordinances.

Index

A

Access roads ... 32
Adhesive
- drywall ... 217
- paneling ... 275

Air conditioning ... 186
Air infiltration ... 205
Aluminum siding ... 225
Anchor bolts ... 69
Appliances ... 290, 293
- cost estimate worksheet ... 295

Appraiser ... 5
Architect fees ... 13
Area formulas ... 145
Area of walls and ceiling
- worksheet ... 215-216

Asphalt paving ... 301
Asphalt shingles ... 174
- estimating ... 175-176

Attic ventilation ... 170

B

Backing, ceiling joist ... 126
Balusters ... 267-268
Base shoe ... 261-262, 274
Baseboard ... 261-262, 273
Basement
- calculating block walls ... 57
- excavation ... 30-32
- floor drain ... 76
- openings ... 67
- sump pit and pump ... 76

Bathroom
- accessories ... 269
- trim ... 269

Batt insulation ... 204
Beams ... 69
Bearing walls ... 109
Bench mark ... 16
Bird's mouth cut ... 131
Bituminous coating ... 75
Block flooring ... 259
Blueprints ... 13
Board insulation ... 205
Board measure ... 81-83
Board sheathing ... 118
Brick
- chimneys ... 193-195
- cleaning ... 199
- courses, calculating ... 192
- estimating ... 196-199
- fireplace ... 193-194
- mortar ... 190
- mortar joints ... 194
- mortar mix factor ... 190-191
- retaining walls and steps ... 197
- standard brick factor ... 191
- types ... 189
- wall ties ... 198
- window and door openings ... 196

Brickwork ... 189-204
- cost estimate worksheet ... 202
- labor costs ... 200-201

Bridging ... 93-94
Broker commissions ... 310
BTU ... 203
Building permits ... 16-20
- sample ... 18-19

Building plan services ... 13
Building site ... 5-11
- checklist before buying ... 9
- cost estimate worksheet ... 12
- maintenance ... 11
- neighborhood ... 7-8
- population trends ... 6
- purchase price ... 8
- zoning ... 7

Business expenses ... 307

C

Calcium chloride ... 37
Calculations
- common rafter length ... 130-131
- crushed stone ... 53-54
- displaced block ... 60-63
- excavation ... 30-35
- hip and valley rafters ... 135-137
- house footings ... 39-41
- jack rafters ... 138-139
- linear feet to board measure ... 82
- lumber pricing ... 83
- masonry block ... 55-56
- roof area ... 145-147
- setting up proportions ... 101-103
- stepped footings ... 41
- stud factor ... 112

Carpet ... 288, 293
- padding ... 289, 293

Carriage, stair ... 267-268
CD ROM, cost estimate
- worksheets ... 9, 310, 312

Cedar sidewall shingles ... 226
Ceiling backing ... 126
Ceiling joists
- backing ... 126
- estimating ... 123-128
- stiffeners ... 127

Ceilings
- finishing ... 213-222
- insulating ... 206
- painting ... 287-288

Cellulose fiber insulation ... 204

Ceramic tile flooring 290, 292
Chair rail 263
Chimney 193-195
  estimating brick 197
  flashing 169
  footings 43
Cleanup, site 36
Closet shelves 267, 278
Code requirements, footings 37, 43
Collar beam 142
Column posts 69, 72, 229
Commission, real estate 310
Common rafter 128-129
  common rafter worksheet 130-131
  computing length 129-133
  pitch ratios 133
  shed roof worksheet 156-157
  worksheet 130-131
Compaction, fill dirt 33
Comparable sales 6
Concrete 239-250
  adding calcium chloride 37
  additives 245-246
  cold weather pours 245-246
  compression strength 239
  cost estimate worksheet 256
  expansion joints 243
  finishing 246
  labor costs 253-255
  lintels 198-199
  multiplying factors 242
  quantity factors 241
  reinforcement 243-245
  vapor retarder 242
  volume factors 42
Conservation, energy 203
Contingency 308
Conversion table, decimals to fractions 31
Corner bracing 117-118
Corner post 112
Cost estimate worksheets
  brickwork 202
  building site 12
  CD ROM 9
  concrete floors, walks and terraces 256
  electrical, plumbing, and HVAC 187
  energy-saving materials 212
  exterior trim 236-237
  floor system 105
  footings 51
  foundation 80
  gutters, on-site improvements, and misc. 303
  home building costs 311
  interior trim 283-286
  interior wall and ceiling finish 222
  painting, floor covering and appliances 295
  preliminary costs 23
  roofing 179
  site clearing, excavations and fill dirt 36
  superstructure 163-165
Counterflashing 169
Countertops
  bathroom 266-267
  kitchen 266
Course factor, block 55
Crawl space vents 67
Cripple rafter 128
Cripple stud 109
Cross bridging 93-94
Crushed stone
  calculating weight 53-54
  driveways 240, 242
  foundation drain pipe fill 77
Cuts and fills, foundation plan 29
Cuts, rafter 131

D

Decimal equivalent tables
  fractional part of a foot 193
  fractional part of an inch 132
Direct overhead 306-307
Direction of growth 6
Displaced block, calculating 60-63
Door openings 196
  lintels 199
Doors
  basement 68
  exterior 224
  heat loss comparison 209
  insulating 209
  interior 260, 271
  tub and shower 269
Downspouts 297-298
  estimating 300
Drain pipe, footing 76
Drip edge 173
Driveways 298
  crushed stone 301
  paving 301
Drywall 214
  area of offset rooms 218-219
  area of walls and ceiling 215
  calculating costs 216
  estimating square feet 214-215
  installation materials 215
  weight and bending radii 215

E

Easement 9
Electrical 181-182
  cost estimate worksheet 187
Elevations 14-16
  grading 29
Employee benefits 306
Employer taxes 306
End connector plugs 171
Energy saving materials 203-205
  cost estimate worksheet 212
Engineering fees 10
Estimating
  basement openings 67-68
  block foundation 54-56
  brick 196-199
  CD-ROM instructions 312
  ceiling joists 123-128
  common rafter lengths 129-134
  common rafter quantities 134
  concrete for footings 38-44
  concrete materials 246-253
  concrete quantities 240
  crushed stone 53-54
  drywall 214-218
  electrical 181-182
  excavation 28-35
  exterior trim 223-227
  floor system labor 101-104
  footing forms 44-47
  footings 37
  foundation 54-56
  girders 86
  gutters 300
  HVAC 184-186
  insulation materials 211
  interior trim 270-281
  labor for footings 48
  ledger strip 87
  masonry blocks 54-56
  masonry labor 79
  miscellaneous materials 298-299
  mortar 65-66
  nails, flooring system 95
  overhead expenses 308, 310
  painting 288
  plumbing 183-184
  porch shed roof framing 154
  reinforcing rod 43-44
  roof framing 129-151
  roof sheathing nails 152
  roofing materials 167-178

sill plate 84-85
wall framing 109-113
wall opening headers, form 118
Excavation 28
basement 30-32
foundation plan 29
how to compute quantity 29-30
Expenses, overhead 305-308
Exterior trim 223-235
cost estimate worksheets 236-237
estimating labor 230
fascia, frieze and rake board 226-228
molding 230
porch 228-229
siding 224-226
soffit 228-229
Exterior walls 107
lintels 199
openings 196

F

Fascia 226-228
Fees
architect 13
building permit 17, 20
engineering 10
water connection 20
Felt underlayment 168, 173
Fiberglass shingles 175
estimating 175-176
Fill dirt
cost to spread 35
foundation plan 29
how to compute quantity 33-35
storing 32
Finishing concrete 246
Fireplaces 193
chimneys 194-195
estimating brick 198
prefabricated 193-194
Flashing 169, 229
Floor area waste factor 97
Floor covering
carpet 288, 291
cost estimate worksheet 295
masonry tile 290, 291
resilient tile 289
Floor plan 108
Floor system 81-104
board measure, lumber 81-83
bridging 93-94
estimating labor 101-104
floor joists 88-92
foundation plan 85
girders 86-87
headers 93
ledgers 87
nailing 95
sheathing 96-101
sill plate 84-85
subfloor 96-101
trimmers 93
Flooring 259
underlayment 258, 270
Floors, insulating 208, 210
Flues 195
Footing drain 76-78
Footings 37-51
calculating displaced block 60-63
concrete 38-44
estimating block for stepped footing 58-64
estimating labor 48
factors for estimating concrete 42
forms 44-47
grade stakes 46
layout 38
reinforcing rod 43-44
stepped 41
stepped, forms 47
Form 100 310-311
Forms
concrete 245
footing 44-46
stepped footings 47
Foundation
footing plan 39
layout 38
plan 29
Foundations 53-79
anchor bolts 69-71
basement door 68
basement windows 67
beams 69
bituminous coating 76
calculating displaced block 60-63
calculating stepped footing 58-63
column posts 69, 72
cost estimate worksheet 80
crushed stone 77
estimating labor 78
footing drain 76-78
footing drain labor 78
lintels 68
masonry blocks 54-56, 64-65
masonry labor 79
miscellaneous materials 78
mortar 65-66
parging 75
vents, estimating 67
waterproofing 72-75
Framing, exterior walls 107
Framing nails 151
Frieze 226-228
Fringe benefits 306

G

Gable ends 113-114, 120
Gable roof
calculating area 146, 148
offset 140
Garage and utility area
calculating block for stepped footing 58-63
grade calculations 33
Geometry formulas 145
Girders 69, 86
Glass fiber insulation 204
Glossary 313-324
Grade
calculating cuts and fills 30-35
elevations 29
Grade beam 244
Grade stakes 46
Grantee 9
Grantor 9
Gutters 297-298
connection kits 300
cost estimate worksheet 303
estimating 300
labor 302
Gypsum board 214
area of offset rooms 218-219
area of walls and ceiling 215
calculating costs 216
estimating square feet 214-215
installation materials 215
weight and bending radii 215

H

Handrails 267-268
Hardboard
paneling 264
siding 224
Headers
exterior openings, form 117
interior openings, form 117
joist 93
wall openings 115-116
Heat loss comparison
ceilings 207
doors 209
windows 209
Heat pump 186

Heating system 184-185
High water table 72
Hip caps 175
Hip rafter 128
  calculating length 135-137
Hip roof area 141
  calculating 146-149
Horizontal rods, estimating 244, 250
Hourly wage, loaded 307
How to use CD-ROM 312
HVAC 184-186
  cost estimate worksheet 187
Hydrostatic pressure 76

I

Indirect overhead 307
Infiltration 205, 207
Instructions, CD-ROM 312
Insulation 203-205
  batt 203
  ceiling 206
  floor 208
  labor 211-212
  loose fill 203
  problem areas 210-211
  rigid board 205
  vapor retarders 205
  wall 207
  windows and doors 209
Insurance 306
  liability 12
Interest payments 10
Interior doors 260
Interior trim 257-265
  baseboard 261-262
  bathroom 269
  cost estimate worksheet 283-286
  labor 282
  perimeter of rooms table 262
  square feet of floors table 258
  windows 260-261
Interior wall and ceiling finish 213-222
  cost estimate worksheet 222
  painting 287-288
  square feet of walls and ceiling table 214
  worksheet for area 215
  worksheet for offset rooms 219

J

Jack rafter 128, 137-139
Job costs, indirect overhead 307
Joint compound 217
Joint straps 171
Joist factor 89, 124
Joists
  ceiling 123-128
  floor 88-92
  spacing 89

K

Kitchen
  appliances 290, 294
  cabinets 266, 276

L

Labor costs
  brickwork 200-201
  concrete 253-255
  drywall 221
  exterior trim 235
  floor systems 101-104
  footings 48
  foundation 79
  gutters 297-298
  insulation 211-212
  interior trim 282
  masonry 79
  on-site improvements 302
  painting 288, 291
  roofing 177-178
  superstructure 160-162
Labor tracking, time sheets 49
Land purchase 8-9
Landscaping 302
Lapped siding 224
Lath and plaster 213
Layout
  foundation 38
  stepped footing 59
  subfloor 99
Leaders 297-298, 300
Ledger 87
Legal fees 9
Let-in bracing 118
Liability insurance 12
Linen closet 267
Lintels 68
  brickwork 198-199
Loan, interest on 10-11
Lookouts 143-144
Loose-fill insulation 204-205
Lumber pricing 83
Lumber, board feet 81-83

M

Maintenance, building site 11
Masonry block
  course factor 55
  displaced by stepped footing 60-63
  estimating labor, foundation 79
  foundation 54-55
  sizes 55
  square-foot factor 55-56
Masonry flooring 290
Maximum clear span 129
Medical and pension costs 306
Medicine cabinets 269
Metal corner bead 216
Metal drip edge 173
Metes and bounds 14, 16
Mirrors 269
Molding
  exterior 230
  interior trim 260-263, 275
  stairs 267
Mortar joints 193-194
Mortar mix 65-66
  estimating 197
  factors 190-191
Multiplying factors
  brick and mortar mixes 190-191
  brick wall height 192
  concrete 242
  concrete for footings 42
  crushed stone 242
  rafter factors per foot run 133
  ratio of common rafters 133
  ratio of hip and valley rafters 137
  roof starter course 175
  shingles for ridge and hips 175
Muriatic acid 199

N

Nailers 143-144
Nails
  flooring system 95
  siding 231
  subfloor 100
  wall framing 121-122
Net rafter length 130
Newel post 267-268
Nominal size
  lumber 82
  masonry blocks 55
Norman brick 189

O

Offset rooms, area of 218-219
On-site improvements 298
  labor 302
Operating expenses, overhead 307-308

OSB
  sheathing 118
  subfloor 98
Overhang, roof 130
Overhead 305
  direct 306-307
  indirect 307-308

P

Padding, carpet 289
Painting 287-288, 290
  cost estimate worksheet 295
Panel molding 263
Panel siding 225
Paneling 264-266, 274
  adhesive 275
Parallelogram, area of 145
Parging 73-74
Parquet flooring 259
Pension costs 306
Perlite insulation 204
Permit fees 17, 20
Pipe, footing drain 76
Plan check fee 20
Plank flooring 259
Plans 13
  foundation 29
  foundation footing 39
  grade and floor elevations 30
  plot plan 14-16
Plaster 213
Plot plans 14-16
  drawing 15
Plumb cut 131
Plumbing 183-184
  cost estimate worksheet 187
Plywood
  paneling 265
  sheathing 118
  subfloor 98
Population trends 6
Porch
  ceiling 228-229
  column posts 229
Porch shed roofs
  framing 154-158
  trusses 158-159
Posts, porch 229
Preliminary costs 13-22
  architect fees 13
  building permits 16
  cost estimate worksheet 23
  plan costs 13-15
  sewer connection 21
  temporary utilities 21-22
  water connection 20
Private road assessments 9
Private sewage system 21
Profit 309-310
Property
  appraisal 5
  sales price 310
  taxes 11
  valuation 17, 20
Proportions, calculations
    using 101-103

R

R-value 203
  house insulation
    requirements 206-210
Rafter length factor 133
Rafters 128-140
  common 128-134
  common cuts 131
  hip 135-136
  jack 137-139
  maximum clear span 129
  support 142-144
  types 128
  valley 135-136
  worksheet for common
    rafters 131
  worksheet for common rafters
    (shed roof) 156
Rake board 226-227
Real estate commission 310
Recording fees 9
Rectangle, area of 145
Reinforcing 72
Reinforcing rods 243-245
  estimating worksheet 244, 250
  footing 43-44
  number and weight chart 243
Resilient flooring 289
Retaining wall 197
Ridge 128, 140
Ridge vents 170
Rigid insulation boards 205
Rise, roof 130
Risers 267-268
Rockwool insulation 204
Roman brick 189
Roof
  area, calculating 145-147, 150
  pitch factor 129-130
  rise 130
Roof framing
  collar beam 142
  common rafter lengths 129-134
  hip and valley rafters 135-137
  jack rafters 137-139
  lookouts and nailers 143-144
  nailing 151
  pitch factor 129-130
  rafter supports 142-144
  rafter types 128
  ridge length 140
Roof sheathing
  calculating area 145-150
  materials 144
  nails 152
  porch shed roof 158-159
Roof ventilating louvers 171
Roofing 167-178
  asphalt shingles 174
  cost estimate worksheet 179
  end plugs 170-171
  felt underlayment 168
  fiberglass shingles 175
  flashing 169
  joint straps 170-171
  labor costs 177-178
  metal drip edge 173
  nails 176
  ridge vents 170-171
  soffit vents 172
  square 168
Run 130
Run-off water 10, 12

S

Saddle, chimney 169
Sales price 310
SCR brick 189
Screeds 245
Septic systems 21
Sewer connections 21
Sewer hook-up 183
Sheathing
  OSB 118
  plywood 118
  roof 144-151
  wall 118-121
Shed roof
  framing 154-158
  trusses 158-159
  worksheet for common
    rafters 156-157
Sheet vinyl 289
Shower door 269
Siding
  nails 231
  types 224-226
Sill plate 70-71, 84-85
Site clearing 36
  cost guidelines 27
  estimating 25-26

Site selection ... 5-8
Skirtboard, stair ... 267
Slabs, reinforcing ... 244-245
Soffit ... 228-229
  vents ... 172
Sole plate ... 105, 107, 109-110
Span ... 130
Speculative builder ... 5
Splash blocks ... 298, 300
Springs, underground ... 10
Square-foot factor, block ... 55-56
Squares, roofing ... 168
Stair ... 267-273, 279
  components ... 267-268
  floor-to-floor rise ... 268
  labor ... 268
  stringer lengths ... 159-160
Standard brick ... 189
  factor ... 191
State surcharge fee ... 20
Stepped footings ... 41
  forms ... 47
Steps, brick ... 197
Stiffener, ceiling joist ... 127
Storm pipe size ... 10
Stringer, stair ... 267-268
Strip flooring ... 259
Stud factor ... 112
Studs ... 111-115
  gable ... 113
  plain wall ... 111
  wall with openings ... 112
Subfloor ... 96-101
  area waste factor ... 97
  layout ... 99
  nailing ... 100
  OSB ... 90
  plywood ... 98
Sump pit ... 76
Superstructure ... 107-163
  ceiling joists ... 123
  cost estimate
    worksheets ... 163-165
  exterior walls ... 107
  framing nails ... 151
  interior walls ... 107
  labor costs ... 160-162
  porch shed roofs ... 154
  rafter supports ... 143
  rafters ... 129
  roof sheathing ... 144
  stair stringers ... 159
  trusses ... 152

T

Tail cut ... 131
Tape, drywall ... 217
Taxes, property ... 11
Temporary
  electric service ... 22
  wall braces ... 117-118
  water service ... 21
Thin-coat plaster ... 213
Tie wire ... 245
Time sheets ... 49
Title check ... 9
Tooled mortar joints ... 193
Top plate ... 105, 109-110
Trapezoid, area of ... 145
Traverse rods, estimating ... 244, 250
Treads ... 267
Triangle, area of ... 145
Trim, window and door ... 264-265
Trimmer ... 93, 109
True rafter length ... 130
Trusses
  porch roof ... 158-159
  roof ... 152-153
Turnout, stair ... 267-268

U

Underground springs ... 10
Underlayment, roofing ... 168
Using CD-ROM ... 312

V

Valley rafter ... 128
  calculating length ... 135-137
Vanities ... 266-267, 277
Vapor barriers ... 205
Vapor retarder, under concrete ... 242
Veneer, brick ... 196
Vents
  foundation ... 67
  ridge ... 169
  soffit ... 172
Vermiculite insulation ... 204
Vertex, roof ... 130
Vinyl flooring ... 289, 292
Vinyl siding ... 225

W

Wage, loaded ... 307
Wall
  coverings ... 288
  finishes ... 213
  flashing ... 229
  insulation ... 207, 210
  molding ... 262-264, 274
  sheathing ... 118-121
  ties ... 198
Wall framing ... 109
  estimating nails ... 121-122
  gable ends ... 113, 120
  headers ... 115-116
  sheathing ... 118-121
  studs ... 111-115
  temporary braces ... 117-118
  wall intersections ... 112
  window and door openings ... 112
Wallboard, gypsum ... 214
  area of offset rooms ... 218-219
  area of walls and ceiling ... 215
  calculating costs ... 216
  estimating square feet ... 214-215
  installation materials ... 215
  weight and bending radii ... 215
Water connection ... 20
Water run-off ... 10, 12
Water supply ... 183
Waterproofing, foundation ... 72-75
Weatherstripping ... 205
Weekly time sheet ... 49
Welded wire fabric ... 243
Window openings ... 196
  lintels ... 199
Windows
  basement ... 67
  heat loss comparison ... 209
  insulating ... 209
  trim ... 260-261, 272
  types ... 223
Wire mesh ... 243
Wood flooring ... 259
  finishing and installation ... 260
Wood paneling ... 265
Wood siding ... 225
  estimating factors ... 225
Worksheets
  area of walls and ceiling ... 215-216
  area of walls and ceiling,
    offset rooms ... 219
  CD-ROM instructions ... 312
  common rafters ... 131
  common rafters (shed roof) ... 156
  estimating reinforcing rods ... 244
  headers for exterior openings ... 117
  headers for interior wall
    openings ... 117

XYZ

Yard, grade calculations ... 34
Zoning ordinances ... 7-8

Practical References for Builders

National Construction Estimator

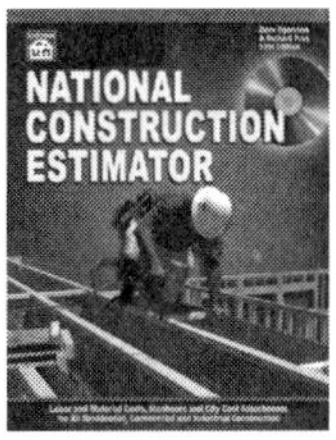

Current building costs for residential, commercial, and industrial construction. Estimated prices for every common building material. Provides manhours, recommended crew, and gives the labor cost for installation. Includes a CD-ROM with an electronic version of the book with *National Estimator*, a stand-alone *Windows*™ estimating program, plus an interactive multimedia video that shows how to use the disk to compile construction cost estimates.
672 pages, 8½ x 11, $62.50. Revised annually

Getting Financing & Developing Land

Developing land is a major leap for most builders — yet that's where the big money is made. This book gives you the practical knowledge you need to make that leap. Learn how to prepare a market study, select a building site, obtain financing, guide your plans through approval, and then control your building costs so you can ensure yourself a good profit. Includes a CD-ROM with forms, checklists, and a sample business plan you can customize and use to help you sell your idea to lenders and investors. **232 pages, 8½ x 11, $39.00**

Construction Forms for Contractors

This practical guide contains 78 practical forms, letters and checklists, guaranteed to help you streamline your office, organize your jobsites, gather and organize records and documents, keep a handle on your subs, reduce estimating errors, administer change orders and lien issues, monitor crew productivity, track your equipment use, and more. Includes accounting forms, change order forms, forms for customers, estimating forms, field work forms, HR forms, lien forms, office forms, bids and proposals, subcontracts, and more. All are also on the CD-ROM included, in Excel spreadsheets, as formatted Rich Text that you can fill out on your computer, and as PDFs. **360 pages, 8½ x 11, $48.50**

Land Development, Tenth Edition

The industry's bible. Nine chapters cover everything you need to know about land development from initial market studies to site selection and analysis. New and innovative design ideas for streets, houses, and neighborhoods are included. Whether you're developing a whole neighborhood or just one site, you shouldn't be without this essential reference. **360 pages, 6 x 9, $55.00**

Residential Property Inspection Reports on CD-ROM

This CD-ROM contains 50 pages of property inspection forms in both Rich Text and PDF formats. You can easily customize each form with your logo and address, and use them for your home inspections. Use the CD-ROM to write your inspections with your word processor, print them, and save copies for your records. Includes inspection forms for grounds and exterior, foundations, garages and carports, roofs and attics, pools and spas, electrical, plumbing, and HVAC, living rooms, family rooms, dens, studies, kitchens, breakfast rooms, dining rooms, hallways, stairways, entries, laundry rooms. **$79.95**

CD Estimator

If your computer has *Windows*™ and a CD-ROM drive, CD Estimator puts at your fingertips over 150,000 construction costs for new construction, remodeling, renovation & insurance repair, home improvement, framing & finish carpentry, electrical, concrete & masonry, painting, earthwork and heavy equipment, and plumbing & HVAC. Monthly cost updates are available at no charge on the Internet. You'll also have the *National Estimator* program — a stand-alone estimating program for *Windows*™ that *Remodeling* magazine called a "computer wiz," and *Job Cost Wizard*, a program that lets you export your estimates to QuickBooks Pro for actual job costing. A 60-minute interactive video teaches you how to use this CD-ROM to estimate construction costs. And to top it off, to help you create professional-looking estimates, the disk includes over 40 construction estimating and bidding forms in a format that's perfect for nearly any *Windows*™ word processing or spreadsheet program. **CD Estimator is $98.50**

National Home Improvement Estimator

Current labor and material prices for home improvement projects. Provides manhours for each job, recommended crew size, and the labor cost for removal and installation work. Material prices are current, with location adjustment factors and free monthly updates on the Web. Gives step-by-step instructions for the work, with helpful diagrams, and home improvement shortcuts and tips from experts. Includes a CD-ROM with an electronic version of the book, and *National Estimator*, a stand-alone *Windows*™ estimating program, plus an interactive multimedia tutorial that shows how to use the disk to compile home improvement cost estimates.
520 pages, 8½ x 11, $63.75. Revised annually

Drafting House Plans

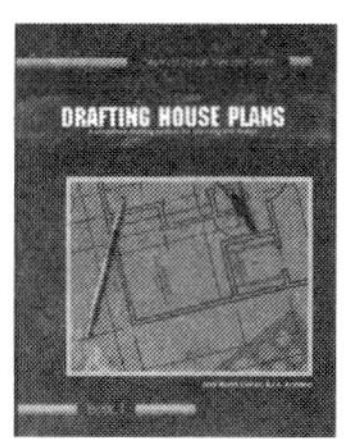

Here you'll find step-by-step instructions for drawing a complete set of house plans for a one-story house, an addition to an existing house, or a remodeling project. This book shows how to visualize spatial relationships, use architectural scales and symbols, sketch preliminary drawings, develop detailed floor plans and exterior elevations, and prepare a final plot plan. It even includes code-approved joist and rafter spans and how to make sure that drawings meet code requirements. **192 pages, 8½ x 11, $34.95**

Contractor's Guide to Change Orders

This book gives you the ammunition you need to keep contract disputes from robbing you of your profit. You'll learn how to identify trouble spots in your contract, plans, specifications and site; negotiate and resolve change order disputes, and collect facts for evidence to support your claims. You'll also find detailed checklists to organize your procedures, field-tested sample forms and worksheets ready for duplication, and various professional letters for almost any situation. **382 pages, 8½ x 11, $79.00**

Stained Concrete Interior Floors

Turn concrete slabs into works of art. This book shows you concrete design options, how to estimate costs, and even how to sell your services and write contracts for floor staining. Large color photos show how to correct problem floors, test your stain before beginning, apply dyes, grouts and sealers, and maintain stained concrete floors. In addition, you'll find information on specialty techniques and the tools you'll need for the job. Includes resources and websites for tools, supplies, equipment and training to help you become a concrete floor staining "pro." Full of charts, helpful hints and step-by-step instructions that should help you make extra income as a concrete floor staining expert. **100 pages, 8½ x 11, $35.00**

DeWalt Construction Estimating Complete Handbook

This new edition is loaded with tips, checklists, worksheets, data tables, and tutorials that will provide you tools to effectively navigate through every step of the estimating process. Focus is on the construction estimating "how-to" essentials, with on-the-spot answers backed up with graphic examples. With a thorough understanding of the important profit-making factors within construction, its comprehensive coverage includes marketing, bid planning, scope review, quantity take-off for all trades and divisions, cost analysis, value engineering, *Excel* spreadsheet estimating, and more. Includes on-line tutorials.
256 pages, 8 x 10, $24.95

Craftsman's Construction Installation Encyclopedia

Step-by-step installation instructions for just about any residential construction, remodeling or repair task, arranged alphabetically, from *Acoustic tile* to *Wood flooring*. Includes hundreds of illustrations that show how to build, install, or remodel each part of the job, as well as manhour tables for each work item so you can estimate and bid with confidence. Also includes a CD-ROM with all the material in the book, handy look-up features, and the ability to capture and print out for your crew the instructions and diagrams for any job. **792 pages, 8½ x 11, $65.00**

Contractor's Plain-English Legal Guide

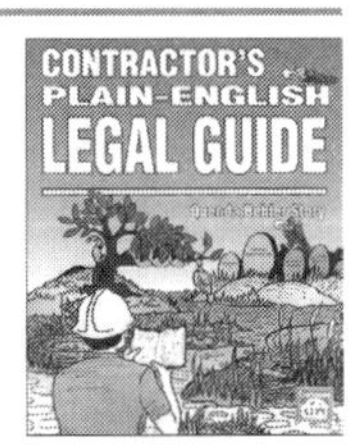

For today's contractors, legal problems are like snakes in the swamp — you might not see them, but you know they're there. This book tells you where the snakes are hiding and directs you to the safe path. With the directions in this easy-to-read handbook you're less likely to need a $200-an-hour lawyer. Includes simple directions for starting your business, writing contracts that cover just about any eventuality, collecting what's owed you, filing liens, protecting yourself from unethical subcontractors, and more. For about the price of 15 minutes in a lawyer's office, you'll have a guide that will make many of those visits unnecessary. Includes a CD-ROM with blank copies of all the forms and contracts in the book. **272 pages, 8½ x 11, $49.50**

Construction Estimating

This unusually well-organized book shows the best and easiest way to estimate materials for room additions or residential structures. It gives estimating tables and procedures needed to make a fast, accurate, and complete material list of the structural members found in wood- and steel-framed buildings. This book is divided into 72 units, each of them covering a separate element in the estimating procedure. Covers estimating foundations, floor framing, wall framing, ceiling framing, roof framing, roofing materials, exterior and interior finish materials, hardware, steel joist floor framing, steel stud framing, and steel ceiling joist and rafter framing. **496 pages, 8½ x 11, $49.50**

Scale MasterPro XE w/PC Cable

If you're looking for a simple scaler for linear, area, and volume takeoffs with a PC interface, the Scale MasterPro XE offers a value-priced way to transfer measurements from plans, blueprints, and maps of any scale, into your computer for estimating with your *National Estimator* program. The PC interface is fully compatible with *Windows* XP or higher. Transfer any rolled values directly into Excel spreadsheets or the *National Estimator*. Includes Scale MasterPro XE, and connectors — no software needed. **$149.95**

Residential Construction Performance Guidelines

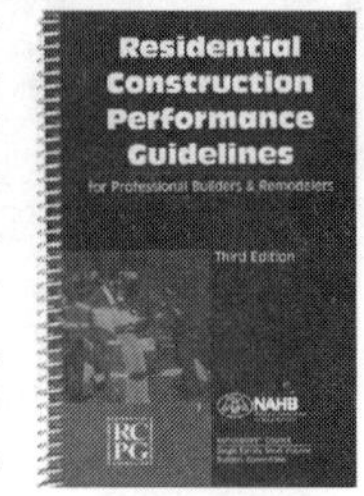

Created and reviewed by more than 300 builders and remodelers, this guide gives cut-and-dried construction standards that apply to new construction and remodeling. It defines corrective action necessary to bring all construction up to standards. Standards are listed for sitework, foundations, interior concrete slabs, basement and crawl spaces for block walls and poured walls, wood-floor framing, beams, columns and posts, plywood and joists, walls, wall insulation, windows, doors, exterior finishes and trim, roofs, roof sheathing, roof installation and leaks, plumbing, sanitary and sewer systems, electrical, interior climate control, HVAC systems, cabinets and countertops, floor finishes and more.
120 pages, 6½ x 8½, $39.95

Construction Contract Writer

Relying on a "one-size-fits-all" boilerplate construction contract to fit your jobs can be dangerous — almost as dangerous as a handshake agreement. *Construction Contract Writer* lets you draft a contract in minutes that precisely fits your needs and the particular job, and meets both state and federal requirements. You just answer a series of questions — like an interview — to construct a legal contract for each project you take on. Anticipate where disputes could arise and settle them in the contract before they happen. Include the warranty protection you intend, the payment schedule, and create subcontracts from the prime contract by just clicking a box. Includes a feedback button to an attorney on the Craftsman staff to help should you get stumped — *No extra charge*. **$99.95**. Download *Construction Contract Writer* at http://www.constructioncontractwriter.com

Construction Estimating Reference Data

Provides the 300 most useful manhour tables for practically every item of construction. Labor requirements are listed for sitework, concrete work, masonry, steel, carpentry, thermal and moisture protection, doors and windows, finishes, mechanical and electrical. Each section details the work being estimated and gives appropriate crew size and equipment needed. Includes a CD-ROM with an electronic version of the book with *National Estimator*, a stand-alone *Windows*™ estimating program, plus an interactive multimedia video that shows how to use the disk to compile construction cost estimates. **432 pages, 11 x 8½, $39.50**

Home Builders' Jobsite Codes

A spiral-bound, quick reference to the 2009 *International Residential Code* that's filled with easy-to-read and understand code requirements for every aspect of residential construction. This user-friendly guide through the morass of the code is packed with illustrations, tables, and figures, to illuminate your path to inspection and approval. **281 pages, 5½ x 8½, $28.95**

National Repair & Remodeling Estimator

The complete pricing guide for dwelling reconstruction costs. Reliable, specific data you can apply on every repair and remodeling job. Up-to-date material costs and labor figures based on thousands of jobs across the country. Provides recommended crew sizes; average production rates; exact material, equipment, and labor costs; a total unit cost and a total price including overhead and profit. Separate listings for high- and low-volume builders, so prices shown are specific for any size business. Estimating tips specific to repair and remodeling work to make your bids complete, realistic, and profitable. Includes a CD-ROM with an electronic version of the book with *National Estimator*, a stand-alone *Windows*™ estimating program, plus an interactive multimedia video that shows how to use the disk to compile construction cost estimates. **496 pages, 8½ x 11, $63.50. Revised annually**

Estimating With Microsoft *Excel*

Most builders estimate with *Excel* because it's easy to learn, quick to use, and can be customized to your style of estimating. Here you'll find step-by-step instructions on how to create your own customized automated spreadsheet estimating program for use with *Excel*. You'll learn how to use the magic of *Excel* to create detail sheets, cost breakdown summaries, and links. You'll put this all to use in estimating concrete, rebar, permit fees, and roofing. You can even create your own macros. Includes a CD-ROM that illustrates examples in the book and provides you with templates you can use to set up your own estimating system.
148 pages, 7 x 9, $39.95

Estimating & Bidding for Builders & Remodelers

This 5th edition has all the information you need for estimating and bidding new construction and home improvement projects. It shows how to select jobs that will be profitable, do a labor and materials take-off from the plans, calculate overhead and figure your markup, and schedule the work. Includes a CD with an easy-to-use construction estimating program and a data-base of 50,000 current labor and material cost estimates for new construction and home improvement work, with area modifiers for every zip code. Price updates on the Web are free and automatic.
272 pages, 8½ x 11, $89.50

Estimating Home Construction Costs 2nd edition

This book walks you step-by-step through the process of estimating costs for new home construction. It discusses the different types of estimates and when to use each, how to integrate estimating into other functions, and describes the benefits of computer estimating. Includes forms, checklists and conversion tables to help your estimating go more easily and be more accurate. **116 pages, 8½ x 11, $29.95**

Residential Wiring to the 2008 *NEC*

This completely revised manual explains in simple terms how to install rough and finish wiring in new construction, alterations, and additions. It takes you from basic electrical theory to current wiring methods that comply with the 2008 *National Electrical Code*. You'll find complete instructions on troubleshooting and repairs of existing wiring, and how to extend service into additions and remodels. Hundreds of drawings and photos show you the tools and gauges you need, and how to plan and install the wiring. Includes demand factors, circuit loads, the formulas you need, and over 20 pages of the most-needed 2008 *NEC* tables to help your wiring pass inspection the first time. Includes a CD-ROM with an Interactive Study Center that helps you retain what you've learned, and study for the electrician's exam. Also on the CD is the entire book in PDF format, with easy search features so you can quickly find answers to your residential wiring questions. **304 pages, 8½ x 11, $42.00**

Construction Forms & Contracts

125 forms you can copy and use — or load into your computer (from the FREE disk enclosed). Then you can customize the forms to fit your company, fill them out, and print. Loads into *Word for Windows*, *Lotus 1-2-3*, *WordPerfect*, *Works*, or *Excel* programs. You'll find forms covering accounting, estimating, fieldwork, contracts, and general office. Each form comes with complete instructions on when to use it and how to fill it out. These forms were designed, tested and used by contractors, and will help keep your business organized, profitable and out of legal, accounting and collection troubles. Includes a CD-ROM for *Windows*™ and *Mac*™. **432 pages, 8½ x 11, $41.75**

Standard Estimating Practice

Estimating isn't always an easy job. Sometimes snap decisions can produce negative long-term effects. This book was designed by the American Society of Professional Estimators as a set of standards to guide professional estimators. It's intended to help every estimator develop estimates that are uniform and verifiable. Every step that should be included in the estimate is listed, as well as aspects in the plans to consider when you're estimating a job, and what you should look for that may not be included. The result should help you produce more consistently accurate estimates. **506 pages, 8½ x 11, $89.00**

Defensive Estimating

More than a few residential builders and remodelers have walked away from closings with less profit than they deserve. You may get the job, but find out later you needed to make a lot more money on it. Performed correctly, estimating is a small builder's or remodeler's key to profit. Here you'll find how to estimate based on fiscal goals while protecting your company's bottom line. This unique approach to estimating gives readers user-friendly tips and methods for improving the process and providing hundreds of ideas and simple suggestions. Explains how to defend each line of an estimate so that your system of planned profit is consistent and bankable. **142 pages, 8½ x 11, $29.95**

2008 *National Electrical Code*

This new electrical code incorporates sweeping improvements to make the code more functional and user-friendly. Here you'll find the essential foundation for electrical code requirements for the 21st century. With hundreds of significant and widespread changes, this 2008 *NEC* contains all the latest electrical technologies, recently-developed techniques, and enhanced safety standards for electrical work. This is the standard all electricians are required to know, even if it hasn't yet been adopted by their local or state jurisdictions.
784 pages, 8½ x 11, $75.00

2009 *International Residential Code*

Replacing the *CABO One- and Two-Family Dwelling Code*, this book has the latest technological advances in building design and construction. Among the changes are provisions for steel framing and energy savings. Also contains mechanical, fuel gas and plumbing provisions that coordinate with the *International Mechanical Code* and *International Plumbing Code*. **868 pages, 8½ x 11, $88.00**

Also available:

2006 *International Residential Code* $81.50
2003 *International Residential Code*, $72.50
2000 *International Residential Code*, $59.00
2000 *International Residential Code* on CD-ROM, $48.00

2008 Ugly's Electrical Reference

The most popular pocket-sized electrical book in America used by electricians, engineers, designers and maintenance workers. This unique book explains everything from bending conduit to complex electrical formulas. This 2008 edition contains all the electrical material that has made this reference famous, but also reflects 2008 *NEC* changes and new color-coded wiring diagrams. Also includes a Basic Math Review and a General First Aid Section. **162 pages, 5 x 7, $16.95**

Builder's Guide to Accounting Revised

Step-by-step, easy-to-follow guidelines for setting up and maintaining records for your building business. This practical guide to all accounting methods shows how to meet state and federal accounting requirements, explains the new depreciation rules, and describes how the Tax Reform Act can affect the way you keep records. Full of charts, diagrams, simple directions and examples to help you keep track of where your money is going. Recommended reading for many state contractor's exams. Each chapter ends with a set of test questions, and a CD-ROM included FREE has all the questions in interactive self-test software. Use the Study Mode to make studying for the exam much easier, and Exam Mode to practice your skills.
360 pages, 8½ x 11, $35.50

Basic Engineering for Builders

This book is for you if you've ever been stumped by an engineering problem on the job, yet wanted to avoid the expense of hiring a qualified engineer. Here you'll find engineering principles explained in non-technical language and practical methods for applying them on the job. With the help of this book you'll be able to understand engineering functions in the plans and how to meet the requirements, how to get permits issued without the help of an engineer, and anticipate requirements for concrete, steel, wood and masonry. See why you sometimes have to hire an engineer and what you can undertake yourself: surveying, concrete, lumber loads and stresses, steel, masonry, plumbing, and HVAC systems. This book is designed to help you, the builder, save money by understanding engineering principles that you can incorporate into the jobs you bid. **400 pages, 8½ x 11, $39.50**

Basic Lumber Engineering for Builders

Beam and lumber requirements for many jobs aren't always clear, especially with changing building codes and lumber products. Most of the time you rely on your own "rules of thumb" when figuring spans or lumber engineering. This book can help you fill the gap between what you can find in the building code span tables and what you need to pay a certified engineer to do. With its large, clear illustrations and examples, this book shows you how to figure stresses for pre-engineered wood or wood structural members, how to calculate loads, and how to design your own girders, joists and beams. Included FREE with the book — an easy-to-use limited version of NorthBridge Software's *Wood Beam Sizing* program.
272 pages, 8½ x 11, $38.00

Planning Drain, Waste & Vent Systems

How to design plumbing systems in residential, commercial, and industrial buildings. Covers designing systems that meet code requirements for homes, commercial buildings, private sewage disposal systems, and even mobile home parks. Includes relevant code sections and many illustrations to guide you through what the code requires in designing drainage, waste, and vent systems.
202 pages, 8½ x 11, $29.95

Easy Scheduling

Easy Scheduling presents you with a complete set of "real world" scheduling tools that are specifically tailored to meet the needs of small- to medium-sized construction businesses. Step by step, it shows you how to use *Microsoft Project* to build a schedule that will synchronize everyone's efforts into an organized system that becomes the foundation of all planning and communication for all your jobs. You'll see how to establish realistic project goals, set checkpoints, activities, relationships and time estimates for each task, as well as establish priorities. You'll learn how to create a project flowchart to keep everyone focused and on track, and see how to use CSI (Construction Specification Institute) coding to organize and sort tasks, methods, and materials across multiple projects. If you want an easy way to schedule your jobs, *Microsoft Project* and *Easy Scheduling* is the answer for you. (Does not include Microsoft Project.) **316 pages, 8½ x 11, $59.95. Published by BNI.**

Commercial Metal Stud Framing

Framing commercial jobs can be more lucrative than residential work. But most commercial jobs require some form of metal stud framing. This book teaches step-by-step, with hundreds of job site photos, high-speed metal stud framing that works in both residential and commercial construction. It describes the special tools you'll need, how to use them effectively, and the material and equipment you'll be working with. You'll find the shortcuts, tips and tricks-of-the-trade that take most steel framers years on the job to discover. Shows how to set up a crew to maintain a rhythm that will speed progress faster than any wood framing job. If you've framed with wood, this book will teach you how to be one of the few top-notch metal stud framers in both commercial and residential construction.
208 pages, 8½ x 11, $45.00

Steel-Frame House Construction

Framing with steel has obvious advantages over wood, yet building with steel requires new skills that can present challenges to the wood builder. This book explains the secrets of steel framing techniques for building homes, whether pre-engineered or built stick by stick. It shows you the techniques, the tools, the materials, and how you can make it happen. Includes hundreds of photos and illustrations, plus a FREE download with steel framing details and a database of steel materials and manhours, with an estimating program. **320 pages, 8½ x 11, $39.75**

Concrete Manual

Filled with illustrations, diagrams, and photographs, this essential code resource covers every aspect of concrete. Shows how to pour, how to affect strength, how to affect durability, how to avoid cracks and blemishes, types of aggregates, types of sealants, formwork requirements, proportioning the concrete mixture, how to test and control concrete, slabs on ground, finish and curing, precast and prestressed concrete, steel reinforcing, and special concrete techniques. Includes a FREE CD-ROM for searching text and copying images. **340 pages, 8½ x 11, $73.00**

A Roof Cutter's Secrets to Custom Homes

A master framer spills his secrets to framing irregular roofs, jobsite solutions for rake walls, and curved and two-story walls. You'll also find step-by-step techniques for cutting bay roofs, gambrels, and shed, gable, and eyebrow dormers. You'll even find instructions on custom work like coffered ceilings, arches and barrel vaults, even round towers, hexagons, and other polygons. Includes instructions for figuring most of the equations in this book with the keypad of the Construction Master Pro calculator.
342 pages, 8½ x 5½, $32.50

National Concrete & Masonry Estimator

Since you don't get every concrete or masonry job you bid, why generate a detailed list of materials for each one? The data in this book will allow you to get a quick and accurate bid, and allow you to do a detailed material takeoff, only for the jobs on which you're the successful bidder. Includes assembly prices for bricks, and labor and material prices for brick bonds, brick specialties, concrete blocks, CMU, concrete footings and foundations, concrete on grade, concrete specialties, concrete beams and columns, beams for elevated slabs, elevated slab costs, and more. Includes a CD-ROM with an electronic version of the book with *National Estimator*, a stand-alone *Windows*™ estimating program, plus an interactive multimedia video that shows how to use the disk to compile construction cost estimates.
608 pages, 8½ x 11, $64.00. Revised annually

Business Letters for the Construction Industry

Here you'll find over 100 professionally-written model letters for virtually every situation. Put your business in the best possible light with a well-written letter — especially when you can use that letter to improve a bad situation. Use these letters over and over again to resolve disputes, win new clients, clarify proposals, coordinate with architects, subcontractors, owners, and insurers, schedule meetings and inspections, and to respond to complaints or difficult situations. Included are letters responding to threats of legal action, of commendation to workers, of job performance, apology for defective or delayed work; letters for justification of change orders and price increases; letters explaining your insurance liability, drug testing, injury at work, overtime, equipment use, and more. Practically every letter you'll have to write is in this book, already written, and available on *MS-Word* on the CD-ROM enclosed. Just load the letter you need, change a few phrases, print out and send, or e-mail, and you're free to spend your time on more productive endeavors.
376 pages, 8½ x 11, $59.95

Estimating Excavation

How to calculate the amount of dirt you'll have to move and the cost of owning and operating the machines you'll do it with. Detailed, step-by-step instructions on how to assign bid prices to each part of the job, including labor and equipment costs. Also, the best ways to set up an organized and logical estimating system, take off from contour maps, estimate quantities in irregular areas, and figure your overhead. **448 pages, 8½ x 11, $39.50**

Contractor's Guide to *QuickBooks Pro* 2010

This user-friendly manual walks you through *QuickBooks Pro's* detailed setup procedure and explains step-by-step how to create a first-rate accounting system. You'll learn in days, rather than weeks, how to use *QuickBooks Pro* to get your contracting business organized, with simple, fast accounting procedures. On the CD included with the book you'll find a *QuickBooks Pro* file for a construction company. Open it, enter your own company's data, and add info on your suppliers and subs. You also get a complete estimating program, including a database, and a job costing program that lets you export your estimates to *QuickBooks Pro*. It even includes many useful construction forms to use in your business.
344 pages, 8½ x 11, $57.00
See checklist for other available editions.

Architects, Contractors, Engineers, Guide to Construction Costs

This unique resource contains both the detailed cost coverage for all construction items as well as Quick Estimating Sections and Square Foot costs for preliminary estimates and conceptual budgets. The metropolitan area cost modifiers have been greatly expanded in this year's edition to cover virtually every region of the country. **8½ x 11, 176 pages, $57.95**

Moving to Commercial Construction

In commercial work, a single job can keep you and your crews busy for a year or more. The profit percentages are higher, but so is the risk involved. This book takes you step-by-step through the process of setting up a successful commercial business: finding work, estimating and bidding, value engineering, getting through the submittal and shop drawing process, keeping a stable work force, controlling costs, and promoting your business. Explains the design/build and partnering business concepts and their advantage over the competitive bid process. Includes sample letters, contracts, checklists and forms that you can use in your business, plus a CD-ROM with blank copies in several word-processing formats for both Mac™ and PC computers.
256 pages, 8½ x 11, $42.00

National Building Cost Manual

Square-foot costs for residential, commercial, industrial, military, schools, greenhouses, churches and farm buildings. Includes important variables that can make any building unique from a cost standpoint. Quickly work up a reliable budget estimate based on actual materials and design features, area, shape, wall height, number of floors, and support requirements. Now includes easy-to-use software that calculates total in-place cost estimates. Use the regional cost adjustment factors provided to tailor the estimate to any jobsite in the U.S. Then view, print, email or save the detailed PDF report as needed.
264 pages, 8½ x 11, $53.00. Revised annually

Markup & Profit: A Contractor's Guide

In order to succeed in a construction business, you have to be able to price your jobs to cover all labor, material and overhead expenses, *and* make a decent profit. The problem is knowing what markup to use. You don't want to lose jobs because you charged too much, and you don't want to work for free because you charged too little. If you know how to calculate markup, you can apply it to your job costs to find the right sales price for your work. This book gives you tried and tested formulas, with step-by-step instructions and easy-to-follow examples, so you can easily figure the markup that's right for *your* business. Includes a CD-ROM with forms and checklists for your use. **320 pages, 8½ x 11, $32.50**

Excavation & Grading Handbook Revised

The foreman's, superintendent's and operator's guide to highway, subdivision and pipeline jobs: how to read plans and survey stake markings, set grade, excavate, compact, pave and lay pipe on nearly any job. Includes hundreds of informative, on-the-job photos and diagrams that even experienced pros will find invaluable. This new edition has been completely revised to be current with state-of-the-art equipment usage and the most efficient excavating and grading techniques. You'll learn how to read topo maps, use a laser level, set crows feet, cut drainage channels, lay or remove asphaltic concrete, and use GPS and sonar for absolute precision. For those in training, each chapter has a set of self-test questions, and a Study Center CD-ROM included has all 250 questions in a simple interactive format to make learning easy and fun.
512 pages, 8½ x 11, $42.00

Building Layout

Shows how to use a transit to locate a building correctly on the lot, plan proper grades with minimum excavation, find utility lines and easements, establish correct elevations, lay out accurate foundations, and set correct floor heights. Explains how to plan sewer connections, level an out-of-level foundation, use a story pole and batterboards, work on steep sites, and minimize excavation costs.
240 pages, 5½ x 8½, $22.00

NO POSTAGE NECESSARY IF MAILED IN THE UNITED STATES

BUSINESS REPLY MAIL

FIRST CLASS MAIL PERMIT NO. 271 CARLSBAD, CA

POSTAGE WILL BE PAID BY ADDRESSEE

Craftsman Book Company
6058 Corte del Cedro
P.O. Box 6500
Carlsbad, CA 92018-9974

NO POSTAGE NECESSARY IF MAILED IN THE UNITED STATES

BUSINESS REPLY MAIL

FIRST CLASS MAIL PERMIT NO. 271 CARLSBAD, CA

POSTAGE WILL BE PAID BY ADDRESSEE

Craftsman Book Company
6058 Corte del Cedro
P.O. Box 6500
Carlsbad, CA 92018-9974

NO POSTAGE NECESSARY IF MAILED IN THE UNITED STATES

BUSINESS REPLY MAIL

FIRST CLASS MAIL PERMIT NO. 271 CARLSBAD, CA

POSTAGE WILL BE PAID BY ADDRESSEE

Craftsman Book Company
6058 Corte del Cedro
P.O. Box 6500
Carlsbad, CA 92018-9974